ELECTRICAL
SAFETY

Systems, Sustainability, and Stewardship

Edited by

MARTHA J. BOSS
GAYLE NICOLL

CRC Press
Taylor & Francis Group
Boca Raton London New York

CRC Press is an imprint of the
Taylor & Francis Group, an **informa** business

CRC Press
Taylor & Francis Group
6000 Broken Sound Parkway NW, Suite 300
Boca Raton, FL 33487-2742

Printed on acid-free paper
Version Date: 20140520

International Standard Book Number-13: 978-1-4822-3017-8 (Paperback)

Library of Congress Cataloging-in-Publication Data

Electrical safety : systems, sustainability, and stewardship / editors, Martha J. Boss and
    Gayle Nicoll.
      pages cm
    Includes bibliographical references and index.
    ISBN 978-1-4822-3017-8 (alk. paper)
    1. Electrical engineering--Safety measures. I. Boss, Martha J. II. Nicoll, Gayle.

TK152.E5435 2014
621.3028'9--dc23
                                            2014017889

Visit the Taylor & Francis Web site at
http://www.taylorandfrancis.com

and the CRC Press Web site at
http://www.crcpress.com

Contents

Preface

This book is dedicated to the concept of moving forward. Forward into a future where electricity more efficiently powers our industries, lights the darkness, and sends our voices around the world and beyond. As we find new applications for electric and magnetic fields, our hope is that we will continue to use them to enhance our lives and to provide sustainable solutions for energy production and use, thus becoming better stewards of this resource—electricity. We will make mistakes, and we will misunderstand; however, communication is the key. This book was written to help with that communication, to help us to move forward. These ideas are exemplified by the words of the song "Forward" by Brock Boss of Ra Ra Rasputin:

> The rain starts coming down on us
> We keep moving forward
> Losing all … now
> We keep moving forward
> Taking steps when in doubt
> Towards the places we can't live without
> Loosen up keep it moving
> Keep moving forward
> Don't stop
> The horizon stretches in front of us
> Looking around for something new
> Doesn't matter what you've got
> All you need is a desire
> Forever's on the horizon out there

Acknowledgment

Lloyd A. (Pete) Morley recognized early in his career that understanding electrical power systems and the transmittal of electricity was crucial to safe usage and worker protection. He authored *Mine Power Systems* to provide information relative to the safe use of electricity in a mine environment. As editors, we acknowledge that work and its outstanding contribution to electrical safety education. Both in the mine environment and in other industrial usage areas, providing clear dialog is of primary importance when power systems and electricity are discussed, especially considering the complexity of the subject. Dr. Morley's work inspired us and has inspired many other engineering and safety professionals to enter into discussions that have led to safer working environments. We as editors and authors thank Dr. Morley for his contribution to our profession and for his ongoing outreach efforts to provide sustainable and safe environments.

Editors

Martha J. Boss is a certified industrial hygienist and certified safety professional. During the course of her career, she has provided analyses of electrical safety status for governmental and private facilities worldwide. Her expertise includes National Electric Code (NEC), International Electric Code, and National Fire Protection Association (NFPA) criteria analysis with regard to U.S. Occupational Safety and Health Administration (OSHA) requirements. She has led teams in the inspection, analysis, and risk management of electrical safety components, including electromagnetic field (EMF) analyses. Martha earned BS and BA degrees from the University of Nebraska and the University of Northern Iowa in biology, with a chemistry and physics emphasis. Martha edited and contributed to *Building Vulnerability Assessments*, *Biological Risk Engineering Handbook*, and *Air Sampling and Industrial Hygiene Engineering* (CRC Press). Martha is also the co-author of *Evaluation of Stored Energy Hazards in Underground Coal Mines during and after an Emergency*, provided to the National Institute of Occupational Health (NIOSH).

Gayle Nicoll has extensive experience in a variety of environmental, health, and safety (EHS) endeavors, including industrial facilities, nuclear research facilities, chemical laboratories, and metal-working facilities. Dr. Nicoll has completed EMF studies for facilities and equipment in various settings and conducted electrical safety training internationally. Dr. Nicoll's expertise includes regulatory compliance and industry guidance requirements, including National Electric Code (NEC), National Fire Protection Association (NFPA), Nuclear Regulatory Commission (NRC), International Commission on Non-Ionizing Radiation Protection (ICNIRP), U.S. Occupational Safety and Health Administration (OSHA), Mine Safety and Health Administration (MSHA), World Health Organization (WHO), and International Organization for Standardization (ISO) standards requirements. Dr. Nicoll earned BS degrees in physics and chemistry from Indiana University, a MS in analytical chemistry from Purdue University, and a PhD in chemistry education from Purdue University. She was a contributor to *Building Vulnerability Assessments* (CRC Press) and co-author of the *Evaluation of Stored Energy Hazards in Underground Coal Mines during and after an Emergency*, provided to the National Institute of Occupational Health (NIOSH).

Contributors

Randy Boss is a registered professional engineer. During the course of his career, he has provided professional engineering and project managing expertise for the U.S. Department of Defense. He has served in engineering squadrons dedicated to missile and missile defense systems and in an environmental engineering capacity for both the U.S. Air Force and Army. His expertise includes missile field system survivability and its structural, electromagnetic, electrical, and mechanical components. Critical factors analyzed for survivability during these engineering assignments have included electromagnetic field shielding for buildings, missiles, and their associated communication devices. Randy earned a BS in civil engineering and a MS in environmental engineering from Iowa State University and is a graduate of the U.S. Air Force Air War College. Randy was a contributor to *Building Vulnerability Assessments* (CRC Press).

Dennis Day is a certified industrial hygienist and certified safety professional. He has conducted electrical safety inspections and analyzed electromagnetic fields for industrial and commercial clients worldwide. His expertise includes OSHA, NEC, International Electric Code, and NFPA requirements and guideline interpretation. Dennis earned a BS from the University of Missouri and completed graduate studies at Creighton University. Dennis was an editor of and a contributor to *Building Vulnerability Assessments*, *Biological Risk Engineering Handbook*, and *Air Sampling and Industrial Hygiene Engineering* (CRC Press). Dennis was also the co-author for the *Evaluation of Stored Energy Hazards in Underground Coal Mines during and after an Emergency* provided to the National Institute of Occupational Health (NIOSH).

Reza Tajali is an engineering manager for Schneider Electric Engineering Services. He has over 30 years of experience with electrical power distribution and control and holds two U.S. patents on switchgear products. Reza manages a power system engineering team in the midwestern region of the United States. His team of engineers supports industrial and commercial customers with power system design, analysis, and power quality improvement plans. Reza earned a MS in electrical power systems from Tennessee Technological University and is a registered professional engineer in seven states.

Robert Zweifel is a registered professional electrical engineer with over 25 years of experience in the electrical industry in various roles and with various companies. During the course of his career, he has performed analyses of electrical systems in a broad base of industries. His expertise is primarily in arc flash mitigation strategies and the applications for electrical safety devices in industrial facilities. Robert is a member of the Institute of Electrical and Electronics Engineers (IEEE) and IEEE Industry Applications Society. He graduated as an honors scholar with a BS in electrical engineering from the University of Missouri.

Acronyms

A	Amperes
AC	Alternating current
ACGIH	American Conference of Governmental Industrial Hygienist
AHA	American Hardboard Association
AIC	Available interrupting capacity
AISC	American Institute of Steel Construction
AM (radio)	Amplitude modulated
AP	Access point
APA	The Engineered Wood Association (formerly American Plywood Association)
ASME	American Society of Mechanical Engineers
ASTM	American Society for Testing and Materials
ATS	Acceptance Testing Specifications
AWS	American Welding Society
BIL	Basic insulation level
C	Celsius
cal	Calories
CFR	Code of Federal Regulations
cos	Cosine
cpm	Cycles per minute
CT	Current transformer
cw	Continuous wave
°	Degree
DAR	Dielectric absorption ratio
dB	Decibel
dBV	Decibel volt
DC	Direct current
DOD/DoD	Department of Defense
DVB-T	Digital Video Broadcasting—Terrestrial
EER	Equipment Evaluation Report
EIRP	Equivalent Isotropically Radiated Power
ELF	Extremely low frequency
EM	Electromagnetic
EMF	Electromagnetic field
EMI	Electromagnetic interference
ESA	Electrical surge arresters
ETSI	European Telecommunications Standards Institute
EU	European Union
F	Fahrenheit
FCC	Federal Communication Commission
FM	Factory Mutual

FM (radio)	Frequency modulated
F_{max}	Maximum frequency limit
FMEA	Failure mode and effects analysis
F_{min}	Minimum frequency limit
fps	Feet per second
FR	Flame-resistant
ft	Feet
G	Gauss
GFCI	Ground fault circuit interrupter
GHz	Gigahertz
gps	Gallons per second
GSM	Global System for Mobile Communications
GST	Grounded specimen test
HCI	Hardness critical item
HDBK	Handbook
HEMP	High electromagnetic pulse
Hi-Pot	High-potential
HRC	Hazard risk categories
HT	Heat treatable
HVAC	Heating, ventilation, air conditioning
Hz	Hertz
I	Current flowing through the resistance (i.e., in $V = IR$)
ICES	International Committee on Electromagnetic Safety
ICNIRP	International Commission on Non-Ionizing Radiation Protection
IEEE	Institute of Electrical and Electronics Engineers
IEEE-SA	Institute of Electrical and Electronics Engineers Standards Association
in.	Inch
IRT	Infrared thermography
kA	Killiampere
kg	Kilogram
kHz	Kilohertz
kPa	Kilopascal
kV	Kilovolt
kVA	Kilovolt-ampere
kW	Kilowatt
lb	Pound
LF	Low frequency
LO/TO	Lockout/tagout
LTC	Load tap changer
m/s	Meters per second
MC	Metal clad
MCSA	Motor current signature analysis
mG	Milligauss
MHz	Megahertz
μW	Microwatt
MIL	Military

mm	Millimeter
MOV	Measurement of voltage; metal oxide varistor
MPa	Megapascal
MSHA	Mine Safety and Health Administration
MVA	Megavolt-ampere
mW	Milliwatt
N	Newtons
NC	Normally closed
NEC	National Electric Code
NEMA	National Electrical Manufacturers Association
NETA	InterNational Electrical Testing Association
NFPA	National Fire Protection Association
NO	Normally open
NPLFA	Non-power-limited fire alarm
NPT	National (American) Standard Pipe Taper
NRC	Nuclear Regulatory Commission
NRTL	Nationally recognized testing laboratory
ns	Nanosecond
Ω	Ohm
O&M	Operations and maintenance
OET	Office of Engineering and Technology
OSHA	Occupational Safety and Health Administration
PCB	Polychlorinated biphenyl
pd	Partial discharge
PI	Polarization index
PIV	Peak inverse voltage
PLFA	Power-limited fire alarm
PM	Preventive maintenance
PPE	Personal protective equipment
ppm	Parts per million
PPTCT	Polymeric positive temperature coefficient thermistor
psf	Pounds per square foot
psi	Pounds per square inch
psia	Pounds per square inch atmosphere
psm	Pounds per square meter
PT	Potential transformer
PT&I	Predictive testing and inspection
R	Resistance (i.e., in $V = IR$)
RBW	Resolution bandwidth
RC	Resistance–capacitance
RCM	Reliability-centered maintenance
RF	Radiofrequency
RFCA	Root-cause failure analysis
RMS	Root-mean-square
s	Second
SASD	Silicon avalanche suppression diode

SCCR	Short-circuit current rating
SCR	Silicon-controlled rectifier
SEL	Sensitive earth leakage
SELDS	Shielded enclosure leak detection system
sf	Square foot
sm	Square meter
SPD	Surge protection device
STC	Sound transmission class
STD	Standard
T	Tesla
TEMPEST	Standards addressing methods and/or projects to shield equipment against electromagnetic fields and prevent information leakage
TETRA BOS	Digital radio network being established by BDBOS (German Federal Agency for Digital Radio and Security Authorities and Organizations)
TIG	Tungsten inert gas
TTR	Transformer turns ratio
TVSSs	Transient voltage surge suppressors
UFC	United Facilities Criteria
UL	Underwriters Laboratory
UMTS	Universal Mobile Telecommunications System
UST	Ungrounded specimen test
UVR	Undervoltage release
V	Voltage, potential difference between two points (i.e., in $V = IR$)
VA	Volt-amperes
VAC	Volts alternating current
VAR	Volt-ampere reactive
VCB	Vacuum circuit breaker
VLF	Very low frequency
V/m	Volts per meter
WBC	Waveguide below cutoff
WHO	World Health Organization
WiMax	Worldwide Interoperability for Microwave Access
WLAN	Wireless local area network
XFMR	Type of transformer
X/R	Reactance-to-resistance ratio

Introduction

Science and engineering have their own language—not just the mathematics, where abstraction and symbolism explain concepts, but also the language of disciplines within the sciences. Whether by instrument or design, science can easily present itself with dialects understandable to few. Adding to the confusion is the likelihood that in order not to appear less than knowledgeable questions are less likely to be broached and thus the more complicated the language seems.

Electrical safety is one of those topics. This book intends to provide a start (generation) to finish (motor use) description of electrical use within industrial settings. Reading this text will not make you an electrical engineer, an electrician, or a physicist, but reading this text will give you a common-ground approach to understanding how electrical power is used.

In addition to the relevant descriptions of power generation, transmittal, and usage, the regulatory requirements for electrical safety are provided. This presentation sometimes quotes chapter and verse (subpart and section) from the OSHA electrical safety standards; however, you can find those citations for yourself on the OSHA website (osha.gov). The intent herein is to provide a dialog that is relatively unencumbered by citations in order to present an outline. This outline is interfaced with discussions on the various types of power generation. Arc flash, electromagnetic fields, and ohm resistance studies are discussed in detail as these represent both important design criteria and ongoing electrical hazard assessment critical elements.

Often, electricity (the flow of electrons) is compared to the flow of water. Although both have molecular components, electricity as used in industry is often controlled almost without cognizance by users. If you doubt this assertion, ask yourself how the flow of electricity is stopped in situations where you cannot just pull the plug. Yet, these control devices may be all that stands between a worker and electrocution or burning from arc flash energy transmittal.

Suffice to say that most (yes, *most*) industrial sites have not even tracked through ohm/resistance studies if their grounding systems (both service/supply circuitry and earthing grounds) function correctly. The assumption is that grounding electrodes magically stay intact and attached to grounding conductors; that paths to ground, whether using conductors or building frame intermediaries, are continuous; that fault currents will preferentially take the designed path to ground; that overcurrent devices will initiate in time to protect workers; and that conductors are correctly aligned to make the grounding system protective. Of course, the consequence of this belief is that worker safety is dependent on hope—hope that everything works, even though no proof exists that a ground fault will not go directly toward a worker and cause that worker harm.

On the plant floor or in a hospital basement corridor, mysterious devices are mounted along walls. These devices have various names: disconnect, blade, motor control center (MCC), sometimes even erroneously switch (as in an on/off switch). These disconnects are relied upon to protect workers when power must be discontinued (e.g., a worker is caught in a machine apparatus). These same disconnects are called out in hazardous energy control requirements used when machines and/or electrical equipment are intentionally de-energized. However, how many times do you walk onto a production floor and test whether the workers know the location of the MCCs and whether the workers could in fact use them in an emergency?

Confounding all this confusion is that various remodels and retrofits have caused electrical control features to be all but invisible, hidden behind a new wall or buried under the new concrete slab (a popular place for former grounding electrode attachment sites). Single-line diagrams and plant design features may be lost in time or so outdated as to no longer represent the plant. Systems in place to protect machines and equipment often do not have functional means to similarly protect people, especially if a ground fault cannot even be cleared!

For these reasons, this book was written—to give voice to the regulatory body of work (thank you, OSHA); to demonstrate the interplay between industry standards and regulations and real-world situations (thank you, ANSI, ASTM, NEC, and NFPA); and ultimately to start a new dialog. This dialog is envisioned to be based on concepts that, although not simple, are understandable even without a table of acronyms. The hope is that these concepts can be explored by various disciplines and lead to professional interplay that strengthens electrical safety in our industries and commercial settings. Better understanding is a key component to providing sustainable industrial systems to meet electrical safety and efficacy goals. The concept of stewardship is applicable, as these systems are used and maintained to meet both design intent and regulatory requirements.

The book begins with basic sustainability and stewardship concepts inherent to reliability centered maintenance (RCM). Various electrical components are then presented in terms of their role in providing electricity and tests required to adequately sustain an electrical system. The concept of responsible stewardship is presented by looking at electrical safety, electromagnetic field shields, ohm/resistance study criteria, arc flash hazard analysis, and hazardous energy control.

Integrating electrical safety knowledge in a sustainable fashion is an *a priori* requirement for any real stewardship of electrical safety systems. Ultimately, the question is do the systems in place actually protect the work force? Have training events, safety committee inspections, and all the multiplicity of communication means actually informed the workers of electrical safety issues? This book provides a means to open this dialog.

1 Electrical Sustainability and Stewardship

Martha J. Boss

CONTENTS

Value and cost span the entire life of equipment. Identifying the interrelationships among electrical equipment is important in order to develop holistic use and maintenance routines for facilities. The interrelationships discussed in this book that provide a sustainable system for electrical safety management are illustrated in Figure 1.1.

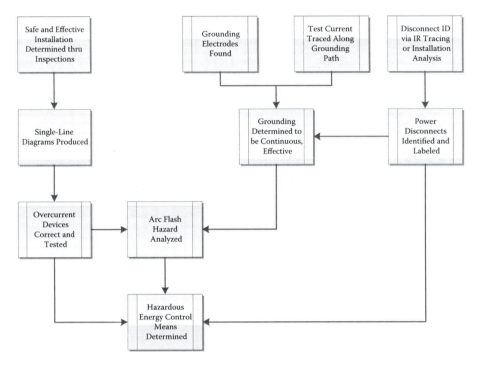

FIGURE 1.1   Electrical safety sustainability and stewardship basic flowchart.

Predictive testing and inspection (PT&I) routines during acceptance testing and throughout the life cycle of equipment usage provide benchmarks for sustaining the entire system. The expected results of applying sustainable management concepts are a quality and safe installation, reduced premature failures, and reduced life-cycle costs. Correctly functioning equipment systems improve overall business success as well as personnel safety. In this regard, stewardship can be defined as initial equipment acceptance and ongoing operation and maintenance (O&M) to provide a safe facility with electrical systems that function within design parameters. A key component of stewardship is proactively identifying hazards before either production or workers are negatively affected. Using the principles of reliably centered maintenance (RCM) as a means to achieve sustainable facility and personnel performance is necessary to achiever proper stewardship.

1.1   COMMISSIONING

Misapplied design, latent manufacturing defects, poor installation practices, and damage incurred during transportation and handling may cause equipment to fail. As an example, recent experience with new construction revealed that 85 to 100% of the rotating equipment was misaligned, was out of balance, or contained defective bearings. In most cases, the faulty equipment would have passed the specified acceptance criteria. This same equipment would most likely experience premature failure during actual operation if the problem conditions were not corrected (NASA, 2004). Premature failures decrease system safety, reliability, and efficiency and may disrupt ongoing critical operations.

Functionality and life-cycle criteria are aspects of both sustainable design and maintenance routines that are developed as electrical systems are commissioned. Often, however, the emphasis is on installation, with future maintenance routines ignored during the commissioning phase. This lack of foresight has resulted in equipment installations that are not reachable for ready maintenance or operational checks and an overall lack of information with regard to equipment manufacturers' requirements. For this reason, major systems and equipment installations are now often turnkey, and the installation contractors are on call for a given time interval to operate the system and work out any issues that may arise.

1.2 TRANSMITTAL OF INFORMATION

Transmittal of information is critically important during commissioning and at all intervals thereafter. The information contained in this text was primarily assembled to aid in that communication so questions are asked at the right time throughout the equipment life-cycle process. Design, construction, and maintenance documentation and testing activities that verify functional operation are often not coordinated between all facility stakeholders. The parameters of pressure, temperature, minimum and maximum airflow, lighting levels, electrical amperage, voltage, torque, fluid volumes, and other thermodynamic measures are checked to confirm that the design intent has been met. However, these measurements and the means by which design intent has been verified may not be sufficiently documented. This lack of sufficient documentation can result in future operation and maintenance checks being performed without knowledge of design intent baselines.

Latent manufacturing and installation defects that may result in premature equipment or system failure are often not apparent until systems have been operational. The maintenance required when these defects become apparent will benefit from knowledge of initial baseline testing methodologies and results. For sustainable systems, maintaining ongoing documentation from system inception is therefore crucial. Realistically, in aged systems, finding design and commissioning documentation may be overly burdensome or impossible; therefore, baseline testing may be the last known and documented successful operational or maintenance test.

1.2.1 TEAM CONCEPT

Team involvement should begin at the earliest stages of project planning, where expertise in system sizing, code compliance, maintainability, user friendliness, maintainability, product quality and reliability, ergonomics, and projected life-cycle costs is applied to the design. It is important to monitor the quality of construction in terms of workmanship and specification and code compliance throughout all of the stages of construction. Monitoring of installed systems should continue to detect any latent installation or operational defects or degradation of system performance. This rigorous process is intended to

- Ensure that a facility has systems at optimal productivity.
- Improve the level of performance.
- Maintain and, as needed, restore high productivity.
- Ensure that facility renovations and equipment upgrades function as designed.

1.3 RELIABILITY CENTERED MAINTENANCE

Reliability centered maintenance (RCM) is an ongoing process that determines the most effective approach to maintenance in support of the mission The RCM approach takes a life-cycle view of facilities and collateral equipment to

- Evaluate equipment optimal design for the highest level of performance throughout the life cycle of the equipment.
- Maximize equipment performance through maintenance actions.
- Identify the optimum mix of applicable and effective maintenance tasks to realize the inherent design reliability of equipment and preserve the safety of systems and personnel.
- Provide a systematic, objective, logic-based approach for selecting the most appropriate maintenance routines and tasks.
- Generate sound technical rationale and economic justification on which maintenance decisions are based with consideration of operational experience and failure history to validate and support those decisions.

This technique provides sustainable O&M decision logic, takes a life-cycle view of facilities and collateral equipment, and recognizes the importance of

- Failure mode and effects analysis (FMEA) to better understand the consequences of equipment failure with regard to efficacy, safety, and facility costs
- Unified acceptance and ongoing use criteria methodologies
- Proactive monitoring and assessment of equipment condition during operation
- Modern testing and inspection technologies

The acceptance criteria and the associated documents provide significant benefit toward integrating design and the O&M professional's knowledge base. Baseline performance data for condition assessment when used to derive O&M programs will eliminate redundant efforts. These criteria will define what technologies will be required to verify a defect-free acceptance, what technologies will be required for equipment condition assessment and maintenance, and when during the entire process those technologies will be required (NASA, 2004).

1.3.1 Reliability

Reliability centered maintenance improves system and equipment reliability, principally through the documentation of initial baseline readings, maintenance experience, and equipment condition data for facility planners, designers, maintenance managers, and manufacturers. The increased reliability that comes from RCM leads to fewer equipment failures and, therefore, lower maintenance costs.

1.3.2 Scheduling

The ability of RCM to forecast maintenance requirements provides the necessary time for planning, obtaining replacement parts, and arranging environmental and operating conditions before the maintenance is done. RCM facilitates obtain maximum use from equipment. With RCM, decisions for equipment replacement consider condition as well as the calendar.

1.3.3 Life-Cycle Costs

A facility's life cycle is often divided into two broad stages: (1) acquisition (planning, design, and construction) and (2) operations. RCM affects all phases of the acquisition and operations stages to some degree. Decisions made early in the acquisition cycle profoundly affect the life-cycle costs of a facility. Even though equipment expenditures may occur later during the acquisition process, their cost is committed at an early stage. These early decisions will have a major impact on equipment life-cycle costs. Ensuring that facilities meet acceptable RCM criteria and obtain and document critical baseline data is extremely important during the construction phase (NASA, 2004).

1.3.4 Philosophies

Reliability centered maintenance is a combination of four distinct philosophies (Figure 1.2):

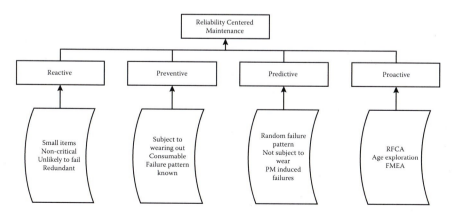

FIGURE 1.2 Components of an RCM program.

- *Reactive maintenance*—Repairing actions after failure
- *Preventive maintenance (PM)*—Restorative actions taken prior to failure; also known as time-directed maintenance
- *Predictive maintenance*—Monitoring actions that are predictive of failure; also known as predictive testing and inspection (PT&I) or condition-based maintenance (CBM)
- *Proactive maintenance*—Includes root-cause failure analysis (RCFA), age exploration, and development of failure mode and effects analysis (FMEA) and incorporates this knowledge into new design to continually improve performance and extend equipment life

1.4   REACTIVE MAINTENANCE STRATEGY

A reactive maintenance strategy is performed when equipment performance is unacceptable. Reactive maintenance allows for a lower skill set, as the failures are well defined and training in equipment evaluation or predictive technologies is not needed.

Advantages
- No downtime between failures
- Traditionally accepted by maintenance personnel
- Easy to justify to outside groups

Disadvantages
- Large spare parts inventory
- Quick response required from trained personnel
- Unscheduled work outages
- Longer restoration time
- Higher restoration costs
- Low manageability of budget, personnel, and parts
- Disregard of safety
- Possible collateral damage

A negative aspect of reactive maintenance is that failures often occur unexpectedly, severely disrupting operations. Emergency repairs can be expensive; labor, parts, and supplies may not be immediately available; costs for returning the equipment back to service may be high; and operational impacts can be far more significant than the mere cost to repair the system. Another disadvantage is having to stock an overly extensive spare parts inventory to be prepared for any possible failure (NASA, 2004).

1.5   PREVENTIVE MAINTENANCE STRATEGY

A preventive maintenance (PM) strategy consists of restorative-type maintenance actions intended to improve equipment condition and prevent or delay failure. PM applies experience and failure history to identify a pattern of degradation and then attempts to apply specific maintenance actions to return equipment to a desirable level of performance. PM tasks include lubrication, servicing and overhaul, and inspections that require the equipment to be shut down. PM is successful in

- Reducing the risk of catastrophic failures
- Extending the interval between failures
- Maintaining equipment at high performance levels
- Overcoming some of the disadvantages of a reactive strategy

The largest disadvantage of PM is that improper execution of maintenance tasks often creates more problems than if the maintenance had never been performed. Many problems may exist immediately after returning equipment into service following maintenance.

Advantages
- Reduces risk of catastrophic failure
- Prevents equipment failure
- Overcomes some disadvantages of reactive maintenance

Disadvantages
- Operating time per cycle is reduced.
- Costly unneeded maintenance is performed.
- Operational restrictions result in deferred maintenance.
- Frequency intervals are based on limited data and vendor recommendations.

One negative aspect of PM is that equipment availability is reduced by intentionally taking equipment offline in order to perform many of the PM actions. PM strategies also tend to perform more maintenance than the other type of strategies, which in turn increases the requirements for labor, spare parts, and supplies. Another disadvantage of a PM strategy is that execution requires very good coordination between operation and maintenance departments. Before equipment can be removed from service for maintenance, the operational schedule must be able to support that period of unavailability. If such periods of unavailability are not immediately obtainable (i.e., operations demand that the equipment continue running), then the maintenance is deferred. For complex equipment, the impacts of deferred maintenance on performance may be difficult to assess (NASA, 2004).

1.6   PREDICTIVE MAINTENANCE STRATEGY AND PREDICTIVE TESTING AND INSPECTION

A PT&I maintenance strategy monitors equipment performance to recognize the onset of failure, determine degradation rate, and forecast failure. Maintenance actions can be performed at the optimum time before failure. PT&I may, however, be expensive to implement, as skilled technicians and therefore increased training are required. PT&I program costs include the one-time costs of acquiring the PT&I

equipment (e.g., infrared camera, vibration analysis equipment, borescope, computers, software, online sensors) and the initial and annual costs to train technicians and use this equipment.

Advantages
- Provides for continuous risk assessment
- Integrates with total resource planning
- Overcomes most of the disadvantages of reactive and preventive maintenance

Disadvantages
- High acquisition and implementation costs
- High training and certification requirements
- Additional maintenance on testing equipment
- May be limited due to non-standard equipment

1.7   PROACTIVE MAINTENANCE STRATEGY

Reliability centered maintenance seeks the optimal mix of PT&I actions, preventive maintenance-based actions, and corrective maintenance actions to form a comprehensive program. The methodology used in RCM to determine this optimum blend of maintenance actions is based on employing these additional proactive maintenance techniques:

- Failure mode and effects analysis (FMEA)
- Root-cause failure analysis (RCFA)
- Age exploration
- Enhanced specifications and acceptance criteria for new/rebuilt equipment
- Precision rebuild and installation, verified with certification
- Failed part analysis
- Reliability engineering
- Recurrence control

These techniques improve maintenance through better design, installation, maintenance procedures, workmanship, and scheduling. The characteristics of proactive maintenance are

- Use feedback and communications to ensure that changes in design or procedures are rapidly made available to designers and managers.
- Employ a life-cycle view of maintenance and supporting functions.
- Ensure that nothing affecting maintenance occurs in isolation.
- Employ a continuous process of improvement.
- Optimize and tailor maintenance techniques and technologies to each application.
- Integrate functions that support maintenance into maintenance program planning.

- Use root-cause failure analysis and predictive analysis to maximize maintenance effectiveness.
- Adopt an ultimate goal of fixing the equipment forever.
- Periodicly evaluate the technical content and performance interval of maintenance tasks (PM and PT&I).

A proactive maintenance program is the capstone of the RCM philosophy. This program's most essential elements are a thorough understanding of the failure modes associated with a system or equipment failure and the accurate assessment of the effects or consequences should a failure occur. A standard decision logic tree may be used to support consistent analysis and determination of the types of maintenance action that are the best solution for any general situation. This decision logic tree is shown in Figure 1.3.

Failure mode and effects analysis provides an evaluation of each function of the system or equipment that may lead to a functional failure. Functional failures are the various ways in which the functional requirements will not be met. Each functional failure is broken down into dominant failure modes, which are observations as to why and how functions will not be met. Each dominant failure mode is then analyzed to determine specific reasons, or failure causes, that will lead to an occurrence of the dominant failure mode. Whereas dominant failure modes address only overall observations without identifying specific failure mechanisms, failure causes will address the failure mechanisms.

1.8 LINK FROM RCM TO DESIGN

Reliability centered maintenance is an ongoing process that continuously generates performance information to measure the success and effectiveness of the program. A constant effort is needed to evaluate the effectiveness of the current maintenance program and to continually make improvements. RCM is often defined as a "living" program to recognize that continuous adjustments are made to incorporate lessons learned. Reliability centered maintenance increases the probability that a machine or component will function as required over the design life cycle. The maintenance decisions must be based on function requirements supported by sound technical and economic justification. In other words, an expensive maintenance task (e.g., infrared thermography, vibration analysis) should not be performed if little or no consequence is associated with the failure. Maintenance actions must be both applicable and cost effective.

When maintenance actions are neither applicable nor cost effective toward eliminating a failure, then the risks associated with that failure can only be mitigated by investigating the design. Any collected knowledge and lessons learned from these adjustments and design considerations can and should be incorporated into new designs. When an organization proactively integrates RCM with the design process, then that organization has effectively bridged the gap between design and O&M to create an optimum solution for system productivity (NASA, 2004).

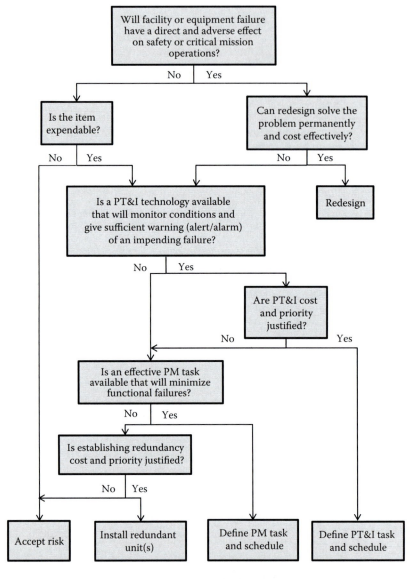

FIGURE 1.3   Reliability centered maintenance (RCM) decision logic tree.

1.9   PREDICTIVE TESTING AND INSPECTION

The process of predictive testing and inspection uses advanced technology to assess machinery and equipment condition. The analysis of PT&I data ideally reveals any degradation in equipment performance and provides insight into the degradation rate. The PT&I data allow for effective planning and scheduling of maintenance or repairs so consequences from failure can be minimized or eliminated. For PT&I

data to be effective, initial baseline data, normally taken at inception, are necessary for comparisons and trending. From an equipment acceptance perspective, PT&I tests have become one of the most effective methods for testing new and in-service equipment for hidden defects.

Although the performance of maintenance and operations occurs during the operations stage of the life cycle, some preparatory activities should be conducted during the construction and acceptance stage. These activities can include personnel selection; planning for the training requirements; procedure preparation; review of specifications, design, and nameplate data; labeling; and review of regulatory planning requirements. An inventory database is useful to categorize each machine by nameplate information and hardware descriptors, as well as usage. PT&I involves the use of

- Acceptance and inspection techniques that are neither intrusive nor destructive in order to avoid introducing problems
- Data collection devices with appropriate documentation (manufacturer, model number, calibration date, certificate of calibration, and serial number)
- Data analysis and computer databases to store and trend information

The tests used are discussed in terms of sustainability and stewardship for the various system components being tested. Electrical systems testing is discussed in Chapter 16, and various component testing suggestions are provided in the other chapters, as applicable.

1.9.1 DOCUMENTATION MARKINGS

All documentation should be marked or otherwise identified and should include the following:

- Equipment identification number
- Location of installation, both the geographic location and the location within the facility
- Date of installation (required or actual acceptance date)
- Applicable reference drawing numbers

1.9.2 ACCEPTANCE AND O&M DOCUMENTATION

Required acceptance and O&M documentation in all instances includes the following:

- Material, equipment, and fixture lists
- Shop drawings
  - Connection diagrams
  - Fabrication drawings
- Product data
  - Equipment and performance data
  - Manufacturer's catalog data

- Test reports
  - Operation test results
  - Performance test results
  - Baseline data from verification test
- Certificates
- Manufacturer's instructions
- Manuals
  - Operations manual
  - Maintenance manual
- Warranty information
- Parts list and recommended spare parts list
- Acceptance documentation (dates and signatures)

1.9.3 THERMODYNAMIC PERFORMANCE TESTS

Thermodynamic tests are used to verify that systems meet the required levels of functional performance in terms of heating, cooling, utility service delivery, and power consumption. These tests directly measure pressure, temperature, flow, voltage, current, and power consumption and indirectly measure heat-transfer characteristics, system capacity performance, energy efficiency, and control system response. The performance of thermodynamic tests is essential if system and facility functionality is to be confirmed and baselined. These types of tests typically use installed sensors, supplemented by temporary sensors, for measuring flow.

1.10   TECHNICAL MANUALS

When systems are procured, technical manuals for all constituent components must be obtained and secured. Parts breakdown must be sufficiently detailed to allow for the identification of all replaceable parts within the equipment being procured. Cut sheets from generic catalogs are not sufficient to meet this requirement. All manuals should be edited to limit the data to the models and configuration of equipment actually delivered, including any and all options.

1.11   TESTING

Table 1.1 illustrates PT&I technologies. Additional information is provided within the chapters specific to equipment types and in Chapter 16.

REFERENCE

NASA. (2004). *Reliability Centered Building and Equipment Acceptance Guide*. Washington, DC: National Aeronautics and Space Administration.

TABLE 1.1

Predictive Testing and Inspection (PT&I) Examples

Equipment Item	Highly Suggested PT&I Technologies	Optional PT&I Technologies
Breakers—general	Contact resistance Insulation resistance	Airborne ultrasonic Power factor Insulation oil High voltage Breaker timing Infrared thermography
Breakers—air blast	Contact resistance Insulation resistance	Airborne ultrasonic Power factor High voltage Breaker timing Infrared thermography
Breakers—air magnetic	Contact resistance Insulation resistance	Airborne ultrasonic Power factor High voltage Breaker timing Infrared thermography
Breakers—oil	Contact resistance Insulation resistance Insulation oil	Airborne ultrasonic Power factor High voltage Breaker timing Infrared thermography
Breakers—SF_6 gas	Contact resistance Insulation resistance Vacuum bottle integrity SF_6 gas test results Air compressor performance SF_6 gas leakage	Airborne ultrasonic Power factor High voltage Breaker timing Infrared thermography
Breakers—vacuum	Contact resistance Insulation resistance	Airborne ultrasonic Power factor High voltage Breaker timing Infrared thermography
Cables—general	Insulation resistance	Airborne ultrasonic Power factor High voltage
Cables—low voltage	Insulation resistance	Airborne ultrasonic
Cables—medium voltage	Insulation resistance	Airborne ultrasonic Power factor High voltage
Cables—high voltage	Insulation resistance	Airborne ultrasonic Power factor High voltage

TABLE 1.1 (continued)
Predictive Testing and Inspection (PT&I) Examples

Equipment Item	Highly Suggested PT&I Technologies	Optional PT&I Technologies
Capacitor banks	Capacitor bank acceptance	Airborne ultrasonic
	Capacitor discharge	
Capacitors—dry type	Insulation resistance	Airborne ultrasonic
	Overpotential	
	Capacitor discharge	
Capacitors—liquid-filled	Insulation resistance	Airborne ultrasonic
	Insulation oil	
Electrical automatic transfer switch	Contact resistance	Airborne ultrasonic
	Insulation resistance	Power factor
	Automatic transfer	Infrared thermography
Electrical buss	Contact resistance	Airborne ultrasonic
	Insulation resistance	Infrared thermography
	Overpotential	
Electrical control panels	Contact resistance	Insulation resistance
	Infrared thermography	Power factor
	Airborne ultrasonic	
Electrical distribution panels	Contact resistance	Insulation resistance
	Infrared thermography	High voltage
	Airborne ultrasonic	Power factor
Electrical grounding grid	Fall-of-potential	—
	Point-to-point	
Electrical lightning protection system	Continuity	Insulation resistance
Electrical power centers	Contact resistance	Insulation resistance
	Infrared thermography	High voltage
	Airborne ultrasonic	Power factor
Electrical power supplies	Contact resistance	Insulation resistance
	Infrared thermography	Power factor
Electrical rectifiers	Contact resistance	High voltage
	Insulation resistance	Infrared thermography
	Power factor	Airborne ultrasonic
	Turns ratio	
Electrical relays	Insulation resistance	Contact resistance
Electrical starters	Airborne ultrasonic	Insulation resistance
	Infrared thermography	Contact resistance
Electrical switches, cutouts	Contact resistance	Airborne ultrasonic
	Insulation resistance	Power factor
		High voltage
		Infrared thermography
Electrical switches—low-voltage air	Contact resistance	Airborne ultrasonic
	Insulation resistance	Infrared thermography
Electrical switches—medium- and high-voltage air	Contact resistance	Airborne ultrasonic
	Insulation resistance	Power factor
		High voltage
		Infrared thermography

TABLE 1.1 (continued)
Predictive Testing and Inspection (PT&I) Examples

Equipment Item	Highly Suggested PT&I Technologies	Optional PT&I Technologies
Electrical switches—medium-voltage air, metal enclosed	Contact resistance Insulation resistance	Airborne ultrasonic Power factor High voltage Infrared thermography
Electrical switches—medium-voltage oil	Contact resistance Insulation resistance Insulating oil	Airborne ultrasonic Power factor High voltage Infrared thermography
Electrical switches—medium-voltage SF_6	Contact resistance Insulation resistance Vacuum bottle integrity SF_6 gas results	Airborne ultrasonic Power factor High voltage Infrared thermography
Electrical switches—medium-voltage vacuum	Contact resistance Insulation resistance Vacuum bottle integrity Insulating oil	Airborne ultrasonic Power factor High voltage Infrared thermography
Electrical transformer load tap changers	Contact resistance Insulation resistance Power factor Turns ratio	High voltage Infrared thermography Airborne ultrasonic
Fans	Vibration analysis Balance and measurement Alignment (laser preferred) Lubricating oil Thermodynamic performance	—
Gearboxes	Vibration analysis Hydraulic oil Lubricating oil	—
Motor control centers	Airborne ultrasonic Infrared thermography	Insulation resistance
Motors—general	Vibration analysis Balance and measurement Alignment (laser preferred) Power factor	Infrared thermography Insulation resistance Motor circuit evaluation High voltage
Motors—hydraulic	Vibration analysis Lubricating oil Performance Alignment (laser preferred)	—
Motors—pneumatic	Vibration analysis Noise level acceptance Lubricating oil Performance	—

TABLE 1.1 (continued)
Predictive Testing and Inspection (PT&I) Examples

Equipment Item	Highly Suggested PT&I Technologies	Optional PT&I Technologies
Switchgear	Airborne ultrasonic	Contact resistance
	Insulation resistance	High voltage
	Infrared thermography	Power factor
Transformers	Airborne ultrasonic	Contact resistance
	Power factor	Insulation resistance
	Insulation oil	High voltage
	Infrared thermography turns ratio	

Source: Adapted from NASA, *Reliability Centered Building and Equipment Acceptance Guide*, National Aeronautics and Space Administration, Washington, DC, 2004.

2 Electrical Components

Gayle Nicoll and Martha J. Boss

CONTENTS

Most of the devices that make up a power generation and distribution environment are based on a small number of fundamental electrical components. Understanding these components is critical to being able to understand the operation of electrical substations and generators. In order to provide a common language for a sustainability discussion, this chapter covers the specific components of an electrical supply and distribution system for a common industrial site. The health and safety aspects of each part of the electrical system are discussed from a scientific standpoint, explaining the rationale for the safety implementation. The relevant health and safety regulations are also presented and explained in light of the scientific basis for the regulations.

2.1 DIRECT CURRENT

Direct current (DC) describes the constant flow of electrons from a negative potential to a positive potential. Direct current assumes that the areas of positive and negative potential are static, and, although this potential may change in magnitude, positive potentials will always stay positive and negative potentials will always stay negative.

2.2 ALTERNATING CURRENT

Alternating current (AC) describes a constant flow of electrons from a negative potential to a positive potential, with the positive and negative potential exchanging places on a periodic basis. In alternating current, current will flow in one direction through a circuit and then reverse direction as the potentials exchange places. In most AC applications, the positive and negative potentials change magnitude at the same rate and in the same direction, leading the current to exhibit a smoothly changing waveform. The point at which both potentials have zero magnitude is called a *zero crossing*. The most common waveform exhibited by AC current is a sinusoid.

2.3 CONDUCTORS

In its simplest form, a conductor is any material capable of conducting electricity. An ideal conductor passes along all current without loss of energy from heat. Although a relatively ideal conductor can be created with superconductors, superconductors are impractical for everyday use. The science of creating efficient and resilient conductors can be very complex; however, for the purposes of this discussion, conductors have a common set of characteristics that must be considered.

1. Conductors have a resistance-per-length rating. This rating describes how many ohms of resistance a conductor has per unit of length. (*Note:* Longer wires will generally have more resistance.)
2. Conductors have a capacitance-per-length rating. This rating describes how much current is absorbed and stored over the length of the conductor.
3. Conductors have a self-inductance-per-length rating. This rating describes the effect that the electric field generated by the conductor has on the conductor itself.

These three characteristics are summarized by the term *impedance*. Impedance is defined as a measure of the opposition that a circuit presents to current when voltage is applied. Impedance has two components: *resistance* and *reactance*. The effect of the capacitance and inductance of a circuit (or conductor) is included in the reactance.

2.4 RESISTORS

A resistor is a conductor that presents a specific amount of resistance to current passing through the conductor. Resistance is measured in ohms in honor of George Ohm, who discovered the fundamental concept that current through a conductor is proportional to the voltage applied to the conductor and the resistance of the conductor. Ohm's law is stated as

$$\text{Current } (I) = \text{Voltage } (V) \div \text{Resistance } (R)$$

Alternatively, this equation can be written as $V = I \times R$.

Although an ideal resistor loses no energy while current flows through the ideal resistor, real resistors lose energy in the form of heat. The amount of heat generated by a resistor is directly proportional to the amount of current flowing through the real resistor.

2.5 CAPACITORS

A capacitor is a device used to store and release current in a circuit. A capacitor is typically formed by two parallel plates separated by a material with a high dielectric constant (insulator). When a voltage is placed across the capacitor's leads, current flows into one plate, generating an electric field that induces an equal and opposite charge on the opposing plate. The maximum quantity of current that can be stored by a capacitor is proportional to the size of the plates and the distance between the plates.

In DC circuits, capacitors will continue to draw current until the capacitor has stored the maximum amount of energy possible. At this point, the capacitors will draw no more current and act as an open circuit. When voltage is removed from the circuit, the capacitor will slowly discharge accumulated charge into the circuit. In this way, a capacitor can act as a short-term battery for a circuit.

In AC circuits, capacitors often do not have time to fully charge before the voltage across the capacitor changes orientation. In this situation, capacitors act as temporary storage for energy, augmenting the current in the circuit with current from the capacitor. For low-frequency AC waveforms, this can have the effect of smoothing out the waveform, essentially turning the waveform into a DC voltage. For high-frequency waveforms, the plates do not have sufficient time to charge, causing the capacitor to have little effect on the waveform. Because of this behavior, capacitors are used as filters—filtering out low-frequency signals but allowing high-frequency signals to pass.

All capacitors have a *breakdown voltage*. The breakdown voltage is the voltage that, when applied across the leads of the capacitor, will cause the charge accumulated on the plates to overcome the insulating capacity of the dielectric material, causing current to flow through the dielectric material. This process may damage some dielectric materials.

2.5.1 CAPACITORS AND OSHA

The Occupational Safety and Health Administration (OSHA) requires that all capacitors be provided with an automatic means of draining the stored charge after the capacitor is disconnected from its power source (supply). (*Note:* Surge capacitors or capacitors as component parts of other apparatus are excluded from this requirement.) Capacitors installed on circuits operating at more than 600 volts, nominal, have additional requirements (OSHA, 2007).

- Group-operated switches used for capacitor switching must be capable of
  - Carrying continuously not less than 135% of the rated current of the capacitor installation
  - Interrupting the maximum continuous load current of each capacitor, capacitor bank, or capacitor installation that will be switched as a unit
  - Withstanding the maximum inrush current, including contributions from adjacent capacitor installations
  - Carrying currents due to faults on the capacitor side of the switch
- A means must be installed to isolate from all sources of voltage each capacitor, capacitor bank, or capacitor installation that will be removed from service as a unit. The isolating means must provide a visible gap in the electric circuit adequate for the operating voltage.
- Isolating or disconnecting switches (with no interrupting rating) must be interlocked with the load-interrupting device and provided with prominently displayed caution signs to prevent switching load current.
- For series capacitors, the proper switching must be ensured by use of at least one of the following:
  - Mechanically sequenced isolating and bypass switches
  - Interlocks
  - Switching procedure prominently displayed at the switching location
- Provisions must be made for sufficient diffusion and ventilation of gases from any storage batteries to prevent the accumulation of explosive mixtures.

2.5.2   PREDICTIVE TESTING AND INSPECTION
SUSTAINABILITY AND STEWARDSHIP—CAPACITORS

Tests for capacitor banks and various types of capacitors are provided below.

2.5.2.1   Capacitor Bank
2.5.2.1.1   Required Testing Results
- Capacitor bank acceptance test results
- Airborne ultrasonic test results (optional)
- Capacitor discharge test results

2.5.2.1.2   Acceptance Technologies
Inspect bolted electrical connections for high resistance as follows:

- Use a low-resistance ohmmeter.
- Verify the tightness of accessible bolted electrical connections by the calibrated torque-wrench method in accordance with the manufacturer's published data.

Use airborne ultrasonics to verify the non-existence of electrical arcing and other high-frequency events.

2.5.2.1.3   Capacitor Discharge Test
- Confirm automatic discharging in accordance with National Fire Prevention Association (NFPA) 70 NEC (National Electrical Code).
- *Minimum acceptance criteria:* Better than or equal to the manufacturer's specifications or, in the absence of the manufacturer's specifications, use NFPA 70 NEC, Article 460. The residual voltage of a capacitor should be reduced to 50 volts after being disconnected from the source of supply.

Rated Voltage	Discharge Time
<600 volts	1 minute
>600 volts	5 minutes

- Document the rated voltage and discharge time within these parameters.

2.5.2.2   Capacitors—Dry

2.5.2.2.1   Required Testing Results
- Insulation resistance test results
- Overpotential test results
- Airborne ultrasonic test results (optional)
- Capacitor discharge test results

2.5.2.2.2   Acceptance Technologies
- Overpotential test
  - The AC overpotential test should not exceed 75% of the factory test voltage for a 1-minute duration. The DC overpotential test should not exceed 100% of the factory test voltage for a 1-minute duration.
  - *Minimum acceptance criteria:* Better than or equal to the manufacturer's specifications or, in the absence of the manufacturer's specifications, the insulation should withstand the overpotential test voltage applied.
  - Use airborne ultrasonics to verify the non-existence of electrical arcing and other high-frequency events.
- Capacitor discharge test
  - Confirm automatic discharging in accordance with NFPA 70 NEC.
  - *Minimum acceptance criteria:* Better than or equal to the manufacturer's specifications. The residual voltage of a capacitor should be reduced to 50 volts after being disconnected from the source of supply.

Rated Voltage	Discharge Time
<600 volts	1 minute
>600 volts	5 minutes

- Document the rated voltage and discharge time within these parameters.

2.5.2.3 Capacitors—Liquid-Filled

2.5.2.3.1 Required Testing Results

- Insulation resistance test results
- Insulation oil test results
- Airborne ultrasonic test results (optional)

2.5.2.3.2 Acceptance Technologies

- Use the insulation resistance test to detect the presence of contamination or insulation degradation.
- Use both the dielectric absorption index and the polarization index.
- Perform winding-to-ground insulation resistance tests.
- Apply voltage in accordance with the manufacturer's published data.
- *Minimum acceptance criterion:* Better than or equal to the manufacturer's specifications.
- Use airborne ultrasonics to verify the non-existence of electrical arcing and other high-frequency events.

2.5.2.3.3 Other Tests

Perform the following tests as appropriate to verify a lack of contaminants and that the necessary inhibitors have been added (acceptance criteria are shown in parentheses):

- Water content using Karl Fisher, ASTM D1533-88 (<25 ppm at 20°C)
- Dielectric breakdown strength test, ASTM D877 and D1816 (>30 kV)
- Acidity test, ASTM D974 (neutralization number <0.05 mg/g)
- Visualization examination, ASTM D1524 (clear)
- Dissolved gas analysis, ASTM D3612-90
  - Nitrogen (N_2) (<100 ppm)
  - Oxygen (O_2) (<10 ppm)
  - Carbon dioxide (CO_2) (<10 ppm)
  - Carbon monoxide (CO) (<100 ppm)
  - Methane (CH_4) (none)
  - Ethane (C_2H_6) (none)
  - Ethylene (C_2H_4) (none)
  - Hydrogen (H_2) (none)
  - Acetylene (C_2H_2) (none)

2.6 INDUCTORS

Inductors are a basic passive element in electrical engineering. An inductor is generated by creating a coil of a conductor, or winding, around a central fixed point (near circle winding) or line (spiral winding). The winding has an inner and outer diameter, with the inner diameter being separated from the fixed point by some small distance. In many cases, the windings of an inductor are wrapped around a shaft of material that is not conductive but does allow electromagnetic fields to penetrate. This inner area (air gap or material) is sometimes referred to as the *core* of the inductor.

Inductors create a potential (voltage) difference across the leads of the inductor proportional to the strength and orientation of the electromagnetic field through the center of the winding and the number of complete loops in the winding. Inductors whose cores are placed in parallel with the flow of an electromagnetic field will produce the highest potential difference, while those placed perpendicular to the field will generate the least. Increasing the number of loops around a central point in a winding will also directly increase the potential difference across the leads. Note that the orientation of the high- and low-potential leads will be directly opposite the flow of the electric field.

Inductors may also be used to generate an electromagnetic field. An inductor generates an electromagnetic field proportional to the number of loops in the inductor's winding and the magnitude of the current flowing through the inductor. The higher the current through an inductor, or the higher the number of coils, the stronger the electromagnetic field produced will be.

Like a capacitor, inductors have an effect on AC waveforms passing through them. In DC circuits, the electromagnetic field generated by the inductor is stable, having a constant effect on the flow of current through a circuit; however, the fact that an inductor stores energy in an electromagnetic field gives the inductor an electrical "inertia" when the direction of current is reversed. This inertia is a reflection of the time to create and eliminate the electric field around an inductor as the current through the inductor changes. Because of this inertia, rapid changes in current direction, such as high-frequency AC waveforms, tend to be resisted and smoothed out, in essence becoming a DC signal. Low-frequency waveforms provide more time for the electric field to collapse as current diminishes and are less affected by an inductor. This makes inductors perfect for creating low-pass filters.

2.6.1 WINDINGS

A winding, a form of inductor commonly used in power systems, is a coil of conductive wire, usually forming a near circle or spiral. A spiral winding is often a set of near-circle windings of similar diameter tightly compressed together in a line formation. The conductive wire is thinly insulated against contact, preventing any portion of the winding from forming a short circuit. As current passes through the winding, heat is generated based on the amount of current flowing through the winding and the resistance (in ohms) of the winding per length. Because a coil of wire is often tightly wound around a central core, heat can build up over time due to the lack of a mechanism for dissipation and can potentially melt the insulating material from the conductor. When the insulator is removed, rings of the conductor can come into contact, creating a short circuit that can effectively remove a large number of coils from the winding. In extreme cases, the conductor itself may melt, rendering the winding useless.

2.6.2 USING INDUCTORS TO INDUCE VOLTAGE
(AND CURRENT) IN ANOTHER INDUCTOR

The core principle in both electrical generators and transformers is the use of two (or more) inductors, electrically isolated from one another. One inductor is designated an *input*, or "primary," inductor. This is the inductor to which the input voltage is applied. The other inductors are designated *output*, or secondary, inductors, which use the electric field generated by the input inductor to induce a voltage across their own windings. The magnitude of the voltage induced across the output windings is proportional to the ratio of the windings in the input and output inductors. If the input inductor has more coils in its windings than the output inductor, then the voltage across the output will be proportionally higher and the output current will be proportionally lower. In transformers, this is often referred to as the basis of a *step-up* transformer. If the input inductor has fewer coils in its windings than the output inductor, then the induced output voltage will be lower and the induced output current will be higher. This is the basis for a *step-down* transformer. The actual voltage and current produced in the secondary inductors will be affected by many different factors, including, but not limited to, the resistance of the conductors used to form the coils, the load attached to the output inductor, and the distance between the inductors.

2.7 DIODES

Diodes are basic circuit elements that allow current to flow only one way. A diode has low resistance to current flow in one direction and near-infinite resistance to current flow in the opposite direction. The voltage drop across a diode is usually near constant and varies mostly with temperature, allowing the diode to have a very small effect on current flow in the allowed direction. Diodes are typically used as waveform rectifiers and the electronic equivalent of a check valve, preventing current from going in a direction that may damage a circuit. Diodes have a breakdown voltage. If a sufficient voltage is applied across a diode in the direction opposite the allowed current flow, the diode will break down and allow current to flow in the wrong direction. Some diodes, called *Zener diodes*, are designed to do this intentionally and repeatedly. These special diodes are used to prevent voltage and current surges from affecting a circuit while allowing normal operation when surges are not present. Many different types of diodes are available; however, for medium- to high-voltage systems only the normal and Zener diodes are of significant interest.

2.8 TRANSISTORS

Transistors are circuit elements that typically have three leads: an emitter, a base, and a collector. Transistors allow current to flow from the collector to the emitter (or in some cases from the emitter to the collector) based on a current applied to the base. Transistors are typically used in low-voltage situations, although some transistors, called *power transistors* (PTs), can be used in medium-voltage environments. Transistors may also be used as switches because of their ability to turn current flow off and on and can be configured to act as amplifiers for signals passing through them. Transistors form the core of most solid-state components.

2.9   IMPLICATIONS FOR SAFETY AUDITS AND INSPECTIONS

A basic understanding of electrical components and the differences between them is necessary in order to discern the different types of components within a facility setting. After identification of the components has been achieved, auditors can then determine which regulations apply to the components. When entering a facility and conducting an electrical safety audit, several actions should be taken:

1. Ask for copies of electrical plans or electrical diagrams.
2. Ask for copies of all previous electrical safety inspections.
3. Ask for copies of all previous electrical safety assessments and testing.
4. Request a tour of the facility with the facility's electrician or someone knowledgeable and familiar with the electrical systems within the facility.

If any of these materials are missing, the facility is deficient in its recordkeeping, and these should be called out as action items. A tour of the facility is essential. During the tour, identify the electrical components inside of the building and in exterior locations. Within the building, identify electrical equipment inside designated electrical closets, along the walls, in the ceilings, and running through the conduits. Request a visual inspection of the facility's grounding rods and their placement within the earth.

REFERENCE

OSHA. (2007). Occupational Safety and Health Standards, 29 CFR, Part 1910, Subpart S, Electrical. Baltimore, MD: Occupational Safety and Health Administration.

3 Transformers

Gayle Nicoll and Martha J. Boss

CONTENTS

A transformer is a direct application of the principles of induction. A transformer consists of a primary inductor and one or more secondary inductors. In most literature, the inductors in a transformer are referred to as *windings*; hence, a transformer has one primary winding and one or more secondary windings. The primary winding is used as the input to the transformer and creates the electromagnetic field that powers the secondary windings. The purpose of a transformer is to transform the current and voltage across the primary winding into a magnitude usable by downstream devices. The relationship between a primary winding and a secondary winding is defined as either step-up or step-down, in reference to the voltage across the windings. A step-down relationship indicates that the voltage induced across the secondary winding will be less than that of the primary coil; however, the current through the secondary will be greater than through the primary. A step-up relationship indicates that the voltage across the secondary winding will be greater than across the primary winding; however, the current through the secondary will be less than through the primary. Transformers are used as the primary distribution mechanism for power in a conventional power distribution system.

3.1 TRANSFORMER CONSTRUCTION

The transformer core and windings provide a path of low resistance to the flux produced by the primary windings. In effect, this becomes the magnetic circuit of the transformer.

3.1.1 CORE

Eddy currents are induced by a changing magnetic field and may be a problem within transformers due to unintended heating effects. To reduce eddy currents, the core of a winding is constructed of laminated sheet steel. Eddy current losses vary as the square of the thickness of the laminations. The laminated material is cut from silicon–iron sheets. Silicon reduces the resistance to hysteresis and prevents increased loss with age. Laminations are stacked one upon another to construct a closed magnetic path. Alternate layers are staggered in an interlaminar core-and-gap construction so all joints do not meet at the same place. The effect is to reduce the air gap between joints, thus making the entire structure function more like a solid piece of iron.

3.1.2 WINDINGS

Primary and secondary windings comprise the current circuit of the transformer. Windings are designed to get the required number of turns into the minimum space through the core opening, which provides just enough room for insulation and cooling ducts. Transformer windings are made of copper or aluminum. Aluminum's lower conductivity requires a larger cross-section of conductor and thus a larger opening in the core. Conductors used in windings may be round, square, or rectangular in cross-section. Aluminum secondaries are often wound using sheet metal. All windings may be insulated with enamel, mica paper, Nomex® paper, silicon, glass tape, or a combination.

3.1.3 FARADAY SHIELD

Transformer windings can be provided with a grounded Faraday (or electrostatic) shield to destroy interwinding capacitance between the primary and secondary, thus reducing the danger of transferring distribution transients to utilization, and to prevent interwinding faults between layered windings. Prevention is especially critical when secondaries are isolated above ground, as primary voltage can be impressed on the secondary without detection. Using the shield, a ground fault will occur in this situation. The resulting ground current must be detected by the upstream ground-fault relaying. Faraday shields consist of a layer of nonmagnetic metal placed between primary and secondary windings, insulated from all windings, and connected solidly to ground. The shield should be made of the same material as the main windings (e.g., aluminum, copper) and physically may be a single turn of sheet metal or closely wound single layer of wire. To obtain interwinding fault protection, the shield must have the same ampacity as the grounding conductors leading to the power center. This ampacity equalization allows the shield to carry the maximum available ground-fault current (at least equivalent to one-half the cross-sectional area of the primary-winding wire). Note that the Faraday shield between the primary and secondary windings does eliminate intercapacitance; however, the shield also creates two new capacitances between the shield and the windings on either side. These new capacitances allow high-frequency currents to flow through the grounding systems of both the primary and secondary windings. Thus, the Faraday shield should be bonded to either the primary or secondary ground, which will establish a current path for the high-frequency noise.

3.2 IMPEDANCE VOLTAGE

The impedance voltage of a transformer is the voltage that must be applied across the primary winding to generate a full load current on the secondary winding when the winding has been shorted. (*Note:* Impedance voltage is usually smaller than the rated voltage of the primary.) The difference between the full rated voltage of the primary and the impedance voltage allows calculation of the maximum current a secondary would produce during a short circuit, thus making it possible to determine what kind of overcurrent protections will be needed for a circuit connected to the secondary.

To calculate the impedance of a transformer, three things must be known: (1) the full capacity of the transformer (in volt-amps, or VA), (2) the rated voltage across the secondary winding, and (3) the full load current of the secondary winding. The full load current can be calculated by dividing the capacity of the transformer by the rated voltage of the second winding. To find the impedance voltage, a variable voltage generator is connected across the transformer's primary, and an ammeter is connected in series with a short across the secondary. The voltage applied to the primary is adjusted upward until the full load current is shown on the ammeter. This voltage is the impedance voltage. The impedance of the transformer is the impedance voltage divided by the rated voltage for the primary. To calculate the maximum current that can be produced when the secondary is shorted out, divide the full load current by the impedance.

Lower limits are placed on the impedance of transformers to limit the current that can be produced during a short circuit. This is particularly important for large high-voltage transformers. In addition, the reactance of the input circuit of a transformer may be intentionally increased to limit available fault current.

3.3   THERMAL AND MAGNETIC STRESS

Transformer windings can undergo considerable thermal and magnetic stress due to the high currents passing through them. To protect a transformer, a temperature-sensing device can be placed in the transformer windings to prevent damage from overheating. This device controls a set of contacts located in the pilot circuit of the incoming distribution line. If the transformer temperature exceeds a specified limit, contacts in the pilot circuit are opened, removing the input current. An additional danger from heat is the thermal expansion of the conductors that form the windings. Although windings are usually constructed from metals that have relatively low expansion rates, adding insulated braces across the transformer's windings may be required in order to contain expansion and deformation of the winding.

Transformers are also subject to magnetic stress. This stress is the result of the minor magnetization of the windings by the established electromagnetic field, causing them to expand and contract at the same frequency as the input current. The constant expansion and contraction are the source of the "hum" heard from most large transformers. The maximum length change happens twice per cycle at the positive and negative current maximums. For 60-Hz systems, this produces a 120-Hz "hum."

Repeated expansion and contraction cause stress to the winding's conductors and insulators, which in turn can cause mechanical breakdown due to friction and the expansion action. This breakdown can cause shorts in the winding as insulators are broken down, significantly damaging the winding.

3.4   COOLING

Transformers are cooled by a wide range of technologies. Most are air cooled, using natural convection through louvers in the housing of the transformer. Others use forced air, sometimes backed by a refrigerant to provide adequate cooling. Transformer windings are designed so that heat generated by electrical power (I^2R)

losses is exposed to an adequate amount of cooling given expected loads. Effective cooling areas are inside of the winding, outside of the winding, and cooling ducts within the winding.

3.5 TRANSFORMER CONFIGURATION AND CLASSIFICATION

3.5.1 BASICS

Transformers are generally classified based on the orientation of the windings on both the primary and the secondary and how the windings are connected within the transformer. These connections determine the way the transformer will work. Single-phase transformers are classified as either *delta* or *wye*. Delta transformers have the windings of the three single-phase transformers connected in series with each other, which forms a triangle and looks like the Greek capital letter Delta (Δ). Wye transformers have the windings of the three single-phase transformers connected to a common point, called the *neutral*. This configuration looks like the capital letter Y, hence the name of the transformer. In a three-phase system, both the primary and the secondary transformers can be classified as delta or wye, leading to classifications of transformers based on both the primary and secondary configurations; for example,

- *Delta–delta* is used for large, low-voltage transformers to increase the number of turns per phase; however, the transformer is not grounded, which presents safety concerns.
- *Delta–wye* is the most popular transformer connection in the world because the secondary can be used to provide a neutral point for supplying line-to-neutral power to serve single-phase loads; also, the system can be grounded for safety.
- *Wye–wye* is generally not used, as harmonics can corrupt the current and voltage waveform.
- *Wye–delta* can be used for high-voltage lines to help protect the system from flashover and ground faults.

For specialty applications, the transformer windings can be connected in different ways. One lesser known application is the zigzag connection, which can be used as a grounding transformer or an autotransformer. In a zigzag configuration, the windings on each phase of the transformer are in two halves, and the connected halves are subsequently connected to the other phases at the neutral point.

3.5.2 CLASSIFICATION

The Institute of Electrical and Electronics Engineers (IEEE) uses three items for general transformer classification:

1. Insulation
2. Power or distribution
3. Substation or unit substation

3.5.2.1 Insulation

Transformers are classified by insulation type as wet (liquid) or dry. Dry transformers are ventilated or sealed gas-filled types. Convection cooling and air insulation within dry transformers offer the following advantages:

- Toxic gases cannot be released.
- Dry transformers cannot explode or catch fire.
- Dry transformers have no oil or other liquid to spill, leak, or dispose of.
- Dry transformers are virtually maintenance free, as they have no valves, pumps, or gauges.

For transformers used in power centers, minimum insulation ratings should be used. Basic insulation level (BIL) coordination with surge arrester characteristics must be maintained.

In liquid-immersed transformers, the insulation withstand level is not a linear function given the impulse width. Instead, the insulation withstand level decreases from front-of-wave to chopped-wave to full-wave values. The full-wave value is the BIL rating. The BIL should be compared with the discharge voltage.

With dry-type transformers, the BIL is practically constant given the width of the applied impulse. Dry transformers used in power centers should have class 220°C insulation. Note that transformers have an absolute allowable maximum temperature when operating continuously under full capacity at the rated voltage and current.

3.5.2.2 Power or Distribution

The capacity or kVA rating determines whether a transformer is classed as a power or distribution unit. Power transformers have capacities greater than 500 kVA. Distribution transformers fall in the range of 3 to 500 kVA. The primary voltage rating of a transformer is dictated by the distribution voltage. The voltage rating of the secondary winding must match the voltage of the utilization equipment. In practice, the transformer secondary is rated at a higher voltage (10% is common) than shown on the nameplate of the utilization equipment to allow for a voltage drop in the trailing cables. Secondary windings are most commonly rated at 480 or 600 V, relating to 440- or 550-V equipment, respectively. Secondaries supplying 950-V machines are usually rated at 1040 V, except for machines that cannot be remotely controlled or in states where face voltages over 1000 V are not allowed (in these instances, 995 V is often specified).

Three-winding transformers are necessary in many power centers. The capacity of each transformer winding (primary, secondary, tertiary) must be individually rated. Power-center transformers are often three phase and either (1) three single-phase units where each transformer is rated at one-third of the total required capacity, or (2) integral, three-phase types with construction allowing field replacement of failed windings. The integral three-phase transformers are preferred and have fewer exposed interconnections than three single-phase units.

Power-center transformers should be provided with voltage taps on the primary windings to account for voltage fluctuations and line losses in the distribution system. Delta primary and wye secondary for standard two-winding power transformers are preferred connections commonly used in power centers.

3.5.2.3 Substation or Unit Substation

The substation transformer classification refers to direct or overhead-line termination facilities. A unit substation has an integral connection to the primary or secondary switchgear. Primary has a voltage of 1000 V or higher; secondary has a rating of less than 1000 V. A load-center transformer can be considered a unit-substation class.

3.5.3 Grounding Transformers

A grounding transformer is used to provide a ground path to an ungrounded wye or delta connected system. These transformers provide relatively low impedance to ground, keeping the system at near ground potential; they provide ground-fault current during line-to-ground faults; and they limit the magnitude of transient overvoltages due to repeated line-to-ground strikes. Standard single-phase or integral three-phase transformers arranged in a wye–delta configuration can be used as grounding transformers in lieu of special zigzag types.

3.6 kVA RATING

The kVA rating of a transformer is based on continuous operation at the rated voltage and frequency without exceeding the specified temperature. The rising or limiting temperature depends upon core and conductor losses. Core losses remain constant with load.

3.7 SINGLE-PHASE TRANSFORMERS

Single-phase transformers are used in power centers to supply 120 V to the control circuit and 240/120 V to convenience outlets. The control circuit consists of ground monitoring, undervoltage releases, ground-fault circuitry for each associated machine circuit, and relay connections for other protection devices. Circuit breakers can be used to protect convenience-transformer primaries.

3.8 GROUNDING RESISTORS AND TRANSFORMERS

Limiting ground-fault current results in a reduction in

- Burning and melting of faulted electrical equipment
- Mechanical stress of faulted electrical equipment
- Overvoltages that might cause insulation failure

To limit the ground-fault current, a grounding resistor is inserted between the neutral of the transformer and the power center frame. The ohmic value of the grounding resistor is based on the maximum ground-fault current condition (ground fault at the secondary terminal of the transformer). For this situation, the transformer impedance is neglected.

When sizing a grounding resistor, the time rating must be considered. Normally, an insignificant amount of current flows through the resistor; however, during a worst-case situation 15 A may flow. Power is dissipated by the neutral grounding connection. Usually, the resistor would only be required to dissipate this power until a circuit breaker has time to trip. The possibility of the circuit breaker's failing to trip must be considered. Resistor ratings are based on an extended time rating, defined as 90 days of operation per year. The end of the resistor that is connected to the neutral of the transformer is required to be insulated for line-to-line voltage (Morley, 1990). Over time, the grounding can deteriorate due to weather, corrosion, or soil conditions; consequently, the transformer grounding should be routinely verified.

3.9   TRANSFORMERS AND OSHA

Signs or visible markings on the equipment or structure (transformer housing) are required to indicate the operating voltage of exposed live parts of transformer equipment. This requirement includes situations where live parts are normally protected within the transformer housing but will be exposed during maintenance and opening of the transformer housing. The polychlorinated biphenyl (PCB)-containing status of all liquid transformers should also be determined and PCB status markings placed on the transformers. Combustible material, combustible buildings and parts of buildings, fire escapes, and door and window openings must be safeguarded from fires that may originate in oil-insulated transformers attached to or adjacent to a building or combustible material.

3.10   TRANSFORMER VAULTS

The Occupational Safety and Health Administration (OSHA) requires that the following types of transformers be installed in a vault:

- Dry-type, high-fire-point, liquid-insulated, and askarel-insulated transformers installed indoors and rated over 35 kV
- Oil-insulated transformers installed indoors

OSHA requires that transformer vaults be constructed to contain fire and combustible liquids within the vault and to prevent unauthorized access. Locks and latches are to be arranged so that a vault door can be readily opened from the inside. Any pipe or duct system foreign to the electrical installation may not enter or pass through a transformer vault. Note that piping or other facilities provided for vault fire protection, or for transformer cooling, are not considered foreign to the electrical installation. Materials may not be stored in transformer vaults (OSHA, 2007).

3.11   BUSWAYS OR BUSBARS

A busway or busbar is a strip, or bar, of electrically conductive metal (copper, brass, or aluminum) that conducts electricity within a substation, switchboard, or distribution board. The power-center output requires numerous taps for feeding utilization circuits. A busway (or busbar) provides a convenient and economical means of providing these taps. The busway consists of a flat, bare conductor supported within the power center enclosure by means of insulators. Busways have either copper or aluminum conductors. Aluminum has a lower electrical conductivity and mechanical strength and quickly forms an insulating film when exposed to the atmosphere. Aluminum conductors should have electroplated contact surfaces (tin or silver). Bolting practices that accommodate aluminum's mechanical properties should be used at electrical joints.

The continuous-current rating of a busway is based on a cross-sectional area of the conductor and provides a maximum temperature rise of 65°C from ambient (40°C), as defined by American National Standards Institute (ANSI) C37.20. To allow heat to efficiently dissipate, busbars are typically either flat strips or hollow tubes to maximize surface area for heat dissipation. Busbars are part of a larger power generation or transmission system. Thus, the continuous rating of the main components (i.e., transformers, rectifiers) determines the current carried by the busbars. In most systems, a safety margin on the busbars must be used, based on a 1- to 4-second short-circuit current. The values of these currents can be calculated and can be 10 to 20 times the continuous current rating. Because these values can be so large, the busbars must be able to withstand the transitory heating effect from such a current surge (Morley, 1990).

3.12 PREDICTIVE TESTING AND INSPECTION SUSTAINABILITY AND STEWARDSHIP—TRANSFORMERS

3.12.1 Required Equipment Information

- Transformer type
- Transformer specifications
- Winding resistances
- Current transformer ratios
- Transformer impedance
- Load loss at rated voltage and current
- Current loading
- Equipment foundation drawings and data

Use ultrasonics to verify the non-existence of electrical arcing and other high-frequency events. Minimum acceptance criteria are determined based on a unit test set.

3.12.2 Required Testing Results

- Contact resistance test results
- Insulation resistance test results
- Airborne ultrasonic test results
- Power factor test results
- High-voltage test results
- Infrared thermography (IRT) test results

- Insulation oil test results
- Transformer turns ratio test results
- Temperature rise test results

3.12.3   INSULATION POWER FACTOR

Insulation power factor is applicable to transformers with at least one winding rated 5 kV or higher. (Low-voltage transformers have not been shown to benefit from insulation power factor testing.) Both the high-voltage and low-voltage windings of a transformer should be power factored. The results of this test should be compared to the factory test to confirm that no damage occurred to the unit during shipping and installation. Additionally, these results should be used for the unit's initial baseline for condition monitoring purposes. Insulation power factor is typically performed at rated voltages up to a maximum of 10 kV. For new oil- and gas-filled units the insulation power factor should not exceed 1.0%. Any values obtained over this value should be investigated and the unit not energized until a reason for the excessive reading is found. For dry type units, power factor standards have not been established due to the hydroscopic characteristics of the windings; consequently, the most useful method of evaluation is comparison with the factory test results.

3.12.3.1   Core and Turn Insulation

The condition of the core and turn insulation can be monitored using an excitation current test. This test uses a simple measurement of a single-phase current on one side, usually the high-voltage side of a transformer with the other side left floating (ungrounded). The test should be performed at the highest possible voltage level without exceeding the rating of the windings. For effective comparisons, the same voltage should be used for subsequent tests. Units with load tap changers should have readings taken at the neutral tap position and one position both higher and lower.

3.12.3.2   Phases

The best approach to the analysis is to compare the results with the factory tests or with other identical units. For three-phase units, the normal pattern is two similar high readings on the outer phases and one lower reading on the center phase. The relationship between outer and center phases should remain the same as a percentage at all tap changer positions.

3.12.4   INSULATION RESISTANCE TESTS

Insulation resistance measurements should be performed to verify that the state of dryness of the winding insulation and the core is acceptable. Insulation resistance tests can also reveal information about concealed damage to bushings that can occur during shipment and/or storage. Measurements normally are performed on transformers at rated voltages up to a maximum of 5000 volts. The temperature of the insulation system must be known when performing the test. Insulation resistance is very sensitive to temperature and varies inversely with temperature. Insulation resistance measurements must be corrected to 20°C for comparison purposes. When the test is performed, the core should be grounded and the windings under test short circuited. Those windings that are not being tested should be grounded, and all bushings should be cleaned prior to beginning the test.

The minimum resistance values depend on the voltage and capacity of the unit under test. For acceptance tests, the minimum value should be determined using the following equation:

$$R = \frac{CE}{\sqrt{kVA}}$$

where
   R = Minimum resistance corrected to 20°C.
   C = Constant (0.8 for oil transformers, 16 for dry type).
   E = Voltage rating of the winding under test.
   kVA = Rated capacity of winding under test.

The results of the test should be compared with the factory results and should be within 5% of the factory readings after correction to 20°C.

Large transformers can sometimes have long charging times due to the absorption current. When this is the case, the polarization index (PI) can be used. The PI is a ratiometric test that will help identify the condition of the insulation even if the charging currents have not diminished to zero. The test lasts for 10 minutes. The 1-minute and 10-minute insulation resistance readings are recorded. The PI is the 10-minute reading divided by the 1-minute reading. For a new transformer, the PI should be greater than 2.

3.12.5 INFRARED THERMOGRAPHY SURVEY

Perform a thermographic survey to detect uneven heating indicative of loose or dirty terminal connections, other internal electrical connections to the individual components, and internal corrosion or other flaws. Uneven heating patterns in transformer oil and windings or localized heating may be indicative of flaws in windings or insufficient ventilation of the surrounding area. Temperature variations in cooling fins or tubes may indicate internal cooling problems (e.g., loss of coolant or plugging). A bank of same-type transformers with significantly different temperature readings may indicate unbalanced loading or a defective transformer. Minimum acceptable limits for this test include the following:

- Qualitatively verify good terminal connections, internal connections to the individual components, and relative equivalent temperatures of the adjacent components during operation.
- Acceptable deviations in temperature will vary with the size and type of the electrical panel; however, these limits should not exceed ±5°C.
- The temperature range should be ≤11.3°C ambient for all of the terminal connections or internal connections to the individual components.

3.12.5.1   Transformer Turns Ratio Test

The transformer turns ratio (TTR) test measures the turns ratio of a transformer and is mainly used as an acceptance test. The TTR test can also be used as a trouble-shooting tool when other electrical tests reveal a possible problem. For acceptance tests, a TTR test is performed to identify short-circuited turns, incorrect tap settings, mislabeled terminals, and functional failures in tap changers. To perform a turns ratio test,

1. A voltage is applied to the primary, and the induced voltage on the secondary is measured.
2. The ratio is then calculated and compared to the nameplate data.

A turns ratio measurement can show that a fault exists; however, this measurement cannot determine the reason or location of the fault. Because the TTR does not give information that has value for trending and because a power factor excitation test will also show a shorted turn condition, the TTR test is not an effective predictive testing and inspection (PT&I) test for maintenance. The minimum acceptance criterion is a turns ratio that is ±2% of design specifications.

3.12.5.2   Temperature Rise Test

Verify that the transformer does not exceed temperature limits when the transformer is delivering rated kVA output at rated secondary voltage.

3.12.5.3   High-Voltage Test

Perform the high-voltage test to verify the insulation in a transformer and to ensure that no excessive leakage current is present. This is a potentially destructive test. Minimum acceptance criteria include results better than or equal to the manufacturer's specifications.

3.12.6   OTHER TESTS

Perform the following tests as appropriate for the type of breaker to verify a lack of contaminants and that the necessary inhibitors have been added (acceptance criteria are shown in parentheses):

- Water content using Karl Fisher, ASTM D1533-88 (< 25 ppm at 20°C)
- Dielectric breakdown strength test, ASTM D877 and D1816 (>30 kV)
- Acidity test, ASTM D974 (neutralization number <0.05 mg/g)
- Visualization examination, ASTM D1524 (clear)
- Dissolved gas analysis, ASTM D3612-90
  - Nitrogen (N_2) (<100 ppm)
  - Oxygen (O_2) (<10 ppm)
  - Carbon dioxide (CO_2) (<10 ppm)
  - Carbon monoxide (CO) (<100 ppm)
  - Methane (CH_4) (none)

- Ethane (C_2H_6) (none)
- Ethylene (C_2H_4) (none)
- Hydrogen (H_2) (none)
- Acetylene (C_2H_2) (none)

3.13 ELECTRICAL TRANSFORMER LOAD TAP CHANGER

Electrical transformer load tap changers (LTCs) are mechanical switching devices that are designed to change the tapping connection of a transformer winding while the transformer is energized (on-load) or not energized (off-load). This adjustment of the tap connection provides a capability to maintain downstream voltages at defined levels as a means of compensating for factors that could cause the voltages to vary. The most common example of an on-load tap changer is a voltage regulator. LTCs can be located either inside the transformer main tank or outside in their own compartment. Most LTCs are adjusted manually, and some have remote or automatic actuators. LTCs contain such components as spring drive mechanisms, motor actuators, moving arms, transition resistors/reactors, oil tanks, and monitoring/control interface equipment. LTCs cause more failures and outages than any other component of a power transformer. When occurring early in life, LTC failures are usually mechanical or related to a defect of the moving mechanisms or contact surfaces and connections. After continued use, LTC failures are usually electrical faults associated with carbon buildup due to arcing across contacts or from mechanical failure due to component load stresses. Other failures may also be associated with transition resistors and insulation. Another common term for LTCs is *automatic voltage regulators.*

3.13.1 REQUIRED EQUIPMENT INFORMATION

- Electrical transformer load tap changer type (NEMA enclosure type)
- Electrical transformer load tap changer specifications
- Step-down voltage configuration (number of positions)
- Maximum current loading
- Maximum tapping range (kV)
- Insulation level (to ground and phase-to-phase)
- Arcing time
- Dimensions, weight
- Oil capacity
- UL certification
- Electromagnetic interference (EMI) levels, if applicable

3.13.2 REQUIRED TESTING RESULTS

- Contact resistance test results
- Insulation resistance test results
- Airborne ultrasonic test results
- Power factor test results

- High-voltage test results
- Infrared thermography (IRT) test results
- Insulation oil test results
- Transformer turns ratio test results
- Temperature rise test results

3.13.2.1   Infrared Thermography Survey

Perform a thermographic survey to detect uneven heating indicative of loose or dirty terminal connections or other internal electrical connections to the individual components. This survey may also indicate internal corrosion or other flaws. Localized heating may be indicative of flaws in the windings or insufficient ventilation of the surrounding area. Uneven heating patterns may be present in the oil and windings of transformers and oil breakers that may be integral to LTCs. A bank of same-type LTCs with significantly different temperature readings may indicate unbalanced loading or a defective LTC. LTC loading should be >50% for performing this test. Minimum acceptable limits for this test are as follows:

- Good terminal connections, internal connections to the individual components, and relative equivalent temperatures of the adjacent components during operation should be qualitatively verified.
- Acceptable deviations in temperature will vary with the size and type of the electrical panel; however, these limits should not exceed ±5°C.
- The temperature range should be ≤11.3°C ambient for all of the terminal connections or internal connections to the individual components.

Use airborne ultrasonic testing to verify the non-existence of electrical arcing and other high-frequency events.

3.13.2.2   Electrical Acceptance Tests
- Perform electrical continuity tests on current, potential, and control circuits.
- Perform ratio and polarity tests on current and potential transformers, if applicable.
- Perform a current test on the remainder of the secondary circuit to detect any open or short-circuit connections (NASA, 2004).

3.14   IMPLICATIONS FOR SAFETY AUDITS AND INSPECTIONS

When conducting an electrical safety inspection, the following questions should be asked:

- *Who owns and maintains the transformers?* The transformers may be owned by the facility or by the electrical supply company. If the facility is responsible for maintaining the transformers, routine inspections should be documented. Review copies of all maintenance records.
- *Where is the transformer ground?* Transformers should have separate grounds to earth and should be verified on a periodic basis.

- *Is appropriate signage visible?* High-voltage areas must be conspicuously marked, and access to these areas should be controlled. Only designated and authorized individuals should have access to these areas. Control may be by fence, lock and key, or other barriers preventing the general public from entering the area.
- *Are arc-flash radii posted?* Because of the large voltage associated with most transformers, the transformers must be labeled with warning signs and approach boundaries for arc-flash hazards.
- *Is shielding necessary or adequate?* If shielding is necessary, the shielding must be periodically tested and inspected to verify the integrity of the shielding. Request copies of all inspection reports and testing.

REFERENCES

Morley, L.A. (1990). *Mine Power Systems*, Information Circular 9258, NTIS No. PB 91-241729. Pittsburgh, PA: U.S. Department of the Interior, Bureau of Mines.

NASA. (2004). *Reliability Centered Building and Equipment Acceptance Guide*. Washington, DC: National Aeronautics and Space Administration.

NFPA. (2005). *NFPA 70: National Electrical Code (NEC) Handbook*. Quincy, MA: National Fire Protection Association.

OSHA. (2007). Occupational Safety and Health Standards, 29 CFR, Part 1910, Subpart S, Electrical. Baltimore, MD: Occupational Safety and Health Administration.

4 Generators

Gayle Nicoll and Martha J. Boss

CONTENTS

Generators use primary inductors and an input voltage from an "exciter" to create output current in a number of secondary inductors.

4.1 GENERATOR FUNCTION

Most generators use induction to produce an output voltage. A typical generator is composed of the following:

- Power source, usually a type of exciter and often direct current (DC); generally, the more current supplied, the stronger the induced electromagnetic field around the rotor windings
  - Rotor (components rotate as single unit)
  - Carbon brushes and copper/steel slip rings that transfer power to a copper conductor
  - Copper conductor that receives power via the slip rings and is wound around the rotor core
- Rotor (field/permanent) windings and core (aforementioned copper conductor is wound to encircle the core)
- Stator (components remain stationary)
  - Stator core that encircles the rotor and holds the stator windings in proper orientation to the rotor (circumference orientation is designed to produce induced current with a sine wave characteristic)
  - Stator windings, where the copper conductor is wound to encircle a portion of the stator core ring

In a typical generator, the input voltage/electric field is supplied by the rotor, while the output voltage is induced across the stator windings. The stator windings are wound around the rotor windings, with some air gap in between. This allows the stator windings to be placed in the optimal orientation to catch the electric field generated by the rotor windings.

Typically, the electromagnetic field generated by a rotor winding is created by the application of a DC voltage source. If the rotor were to stay still, the voltage and current generated by the stator windings would also remain constant. In this configuration, the generator would be, in essence, a transformer; however, most generators are used to generate alternating current (AC) and voltages, which are far more efficient to transmit over long distances.

4.1.1 ALTERNATING CURRENT GENERATORS

To generate alternating current, the rotor is set in motion. This motion has the effect of bringing the electric field generated by the rotor windings into, and out of, alignment with the stator windings at a constant speed. As the windings come into alignment, the voltage and current produced by a stator winding increase. As the windings come out of alignment, the voltage and current produced by the stator decrease. The stator and rotor windings come into alignment twice in each rotation of the rotor. Because the electromagnetic field generated by the rotor is a circle, the flow of the field will actually change between the alignments (on the way in and on the way out). This means that the voltages generated by the electric field will change from positive to negative and back again over the course of a rotation. Because the rotation is done in a circle, the output voltage has the look of a sinusoid. The amplitude of the sinusoid is determined by the ratio of the windings in the stator and rotor.

4.1.1.1 Slip Rings

Because the rotor is in motion, maintaining voltage to the winding of the rotator can be difficult. Static wires would get instantly tangled. To address this, generators use a construct called a *slip ring* to provide voltage to the rotor. In a slip ring, two concentric rings are placed around the shaft of the rotor at a short distance. Brushes attached to the rotor leads touch one of the concentric rings, creating a complete circuit through the rotor. As the rotor spins, the brushes maintain contact with the rings, providing constant current and voltage.

4.1.1.2 Generator Output Frequency

The frequency of the alternating current produced by a generator is determined by rotor revolution speed (revolutions per minute, or RPM). Frequencies are denoted in hertz (Hz), which is the number of complete cycles, positive to negative and back again, that the current makes in 1 second. Frequency is not varied during power production, as the downstream power receivers depend on consistency. In the United States, 60 Hz is common. To achieve a 60-Hz signal, the rotor must spin at 3600 rotations per minute (60 cycles per second × 60 seconds per minute).

Varying the rotation speed of the rotor will affect the overall frequency, as will adding additional windings (usually offset by 90°) to the rotor. A typical rotor will have one or two windings, with opposite orientations. Offsetting the windings by 90° makes them perpendicular to one another, so their electric fields will not interact substantially. The opposite orientation provides a "positive–negative–positive–negative" result on the stator for each revolution of the rotor, in effect doubling the frequency of the output while maintaining the magnitude.

4.1.1.3 Phases

Typical generators have more than one set of stator windings. Generators usually have at least three sets of stator windings, but they can have more. The angle separating the stator windings determines the phase of the output voltage. The phase of an output sinusoid is determined by the difference in time between when two different sinusoids of the same frequency reach their maximum voltage. If two stator windings are offset by 60° from each other, the power generated will be 60° out of phase. In a typical three-stator generator, the stator windings are separated by 120°, producing output voltages at three distinct phases. This is the basis for three-phase power, as seen in commercial applications. The advantage of three-phase power is that, using the three phases as a combined output, the highest potential across all three phases can be matched against the lowest potential across all three phases, providing an overall increase in the voltage potential applied to any circuit. Each phase may also be used separately, allowing a single three-phase power connector to power three complete single-phase circuits, which is what most consumer electronics use.

4.1.2 Direct Current Generators

Direct current generators operate on the same principle as alternating current (AC) generators, with one substantial change. Instead of slip rings, a DC generator uses a commutator to provide power to the rotor. Although the rings in a slip ring are unbroken, the rings in a commutator are split into segments that are electrically isolated from one another. This segmentation allows the polarity of the voltages applied to the inductors on a rotor to shift during each rotation, ensuring that only positive voltages are induced on the stators of the generator.

Stators are placed at small intervals around the housing of the generator and are typically grouped. The maximum output voltage from each stator in a group is used as the output voltage for that group. This grouping allows a set of stators to be used to minimize the fluctuations caused by the movement of the rotor's electromagnetic field across many stators, producing a more stable voltage. Some DC generators use other methods to generate a stable current from the rectified waveform generated through the use of a commutator.

4.2 EXCITERS

Exciters are used to provide the current to the inductors on the rotor of a generator. For proper operation, the current to the rotor inductors of a generator must be direct current. If a generator uses an AC exciter, the input current must be passed through a waveform rectifier and a DC voltage regulator in order to produce a stable DC input. The actual excitation current applied to a rotor's inductors may vary depending on the load on the generator. Increased generator loads require increased input current to compensate. The voltage applied to the rotor of a generator during normal operation is called the *terminal voltage*. The terminal voltage is not, however, the same as the peak voltage that may be applied to the rotor's inductors. Similarly, the current supplied to the inductors of a rotor during normal operation is called the *field amperage*.

Most generators have a maximum field amperage rating and maximum stator amperage rating. If the current applied to the rotor or induced in the stator exceeds these maximums, damage to the generator will result due to excessive heat generation. Because current increases from no-load to maximum-load values during normal generation, the amperage readings must be monitored at all times.

4.2.1 Internal (Self-) Exciters

In an internal excitation system, the excitation voltage and current for a generator's rotor are provided by one or more of the generator's stators. Stator voltages and current are run through a step-down transformer, a waveform rectifier (for AC generators), and a voltage regulator. The regulated and stepped-down voltages are then applied to the inductors of the rotor, allowing the system to power itself once activated.

4.2.2 EXTERNAL (SEPARATE) EXCITERS

External exciters use voltage and current produced by a separate system to power the rotor of a generator. The source may be a battery, another generator, or an electromechanical system. Appropriate electrical filters must be applied to input voltages and current to maintain a steady flow at the operational rating of the generator. External exciters that are designed to power an internally excited system until that system reaches full operation are called *pilot exciters*.

4.3 ALTERNATIVE GENERATOR TYPES

The generators described up until this point have been traditional electric generators and have used common inductance principles based on a supplied electric field; however, several other types of generators exist.

4.3.1 PERMANENT MAGNET GENERATORS

Some generators use permanent magnets to generate an electromagnetic field to feed the stators. Instead of an exciter-driven inductor, a permanent magnet is attached to the shaft of a rotor. The alternating poles of the permanent magnet provide the same alternating induction effect as an electrically driven inductor. Stators are attached around the housing of the generator as with the electrically driven generators. Permanent magnet generators are used primarily for converting mechanical energy sources into electricity. For example, wind turbines, steam turbines, gas turbines, and hydroturbines use their respective power sources to turn the rotor of a permanent magnet generator to generate electricity through the stators. These generators typically have maximum stator amperage ratings and do not have field ampere ratings.

4.4 POWER

Power in electrical systems provides a measure of how much work the system is capable of performing. In direct current circuits, power is measured in watts (volts × amperes) and is simply measured by multiplying the voltage applied to a system by the current flowing through the system. For AC circuits, three distinct types of power must be considered, each of which has certain characteristics.

4.4.1 REACTIVE AND TRUE POWER

Reactive power is the result of voltage and amperage in a circuit becoming out of phase. This can happen in AC circuits when a significant inductance or capacitance is present in the system. Reactive power is measured in VARs (volt-amperes reactive). Capacitive circuits tend to cause current to lead the voltage in phase, whereas inductive circuits tend toward the opposite. Many texts refer to reactive power as not doing any actual work in a system; rather, it simply offsets the inductance and capacitance of the system. This is not entirely accurate, though. Reactive power represents the power in the system that is temporarily stored in a form of magnetic or electric fields due to inductive or capacitive elements in the system. In some ways, reactive power also represents the cost of energizing a system so the system can function.

4.4.2 REAL POWER

Real power is the portion of the total power applied to the system that is affected purely by resistive loads and is therefore where the voltage and current are completely in phase. Real power is always a positive value. Real power is measured in watts.

4.4.3 APPARENT POWER IN AC CIRCUITS

Apparent power is a measure of the overall magnitude of power consumed by a system and is often described as the product of the root-mean squares of the voltage and current in a system. If real power in a system is denoted as P and the reactive power as Q, then the apparent power is calculated as follows:

$$\text{Apparent power} = \sqrt{P^2 + Q^2}$$

Apparent power is given in volt-amperes (VA).

4.4.3.1 Power Factor

The power factor is the ratio of real power to apparent power. A power factor of one is defined as unity and is the ultimate goal that will never be attained. A unity power factor indicates that all power was available to do actual work. In some cases, the power factor is stated as the phase difference (in degrees) between the voltage and current. To obtain a number power factor, take the cosine of the phase difference. For example, if the phase difference between voltage and current is 32°, then the power factor of the circuit is cos(32°) = 0.848.

4.4.3.2 Generator Power Ratings (kVA and kW)

Generator power ratings are given in kVA, which represents the maximum apparent power generated by the system. For generators that produce multiple phases of power, each individual phase has an associated corrective factor. The total kVA rating of the generator is the sum of the kVA rating of each phase multiplied by a correction factor.

4.5 HEAT DISSIPATION

Power generation also generates heat. The large amounts of current flowing through the stator and rotor conductors generate heat due to the resistance of the conductors. Additional heat may come from the friction overcome by the turning of the rotor shaft, the passage of brushes over the commutator or slip rings, or external environmental factors. The resistance of a conductor increases as the conductor is heated, making a generator less efficient as the temperature of the generator increases. In extreme cases, excess heat will cause thermal breakdown of the insulating material around the conductor or melt the conductor itself. Either of these cases will cause significant damage to the generator.

To adequately cool generators, sufficient air flow must be provided. In some cases (e.g., large power plant generators), air cooling is insufficient to dissipate the heat loads generated, thus requiring more elaborate cooling mechanisms (e.g., water cooling, gas cooling). In large kVA-rated generators, hydrogen gas cooling is the most common form of cooling used.

4.5.1 HYDROGEN COOLING

Despite the volatile nature of hydrogen when combined with oxygen, hydrogen gas has many desirable properties for heat dissipation. First, hydrogen gas has a high specific heat and a high thermal conductivity. These properties make hydrogen significantly better than air (which is mostly nitrogen) at absorbing and conducting heat. Second, hydrogen has no natural corrosive properties, leading to less wear on generator parts. Third, hydrogen gas is easily generated through a number of different means and is an inexpensive material to use and replace.

In hydrogen cooling, the generator core is sealed and filled with pure hydrogen gas. A pump system moves the hydrogen gas into and out of the sealed core in order to move heat out of the generator and into dissipation units. Due to the requirements for a sealed environment, hydrogen cooling is generally used only for the rotor, as the rotor can be completely encapsulated in the gas. Stator units are typically cooled using standard water coolers. Additional gas-to-water heat exchangers cool the hydrogen gas for reinjection into the core.

4.5.1.1 Hydrogen Cooling Safety

Hydrogen gas systems must be continuously monitored at both the inlet and outlet of the generator core. Care must be taken to ensure that the hydrogen gas is maintained in an overpressure state to prevent the seepage of oxygen into the generator core, which can cause an explosion in a confined space. In addition, temperature sensors must be used to ensure that the temperatures of the inbound and outbound gas do not exceed safety limits. Moisture sensors must be used to detect the presence of water in the system, which can be a critical safety concern. Care must be taken to keep the hydrogen gas pure and free of contaminants. The presence of water in the gas is to be avoided due to the risks of arcing and corrosion.

4.5.2 WATER COOLING

Water cooling uses the passage of water over a heated area to transfer the heat from the area to be cooled to heat dissipation units. Many generator systems use gas-to-water cooling, which passes a gas with a high specific heat through the generator core (air, hydrogen, or helium) and then uses a heat exchanger to pass the heat from the gas into a water cooling system. In larger generators, the stators require significant cooling, leading to the use of water cooling on the stator bars or housings.

Water cooling is generally achieved by attaching a hollow metal block or plate to the area to be cooled, usually attached by a high thermal conductivity paste and a clamp or bolt to hold the block in place. Demineralized water is passed through the block, allowing the water to absorb heat from the block and transfer this heat downstream to a heat exchanger unit. Care must be taken to keep the water as pure as possible to limit the electrical conductance of the water. Conductance of water is typically measured in mhos (the opposite of "ohms"), and the lower the mho rating the better.

4.5.3 Heat Pipes

Heat pipes are used extensively in areas that are difficult to reach with conventional cooling methods and where air flow is restricted. A heat pipe is a closed system, usually a metal block attached to a long copper (or other high heat-conductivity metal) tube. The tube is partially filled with water. The block is attached to the area to be cooled, usually by a high conductivity thermal paste. As heat passes into the block, the water in the tube begins to boil and turn into a gas. The gas then rises through the tube to the far end, where the tube can be cooled by conventional means (e.g., air cooling, water cooling). As the gaseous water is cooled, this gaseous water returns to a liquid state and runs back down the tube toward the block. This has the effect of transferring the heat along the length of the tube to an area that is more readily cooled. Tubes can be coated in a heat-resistant enamel to prevent electrical conductivity. Heat pipes can be used to cool phase conductors, as the conductors are often in areas of the generator housing that are difficult to reach with conventional cooling.

4.6 IMPLICATIONS FOR SAFETY AUDITS AND INSPECTIONS

When conducting an electrical safety inspection, consider the following points:

1. *How frequently are the generators being inspected?* Generators should be on a routine inspection schedule. Request copies of inspections and verify that all safety points, as discussed above, are being addressed as part of the inspection.
2. *Are the generators grounded?* All dedicated generators should be individually and directly grounded to earth. The efficacy of this ground should be verified on a routine schedule.
3. *Is the generator a backup power source?* If the generator is to serve as a backup power source, the generator should be tested and inspected on a regular schedule. The electrical switch-over system should also be tested and inspected on a routine basis.
4. *Is the area around the generator clean?* The area around a generator should be kept clear of any combustible material that could potentially fuel a fire.
5. *Is the area where the generator is housed suitable?* The entire area around the generator should be inspected to verify that the area is used as a dedicated electrical supply area. Determine whether or not the fire suppression

system is water based. Fire suppression in an electrical-supply area should not be water based. The area should also not have potential sources of water, such as water storage tanks. These tanks can rupture, which would cause a significant hazard in the area of the electrical systems.

REFERENCES

Morley, L.A. (1990). *Mine Power Systems*, Information Circular 9258, NTIS No. PB 91-241729. Pittsburgh, PA: U.S. Department of the Interior, Bureau of Mines.

NFPA. (2005). *NFPA 70: National Electrical Code (NEC) Handbook*. Quincy, MA: National Fire Protection Association.

OSHA. (2007). Occupational Safety and Health Standards, 29 CFR, Part 1910, Subpart S, Electrical. Baltimore, MD: Occupational Safety and Health Administration.

5 Circuit Breakers

Gayle Nicoll and Martha J. Boss

CONTENTS

The circuit breaker is primarily an interrupting device. Circuit breakers are designed to open and close a circuit automatically at a specific current level (without causing circuit breaker internal damage). Breakers are available as single pole, double pole, or triple pole.

5.1   CIRCUIT BREAKER CONSTRUCTION

A typical electromechanical circuit breaker is an electromechanical switch designed to actuate when the current passing through the breaker exceeds or drops below a threshold amount. The circuit breaker contains a spring-loaded switch that is locked into place by an armature. The locked switch connects two contacts that complete the circuit. When the input current exceeds a specified amount, the electromagnet is energized to allow the locking armature to release the spring-loaded switch. When the switch is released, the circuit is opened as the contacts are separated. Most circuit breakers have a manual operation switch to allow manual opening and closing of the contacts within the breaker. Manual operation may be mechanically or electrically actuated. Use of manual circuit breakers is only intended where circuit current is not in excess of the rated continuous current.

5.1.1   TERMINAL CONNECTORS

Terminal connectors are for direct connection of one cable connector per terminal. The basic breaker component function is to connect the circuit breaker to a desired power source and load. Connectors are usually made of copper and must be constructed so that each conductor can be tightened without removing another. Many molded-case breakers have provisions for studs and can be used for connecting more than one conductor per terminal and breaker mounting. The type of terminal used on a breaker may change the heat dissipation properties of the breaker, thus lowering the breaker's interrupting rating.

5.1.2 TERMINAL SHIELDS

Terminal shields are an optional part of circuit breakers and do not come standard on most units. Terminal shields are applied to the line or load side of the breaker and protect workers from accidental contact with the energized terminal contacts.

5.1.3 DISCONNECT SWITCH

Manufacturers may incorporate a disconnect switch in their circuit breakers. The operating mechanism for the switch is mechanically interlocked with the circuit breaker mechanism. If the switch is opened when the breaker is closed, the interlock trips the circuit breaker prior to switch-contact parting.

5.1.4 MOLDED-CASE CIRCUIT BREAKERS

Molded-case circuit breakers are interrupting devices with self-contained, current-responsive elements. These breakers are assembled as an integral unit in a supported and enclosed housing of insulating material. Depending on the amount of protection required, these breakers can sense internally and then clear undervoltage, overcurrent, and short-circuit conditions. Some tripping elements are also externally accessible through control wiring, and other circuit protection can be added. Breakers may rely solely on outside information to perform their prime function. Several trip units may be available for a particular frame size. A specific assembled breaker may have a lower continuous-current rating than the current designation of the frame manufacturers. Arc chutes, extinguishers, and quenchers define the interrupting-current capacity of the assembly, in conjunction with the insulating and heat-dissipation properties of the molded-case chutes.

Contacts often begin to part during the first cycle of a fault. The current breaker must be capable of interrupting the maximum allowable first-cycle asymmetrical current. The operating mechanism provides a means of opening and closing the toggle mechanism of quick-make/quick-break type contacts (snap open or closed independent of the speed of handle movement). The breaker is trip free and cannot be prevented from tripping by holding the breaker handle in the "on" position during a fault condition.

Molded-case circuit breaker use is restricted to low-voltage and medium-voltage systems. With alternating current (AC), this breaker provides high interrupting capacity for short circuits in minimum space. On AC or direct current (DC) systems, this breaker is often the first protection device to handle electrical problems. For low-voltage and medium-voltage circuit breakers, interrupting and close-and-latch ratings are usually the same.

5.1.5 TRIP ELEMENTS

Trip elements trip the operating mechanism of a circuit breaker during either a prolonged overload or a short-circuit current. Some circuit breakers have a screwdriver slot located on the front of the trip unit used for adjusting sensitivity. The maximum setting is established by protection of the minimum conductor size in the circuit.

5.1.5.1 Instantaneous Magnetic Trip

Magnetic trips work by using an electromagnet in series with the load current. When the current reaches the set point, the electromagnet instantaneously trips. This type of trip is commonly found in low-voltage breakers (e.g., household circuit breakers).

5.1.5.2 Thermal Trip

Considered the industry standard, these trip elements work using a bimetal heated by the load current. When overheated, indicating an overload, the bimetal will deflect, which causes the operating mechanism to trip.

5.1.5.3 Electronic Trip

Current transformers and solid-state circuits are used to monitor the current. When an overload or short circuit is detected, the monitors initiate a trip. Electronic trip elements can include trip features and flexibility not present in other types of trip elements (e.g., adjustable pickup, time delays, instantaneous pickup, selective interlocking). Solid-state components have replaced electromechanical-magnetic and thermal-magnetic trip elements in some molded-case breakers.

5.1.5.4 Thermal-Magnetic Trip

A thermal-magnetic trip, in addition to providing short-circuit protection, guards against long-term current overloads existing longer than roughly 10 seconds. Because bimetal deflection is dependent on current and time, the thermal unit provides long-time delay for light overloads and fast response for heavy overloads. The thermal-magnetic unit may be ambient-temperature sensitive (breaker trips at a lower current as ambient temperature rises). The National Electrical Code (NEC) defines the current at which a long-time-delay thermal element must initiate circuit-clearing operation. The NEC specifies a point that is 125% of the rated equipment or conductor ampacity. The circuit breaker will take no action below this current. For conventional (noncompensating) thermal-magnetic elements, the thermal portion defines the continuous current rating of the breaker (specified as 100% at 40°C). The thermal element current rating cannot exceed the frame rating. Manufacturers recommend that continuous current through the breaker be limited to 80% of the frame size.

5.1.5.5 Shunt Trips

A shunt trip relay completes the circuit between the control-power source and the solenoid coil. Shunt trips are used to trip a circuit breaker electrically from a remote location and consist of a momentary-rated solenoid tripping device mounted inside a molded case. The shunt trip can remotely trip the breaker but cannot remotely reclose the breaker. To reclose the breaker, the breaker handle must first be moved to the reset position and then to the "on" position.

5.1.6 UNDERVOLTAGE RELAYS IN CIRCUIT BREAKERS

A molded-case breaker alone in an outgoing circuit can provide overload and short-circuit protection, where the undervoltage relay (UVR) adds undervoltage protection. Undervoltage protection is actually *loss-of-voltage* protection, as the dropout level is well outside the recommended operating range of most machinery. UVRs work using a solenoid wired to the line side of the breaker. When incoming power is present, the current passing through the solenoid retracts the switch. When line power is too low or absent, then the solenoid releases and strikes a mechanical trigger, tripping the breaker. Additional types of protection can be applied using shunt trip and UVR combinations. When a UVR is used, the relay removes the control voltage across the solenoid coil. The UVR trip breaker is for use whenever control voltage to the UVR falls below a predetermined level (usually 35 to 70%). A spring is held in the cocked position by a solenoid after closure. If the voltage drops below the required level, the solenoid releases the spring, causing the circuit breaker to trip. The breaker cannot be turned on again until the voltage returns to 80% of normal.

5.2 ARC SUPPRESSION IN HIGH-VOLTAGE CIRCUIT BREAKERS

When the contacts in a high-voltage breaker are separated, a very-high-voltage arc between the contacts is likely as these contacts are pulled apart. Suppressing this arc is extremely important to prevent damage to the breaker. High-voltage arcs can cause the metal of the breaker contacts to melt, deforming the contacts. In addition, the arc generates a large amount of heat, which must be dissipated to prevent damage to the components of the breaker. Several strategies are available for arc suppression.

5.2.1 OIL-BASED SUPPRESSION

Oil-based suppression systems fill the cavity containing the contacts with a nonconductive mineral oil that possesses a high specific heat. The oil absorbs the heat of the arc flash while the non-conductive oil flowing between the open contacts limits the duration of the arc flash.

5.2.2 SULFUR HEXAFLUORIDE SUPPRESSION

Sulfur hexafluoride (SF_6) suppression systems trigger a blast of SF_6 gas as the contacts open. This gas serves as a mechanism to interrupt the arc, and the high specific heat of the gas allows the SF_6 gas to absorb the heat generated by the arc. Some SF_6 systems capture initial portions of the arc energy to power the injection mechanism.

5.2.3 AIR-BLAST SUPPRESSION

Air-blast suppression uses a jet of air to rapidly cool contacts that are heated up by the arc and to disrupt formation of the arc. The suppression capacity of an air-blast suppressor is determined by the air flow through the contact cavity per time. Some air-blast suppressors have special arcing contacts that will draw the arc to them once the blast has been applied.

5.2.4 Vacuum Suppression

Vacuum suppression is typically used only in medium- and low-voltage situations. In vacuum suppression, no gas is present in the breaker cylinder between the electrical contacts; this vacuum essentially removes the medium for electricity to be conducted. Vacuum has a high dielectric strength, allowing arcs to be suppressed in a very small distance. The electrical contacts in a vacuum suppression system are intended to be ionized and heated rapidly. The arc will continue until the next zero point in the alternating current, at which point the arc will extinguish and the vacuum will prevent the arc from being reestablished. When the arc is extinguished, the heated, ionized metal will collect and cool on the contact. Special contacts are used to assist in this heating/cooling process.

5.2.5 Magnetic Suppression

Magnetic suppression uses a fixed magnet or magnets in the contact chamber to establish a magnetic field between the contacts. This magnetic field diverts the flow of electrons in an arc, causing the path of the arc to be substantially longer and requiring more voltage to sustain the arc. The intent of this method is to extinguish the arc before substantial damage is done to the contacts. Air remains in the contact chamber to increase the dielectric strength of the gap between the contacts.

5.2.6 Dead vs. Live Tank Systems

Some suppression systems refer to "dead" and "live" tank systems. The difference between these two is the grounding state of the breaker. In a dead system, the breaker is grounded, whereas in a live system the breaker is electrically isolated from the ground. Incoming and outgoing conductors in a dead tank system are electrically isolated from the tank to prevent accidental grounding.

5.3 CIRCUIT RECLOSERS

Reclosers can eliminate prolonged outages of a distribution system due to temporary faults or transient overvoltage conditions. Circuit reclosers are circuit breakers that detect line overcapacity as well as timing and interrupt overcurrent. The circuit reclosers can reclose automatically and reenergize the line. Either hydraulic or electronically controlled reclosers may be used. Oil circuit reclosers are reliable interrupters on the transformer secondary at distribution voltages up to 15 kV. If line overcurrent is permanent, after a preset number of operations the recloser will lock open and isolate the failure.

5.4 CIRCUIT BREAKERS AND OSHA

According to 29 CFR 1910 (Occupational Safety and Health Standards):

- Handles or levers of circuit breakers and similar parts that may move suddenly must be guarded or isolated to prevent worker injuries.
- Circuit breakers must clearly indicate whether they are in the open (off) or closed (on) position.
- Where circuit breaker handles on switchboards are operated vertically (rather than horizontally or rotationally), the up position of the handle must be the closed (on) position.

Further, according to 29 CFR 1926 (Safety and Health Regulations for Construction),

- A means must be provided to completely isolate equipment for inspection and repairs. Isolating means which are not designed to interrupt the load current of the circuit shall be either interlocked with an approved circuit interrupter or provided with a sign warning against opening them under load.

5.4.1 CIRCUIT BREAKER INSTALLATIONS OVER 600 VOLTS, NOMINAL

The Safety and Health Regulations for Construction state that, when installed indoors, breakers for systems with an operational voltage over 600 V must be either metal-enclosed units or fire-resistant cell-mounted units. Circuit breakers must be located or shielded such that employees will not be burned or otherwise injured by their operation. An exception to this rule is that open mounting of circuit breakers is permitted in locations accessible only to qualified personnel. Levers, handles, or other exposed moving parts of a circuit breaker must be guarded or isolated in order to limit the likelihood of personnel injury when the breaker actuates.

5.4.2 MEDIUM-VOLTAGE AND LOW-VOLTAGE CIRCUIT BREAKERS (AC AND DC)

Medium- and low-voltage circuit breakers must be internally controlled by self-contained current-responsive elements, external protective relays, or a combination of both. All Underwriters Laboratories (UL), National Electrical Manufacturers Association (NEMA), and American National Standards Institute (ANSI) standards apply to molded-case circuit breakers and low-voltage power circuit breakers.

5.4.3 CIRCUIT BREAKERS IN THREE-PHASE CIRCUITS

Circuit breakers used for overcurrent protection of three-phase circuits must have a minimum of three overcurrent relays operated from three current transformers. On three-phase, three-wire circuits, an overcurrent relay in the residual circuit of the current transformers may replace one of the phase relays. An overcurrent relay, operated from a current transformer that links all phases of a three-phase, three-wire circuit, may replace the residual relay and one other phase-conductor current transformer. Where the neutral is not grounded on the load side of the circuit, the current transformer may link all three phase conductors and the grounded circuit conductor (neutral) (OSHA, 2007).

5.4.4 RATINGS AND USAGE

Circuit breakers used as switches in 120-V and 277-V fluorescent lighting circuits must be listed and marked "SWD." Circuit breakers with a straight voltage rating (e.g., 240 V or 480 V) may only be installed in a circuit in which the nominal voltage between any two conductors does not exceed the circuit breaker's voltage rating. A two-pole circuit breaker may not be used for protecting a three-phase, corner-grounded delta circuit unless the circuit breaker is marked "1Ω–3Ω" to indicate such suitability. A circuit breaker with a slash rating (e.g., 120/240 V or 480Y/277 V) may only be installed in a circuit where (1) the nominal voltage of any conductor to ground does not exceed the lower of the two values of the circuit breaker's voltage rating, and (2) the nominal voltage between any two conductors does not exceed the higher value of the circuit breaker's voltage rating.

5.5 LOW-VOLTAGE POWER CIRCUIT BREAKERS

Molded-case circuit breakers cannot handle available short-circuit currents in certain low-voltage applications. Low-voltage power circuit breakers provide an alternative for applications less than or equal to 1000 V. For low-voltage and medium-voltage circuit breakers, interrupting and close-and-latch ratings are usually the same. Electromechanical and mechanical-displacement dashpot units are available for long-time delay tripping. Dashpots allow the long-time-delay pickup current and operation time to be changed. This quality extends the capabilities of the power circuit breaker by providing short-circuit and overload tripping adjustments, allowing a broader range of applications.

5.6 HIGH-VOLTAGE CIRCUIT BREAKERS

Power circuit breakers used in high-voltage applications include air–magnetic (function the same as for low-voltage circuit breakers), oil, minimum-oil, and vacuum circuit breaker (VCB) types. VCBs are the most popular because of their small size and high efficiency. Interrupting and close-and-latch ratings are very important high-voltage parameters. High-voltage circuit breakers rarely terminate current flow until a few cycles after the first cycle peak. The close-and-latch rating must be higher than the interrupting rating. High-voltage (AC only) breakers must contain a sensing device to perform their intended function. High-voltage sensing devices have interconnections *only* through the control wiring. These high-voltage circuit breakers can create transients. *Note:* Power transients are transients on the power system being monitored and transients on the trip circuit control power.

5.7 USE OF MOLDED-CASE CIRCUIT BREAKERS

Molded-case circuit breakers are used to protect AC utilization equipment and associated cables. The power-center circuit breaker compartment must be designed for easy access and protection from mechanical damage or untoward intrusion (accidental exposure to energized terminals and conductors). Dead-front panels are the preferred method. Dead-front panels are installed on most circuit breaker panelboard assemblies. If the panelboard is open, the work on or near this panelboard is considered "live work" by OSHA, because the area around each circuit breaker is not completely sealed. Consequently, arc-flash radii for open panelboards should be based on both the circuit breaker's potential to arc and the potential arc created if the bus is contacted via insertion of a tool into the area that surrounds each circuit breaker housing. To avoid live work and heightened arc-flash radii restrictions, be sure that all panelboards originally equipped with dead-front panels are now so equipped and that the front panels are not damaged or inadvertently left open.

5.7.1 Selection of a Molded-Case Circuit Breaker

Selection of molded-case circuit breakers is based on voltage, frequency, interrupting capacity, continuous-current rating, and trip settings. Molded-case circuit breakers often begin interrupting short-circuit currents during the first cycle after a fault. These circuit breakers must be selected on the basis of the maximum first-cycle asymmetrical fault current. Breakers are usually rated on a symmetrical current basis, which eliminates applying DC offset multipliers. The continuous-current rating for a molded-case circuit breaker is actually the rating of the thermal trip element, which can be less than the breaker frame size (thermal-magnetic breaker). Magnetic-only devices have a continuous-current rating equal to the frame size.

5.7.2 Short-Circuit and Overload Protection

Two kinds of protection can be provided directly by molded-case circuit breakers: short-circuit and overload. Short-circuit protection is required on all outgoing power (ungrounded) conductors. The maximum allowable instantaneous-trip setting is established by federal regulations based on the regulatory applicability and assumed application. Overload protection is mandated in some states.

5.7.2.1 Magnetic and Thermal Elements

Magnetic elements give instantaneous-trip protection and thermal elements afford overload protection. Pickup settings for both are based on the smallest size of cable being protected. Undervoltage release (UVR) is used and is an auxiliary solenoid that trips the operating mechanism of the breaker whenever the coil voltage of the solenoid drops below 40 to 60% of the rated value. Additional external protective relaying can be provided to the outgoing circuit through UVR. A shunt-trip circuit breaker is a low-voltage, molded-case circuit breaker that has an additional trip solenoid to allow for external control logic to trip the breaker.

5.7.3 Ground-Check Monitors for Resistance-Grounded Systems

A low-voltage and medium-voltage resistance-grounded system is required to have a fail-safe ground-check circuit to continuously monitor the continuity of the grounding conductor. The grounding conductor monitor must cause the associated circuit breaker to trip if the grounding conductor or a pilot wire is broken. An indicator lamp on the monitor should indicate a tripped condition. The monitors are usually enclosed in a dead-front package that is mounted near an associated circuit breaker (Morley, 1990).

5.8 MAIN CIRCUIT BREAKER CONSIDERATIONS

The use of a main circuit breaker is recommended when the number of outgoing circuits exceeds three. The bus work of the breaker serves as the breaker's primary protection zone and as a backup to outgoing breakers. The rating is based on the full-load current of the transformer secondary or ampacity of the bus work (whichever is lower). In general, bus work is sized based on the transformer full-load current.

5.8.1 MAGNETIC AND THERMAL

Some main breakers are provided with only magnetic elements. Thermal elements would provide overload protection to the transformer, and bus work magnetic elements can be set to give short-circuit protection to the bus work. The main breaker must be coordinated with the outgoing breakers and high-voltage fuses on the transformer primary.

When a main circuit breaker is used, neutral relaying may be used in power centers. The neutral conductor from the transformer is encircled by a current transformer (CT). The current transformer and ground-trip relay are the same as for zero-sequence relaying. A time delay is introduced by the ground-trip relay to provide selective interruption of the faulted circuit. If the circuit breaker of the faulted circuit fails to trip, then the main circuit breaker will provide backup protection. The prescribed time delay (0.5 seconds) can be achieved by either pneumatic or electronic means. The breaker trip device may be either a shunt trip or UVR (preferred).

5.8.2 POTENTIAL RELAYING

When a main circuit breaker is used, potential relaying can also be used as backup protection. Unlike zero-sequence and neutral relaying, potential relaying will detect a ground fault with the neutral grounding resistor open. A potential-relaying scheme may use either a potential transformer (PT) or a voltage divider to obtain a voltage within the operating range of the relay. For both schemes, the maximum pickup voltage of the relay should correspond to the voltage developed across the relay at 40% of the maximum fault current. When using a voltage divider circuit, the resistors are sized so that the total equivalent resistance of the parallel combination is not reduced significantly. The combined resistor value must be significantly higher than the value of the neutral grounding resistor. The power (I^2R) rating of the resistors should be capable of withstanding currents expected through them under a maximum ground-fault condition.

5.9 PREDICTIVE TESTING AND INSPECTION SUSTAINABILITY AND STEWARDSHIP— AIR–MAGNETIC, VACUUM, AND OIL CIRCUIT BREAKERS

To determine if breakers meet design intent, collect and document information relative to the following:

- Breaker type
- Breaker specifications (including current transformer ratios)

In addition to the documents mentioned in Section 1.9.1 of Chapter 1, maintain the following documents during both initial acceptance and ongoing operations and maintenance (O&M) routines.

5.9.1 REQUIRED TESTING RESULTS FOR AIR–MAGNETIC, VACUUM, AND OIL BREAKERS

- Contact resistance test results
- Insulation resistance test results
- Airborne ultrasonic test results (optional)
- Power factor test results (optional)
- High-voltage test results (optional)
- Infrared thermography (IRT) test results (optional)
- Breaker timing test results (optional)
- Insulation oil test (for oil breakers only)

5.9.2 CONTACT RESISTANCE TEST

The contact resistance test is used to determine the contact condition on a breaker or switch without visual inspection. Most manufacturers of high- and medium-voltage circuit breakers will specify a maximum contact resistance for both new contacts and in-service contacts. The contact resistance is dependent on two things: (1) the quality of the contact area and (2) the contact pressure. The contact quality can degrade if the breaker is called upon to open under fault conditions and the contact pressure can lessen as the breaker's springs fatigue due to age or a large number of operations.

To measure the contact resistance, a DC current, usually 10 or 100 amps, is applied through the contacts. The voltage across the contacts is measured, and the resistance is calculated using Ohm's law. This value can be trended and compared with maximum limits issued by the breaker or switch manufacturer. For oil-filled breakers, using a 100-amp test set is best because oil tends to glaze on contact surfaces and in some cases 10 amps is not enough to punch through the glaze.

5.9.3 BREAKER TIMING TEST

A breaker timing test is a mechanical test that shows the speed and position of breaker contacts before, during, and after an operation using (1) digital contact timers or (2) digital contact and breaker travel analyzers. Digital contact timers are used only for timing contacts where no travel time is required, and they are only appropriate for new breakers prior to being put into service. A digital contact and breaker analyzer measures the contact velocity, travel, over-travel, bounce back, and acceleration as an indicator of the breaker operating mechanism condition. A voltage is applied to the breaker contacts, and a motion transducer is attached to the operating mechanism. The breaker is then closed and opened, and the test set measures the time frame of voltage changes and plots the voltage changes over the motion waveform produced by the motion transducer. The three square waves (C1, C2, and C3) are the contacts, and the curve is the motion of the mechanism. The "wiggle" at the top of the motion waveform shows the amount of over-travel, bounce back, at seating depth of the contacts. The numbers are normally printed out from the test set, and the chart is stored in memory for downloading into a computer. Analyzing and trending this information will indicate if adjustments to the breaker operating mechanism are necessary.

5.9.3.1 Breaker Timing Test Performance

Perform a breaker timing test (mechanical) to verify the speed and position of breaker contacts before, during, and after an operation. The breaker timing test should perform contact timing during breaker close, open, open–close, close–open, and open–close–open. The test should include a minimum of three dry contact inputs and two wet-input channels to monitor breaker secondary contacts, and it should have a minimum resolution of ±0.0001 seconds over a one-second duration. The test should have travel transducers capable of linear and rotary motion, and it should be capable of slow close-contact point measurement. The minimum acceptance criterion is that the results should be better than or equal to the manufacturer's specifications. This test is not applicable to molded-case breakers or low-voltage breakers.

5.9.4 INSULATION RESISTANCE TEST

Perform an insulation resistance test to determine insulation resistance to ground. Use both the dielectric absorption index and the polarization index. The insulation resistance test set should have all of the following minimum requirements:

- Test voltage increments of 500 V, 1000 V, 2500 V, and 5000 V DC
- Resistance range of 0.0 to 500,000 megohms at 500,000 V DC
- A short-circuit terminal current of at least 2.5 milliamps
- Test voltage stability of ±0.1%
- Resistance accuracy of ±5% at 1 megohm

The minimum acceptance criteria are that the results should be better than or equal to the manufacturer's specifications, and insulation resistance (Megger) values, measured for each phase, should be over 25 megohms for molded-case breakers and over 100 megohms for all others.

5.9.5 ULTRASONICS

Use airborne ultrasonics to verify the non-existence of electrical arcing and other high-frequency events.

5.9.6 HIGH-VOLTAGE TEST

Perform the high-voltage test to verify the insulation in a new breaker and to ensure that no excessive leakage current is present. *Note:* This test is a potentially destructive test. The minimum acceptance criteria are that the results should be better than or equal to the manufacturer's specifications, and the limits must be in accordance with ANSI/IEEE Standard 400.

5.9.7 OTHER TESTS

Perform the following tests as appropriate for the type of breaker to verify a lack of contaminants and that the necessary inhibitors have been added (acceptance criteria are shown in parentheses):

- Water content using Karl Fisher, ASTM D1533-88 (<25 ppm at 20°C)
- Dielectric breakdown strength test, ASTM D877 and D1816 (>30 kV)
- Acidity test, ASTM D974 (neutralization number <0.05 mg/g)
- Visualization examination, ASTM D1524 (clear)
- Dissolved gas analysis, ASTM D3612-90
  - Nitrogen (N_2) (<100 ppm)
  - Oxygen (O_2) (<10 ppm)
  - Carbon dioxide (CO_2) (<10 ppm)
  - Carbon monoxide (CO) (<100 ppm)
  - Methane (CH_4) (none)
  - Ethane (C_2H_6) (none)
  - Ethylene (C_2H_4) (none)
  - Hydrogen (H_2) (none)
  - Acetylene (C_2H_2) (none)

5.10 PREDICTIVE TESTING AND INSPECTION SUSTAINABILITY AND STEWARDSHIP— SF_6 CIRCUIT BREAKERS

In addition to the tests described in Section 5.9, perform the vacuum bottle integrity (overpotential) test across each vacuum bottle with the contacts in the open position in strict accordance with the manufacturer's published data. Do not exceed the maximum voltage stipulated for this test. Provide adequate barriers and protection against x-radiation during this test. Do not perform this test unless the contact separation of each interrupter is within manufacturer's tolerance. (Be aware that some DC high-potential test sets are half-wave rectified and may produce peak voltages in excess of the manufacturer's recommended maximum.)

TABLE 5.1
SF$_6$ Gas Test

Test	Method	Serviceability Limits
Moisture	Hygrometer	Per manufacturer or 200 ppm
SF$_6$ decomposition byproducts	ASTM D2685	500 ppm
Air	ASTM D2685	5000 ppm
Dielectric breakdown, hemispherical contacts	0.10-inch gap at atmospheric pressure	11.5–13.5 kV

Source: NASA, *Reliability Centered Building and Equipment Acceptance Guide*, National Aeronautics and Space Administration, Washington, DC, 2004.

5.10.1 REQUIRED TESTING RESULTS

- Contact resistance test results
- Insulation resistance test results
- Airborne ultrasonic test results (optional)
- Power factor test results (optional)
- High-voltage test results (optional)
- Infrared thermography (IRT) test results (optional)
- Breaker timing test results (optional)
- SF$_6$ gas test results
- SF$_6$ gas leakage test results
- Vacuum bottle integrity test results
- Air-compressor performance test results

5.10.2 SF$_6$ GAS TEST AND GAS LEAKAGE TEST

Remove a sample of SF$_6$ gas if provisions are made for sampling and test in accordance with Table 5.1. A test for gas leaks should reveal no leakage.

5.10.3 VACUUM BOTTLE INTEGRITY TEST

Perform the vacuum bottle integrity (overpotential) test across each vacuum bottle with the contacts in the open position in strict accordance with manufacturer's published data. Do not exceed maximum voltage stipulated for this test. Provide adequate barriers and protection against x-radiation during this test. Do not perform this test unless the contact separation of each interrupter is within the manufacturer's tolerance. (Be aware that some DC high-potential test sets are half-wave rectified and may produce peak voltages in excess of the manufacturer's recommended maximum.)

5.11   CIRCUIT BREAKERS (ABOVE 600 VOLTS)

Circuit breakers should first be power factored in the open position using the grounded specimen test. The porcelain surface of bushings should be clean and dry prior to beginning the test. The load and line side of each phase should read within 10% of each other and the factory test results. Any larger difference should be investigated and repaired prior to acceptance. The breaker should then be closed, with the load and line side bushings tied together, and then power factored. As before, each phase should be within 10% of the other two phases, and anything greater should be investigated and repaired. The manufacturer's instructions and factory test data should be consulted for comparisons. Circuit breakers rated 15 kV and above should also have their bushings power factored. Breaker bushings rated 69 kV and above should have the factory power factor and capacitance values stamped on the bushing base (NASA, 2004).

5.11   IMPLICATIONS FOR SAFETY AUDITS AND INSPECTIONS

A good electrical safety inspection will look for more than labels on circuit breakers and clear paths of approach to circuit breaker panels. Points to consider include

1. *Is the circuit breaker being used appropriate for the application?* Determine the type of circuit breaker being used and what it is being used for. Verify that the circuit breaker is of the correct type and rating for the service intended.
2. *Is the location of the circuit breaker panel appropriate?* Not only should the approach to the circuit breaker panel be unobstructed, but all hazards in the immediate area should also be minimized. For example, circuit breakers that have the potential to be exposed to water should be placed in electrical boxes that are rated for a wet environment or should be relocated to a less hazardous area.
3. *Are the circuit breakers on a routine inspection schedule?* Depending on the type of circuit breakers, the material inside can degrade over time. Thus, circuit breakers should be tested as part of the facility's routine inspection.
4. *Are arc-flash boundaries established?* Work on now opened and previously dead-front panelboards is deemed "live work" by OSHA. Thus, arc-flash boundaries should be established.

REFERENCES

Morley, L.A. (1990). *Mine Power Systems*, Information Circular 9258, NTIS No. PB 91-241729. Pittsburgh, PA: U.S. Department of the Interior, Bureau of Mines.

NASA. (2004). *Reliability Centered Building and Equipment Acceptance Guide*. Washington, DC: National Aeronautics and Space Administration.

NFPA. (2005). *NFPA 70: National Electrical Code (NEC) Handbook*. Quincy, MA: National Fire Protection Association.

OSHA. (2007). Occupational Safety and Health Standards, 29 CFR, Part 1910, Subpart S, Electrical. Washington, DC: Occupational Safety and Health Administration.

6 Relays

Gayle Nicoll and Martha J. Boss

CONTENTS

A relay is an electrical component designed to open or close an electrical circuit as the result of an input voltage. Several different types of relays are in use today.

6.1 TYPES OF RELAYS

6.1.1 ELECTROMECHANICAL RELAYS

Electromechanical relays require an input voltage and current to open and close the circuit. Typically, an electromagnetic relay has a spring-loaded armature that places the relay in its default position (open or closed). When the input voltage exceeds the relay's required level to energize, the armature of the solenoid is moved by means of a magnetic field generated by an internal inductor. This movement places the armature in the opposite position (closed or open) until the energizing voltage and current are removed. Electromechanical relays are often used in high-voltage environments. Electromechanical relays may cause sparks or arcing as the high-voltage circuit opens or closes. An audible click or other noise is often heard as the relay armature actuates, and the armature magnetic coil radiates a magnetic field while energized. Electromechanical relays are temperature responsive and may alter operation if a temperature threshold is exceeded. In general, electromechanical relays have a low speed of operation and may cause a "bounce" phenomenon. This bounce is a state induced by arcing as the switch closes or opens, which may appear as several open-and-close cycles to the connected system.

6.1.2 SOLID-STATE RELAYS

Solid-state relays use low-voltage analog input signals to determine whether to open or close a circuit. These relays are typically constructed from power transistors and may include a small static digital-logic array to determine when to open and close the circuit. Solid-state relays are typically very fast to open or close. Solid-state relays are often used in low-voltage/low-current environments and are considered superior to electromechanical relays in that solid-state relays are noiseless, have a longer life span, and do not arc when opening or closing. However, solid-state relays are impractical for high-voltage operation. Some solid-state relays are temperature responsive and may alter operation if a temperature threshold is exceeded.

6.1.3 NUMERIC OR DIGITAL RELAYS

Numeric and digital relays are programmable relays. Electromechanical and solid-state relays are static and their operating parameters will not change; however, a numeric or digital relay may be reprogrammed to have a number of different actuation states and conditions. As with solid-state relays, numeric and digital relays use power transistors to control the flow of power through the system, with the operation of the transistors being controlled by the programmable logic.

Numeric and digital relays are small and quiet and draw relatively little power to operate. Like solid-state relays, numeric and digital relays operate on relatively low voltages and currents and are not suitable for direct use in transmission systems. However, numeric and digital relays may be used to provide the actuation for electromechanical relays in high-voltage systems. The primary advantage of numeric and digital relays is their ease of monitoring. Monitoring logic can be built directly into the programmable logic, making the determination of system operational status quick and easy.

6.1.4 THERMAL RELAYS

A thermal relay uses bimetallic strips to open or close electrical connections. A bimetallic strip is a length of metal that is composed of two layers, each layer containing a different metal. As the bimetallic strip is heated, the individual layers heat up and expand at different rates. This difference in expansion rates causes a straight bimetallic strip to become curved when exposed to sufficient heat.

The curving reaction of a bimetallic strip exposed to heat makes the strips perfectly suited to opening or closing circuits based on the temperature of the relay. A limitation of a thermal relay is that this relay is very sensitive to the current passing through the bimetallic strip. As current passes through the strip, some of the energy is converted to heat by the resistance of the strip's metals. If the heat generated by the current is sufficient, the current itself can cause the strip to bend; however, when the circuit is disconnected, the strip will cool and become straight again. This reestablishes the flow of current, which in turn causes the strip to heat up, beginning the cycle over again.

Thermal relays are used to protect electrical devices from overheating and are commonly used in motors to protect the motor from damage due to overheating. Thermal relays may be built such that the circuit either opens or closes when the bimetallic strip is heated.

6.2   RELAY ACTION

Relays may be constructed in one of two states. Normally closed (NC) relays are closed circuits that allow current to flow until the relay's actuation conditions have been met. Normally open (NO) relays are open circuits that do not allow current to flow until their actuation condition has been met.

6.2.1 PICK UP

When a relay operates to open NC contacts or to close NO contacts, the relay is said to "pick up" the contacts. The smallest actuating quantity to cause contact operation is the *pick-up value*.

6.2.2 RESET/DROP OUT

When a relay operates to close NC contacts or to open NO contacts, the relay is said to "reset" or "drop out." The largest actuating quantity necessary to return the relay to normal operating condition is the *reset value*. The reset value is almost always greater than zero, and it is often specified as a percentage of normal operation.

6.3 PROTECTIVE RELAYING

Relays perform a major role in power system protection. Relays detect voltage and current anomalies and receive information about system conditions. Information is received through transformers or resistors that reduce system parameters down to levels that relays can handle. Relays designed for protective circuits are usually provided with some means of visual indication that the relay has operated.

With larger substation capacities, relay actuation thresholds for overloads and short circuits are normally too high to protect the transformers. The relay thresholds are intentionally set high to allow for large through-faults from one side of the substation to the other. Unfortunately, this does not allow the relays to protect the transformers themselves. Additional protective systems, such as sudden pressure (current surge) relaying, percentage-differential relaying, and thermal overload protection, are used to protect substation transformers.

By providing sudden pressure relays in combination with percentage-differential relays, the sensitivity of the relay system can be customized to allow normal operation while preventing undesirable operation due to inrush current. Inrush current is the transient exciting current that results from a sudden change in the exciting voltage, which occurs at the instant of energizing after a fault has been cleared or during the inrush period of a nearby transformer. The type of relays used will depend on the type of transformers and the type of operation (high or low through current). Many different types of protective relaying are in use.

6.3.1 ALTERNATING CURRENT DIRECT RELAYING

Alternating current (AC) direct relaying is used to sense the magnitude of current flow. A relay may be used to open or close a circuit if the current passing through the circuit is too high or too low.

6.3.2 ALTERNATING CURRENT POTENTIAL RELAYING

Alternating current potential relaying is used to sense the voltage across a circuit. A relay may be used to open or close a circuit if the voltage across the monitored circuit is outside a predetermined operating range. The actuating input of an AC potential relay is usually connected in parallel with a resistor or used to monitor the potential difference between two conductors.

6.3.3 DIRECT CURRENT CONNECTIONS

Current relaying and potential relaying are the two most-used protective relaying connections for direct current (DC) systems. Relays in DC circuits are used to protect the circuit from overcurrent and overvoltage situations that may damage sensitive components (e g., diodes, capacitors).

6.3.4 OVERCURRENT RELAY

An overcurrent relay is actuated when the current flowing through the relay exceeds a specified or designed threshold. The threshold may be instantaneous, time based, or both. The typical reaction to an overcurrent situation is to trip the circuit breaker for the transmission line.

6.3.5 DISTANCE RELAY

Distance relays are used to isolate and identify faults along transmission lines. Transmission lines typically have a set impedance per mile of length, allowing utility companies to calculate transmission line losses over distance. Distance relays are adjustable to specific transmission line characteristics and can be used to identify distances to faults or to detect multi-phase faults.

6.3.6 DIFFERENTIAL RELAY

A current differential relay is used to sense if two or more connected current sources begin to differ in the amount of current these relays are supplying. Differential relays are used as status sensors, checking the input to a transformer against the output or the readiness of multiple redundant power supplies, or to compare current flowing through multiple circuits to a common standard. A differential relay is used to detect faults in current sources connected to the same system. The relay design is based on the principle that all currents into the electrical node must equal the sum of all currents leaving the node. Differences in input and output current may cause the relay to trip the circuit breaker associated with the system.

6.3.7 SUDDEN PRESSURE RELAY

A sudden pressure relay is designed to activate an alarm or open a circuit in the event of a sudden current surge. Sudden pressure relays are typically used to protect transformers from a sudden increase in current, which can cause spikes in the generated magnetic field. The spikes in turn can cause additional surges downstream, in addition to causing thermal breakdown of the transformer windings.

Transformers and generators may also utilize sudden pressure relays designed to detect gas pressure in their chambers. These relays help to detect when the fluid or gas in a chamber has begun to expand rapidly, possibly indicating an arc fault that has vaporized some of the liquid or caused rapid gas expansion.

6.3.8 GROUND OVERCURRENT

Various relay configurations provide ground-overcurrent protection. The magnitude of the ground-fault current depends on the grounding method of the system. Solidly grounded and low-impedance grounding systems can have relatively high levels of ground-fault currents, which require line tripping to remove the fault from the system. High-impedance ground-fault detection is difficult because the relay has to measure the ground-fault current combined with the unbalanced current generated by line phasing, configuration unbalance, and load unbalance. *Note:* Ungrounded systems, which have no intentional ground, have no alternative path for current to flow. The result is that the only path to ground is through the distributed line-to-ground capacitance of the surrounding system, resulting in a highly hazardous situation.

Ground relays for high-impedance grounded systems require high relay sensitivity, as the fault current is low compared to solidly grounded systems. The point of application for direct and potential ground-fault relaying is usually restricted to the system neutral point or grounding resistor. Residual connection, zero sequence, and broken delta can provide protection anywhere in the system in the combination needed for complete assurance of clearing all ground faults.

6.3.9 DIRECT OR NEUTRAL RELAYING

The simplest form of AC ground-fault protection is direct or neutral relaying. The current transformer (CT) is located between the neutral point of the source transformer and the grounding resistor. The grounding conductor acts as a primary winding of the CT. The secondary winding is connected to the overcurrent ground relay. If current through the grounding conductor exceeds a predetermined value, the relay acts to trip the circuit breaker.

Some ground-current flow is normal due to system unbalance, capacitive-charging currents, and inductive-coupling effects. The circuitry must be adjusted to pick up only when the normal level is exceeded. The pick-up point should always be less than the system current level. The disadvantage of this relay is that, if the grounding resistor or grounding conductors become open, the system will never detect any ground-current flow and the system will continue to operate with no abnormal indication. Thus, direct or neutral relaying systems can become essentially ungrounded, posing a hazard. Direct or neutral relaying does find application in some portions of the ground system; however, some states do not allow the use of direct or neutral relaying on substation grounding resistors, even for a second line of defense.

6.3.10 POTENTIAL RELAYING

Potential relaying is often used as the only means of ground-fault protection; however, potential relaying can also be used as a backup to other protection schemes at a unit substation or power center. The primary winding of a potential transformer (PT) is connected across the neutral grounding resistor. The secondary winding of the PT is connected to a voltage-sensing ground-trip relay. If current flows through the grounding conductor, voltage (potential) is developed across the grounding resistor. When the voltage (potential) rises above a preset level, the ground-trip relay causes the circuit breaker to trip. Note that potential relays only detect that a potential difference occurs; these types of relays cannot identify where within the system the fault occurred. Potential relaying has the advantage of being able to detect a ground fault with the neutral grounding resistor in an open mode of failure; however, if the grounding resistor fails in a shorted mode, potential relaying is rendered inoperable.

6.3.11   Zero-Sequence or Balance-Flux Relay

Zero-sequence relays are the most reliable first defense against ground faults. These relays work by responding to the zero-sequence current of the system, which is caused by an unbalanced fault involving the ground. The circuitry consists of a single window-type CT. The three line conductors are passed through the transformer core, forming the CT primary. Zero-sequence relaying is not affected by CT error and thus is a very sensitive tripping scheme. A common misconception is that zero-sequence relaying operates only on faults causing ground-current flow. In fact, zero-sequence current can occur during normal operation; for example, unbalanced system conditions caused by nontransposed transmission lines or unbalanced loading can cause zero-sequence current. Zero-sequence overcurrent relays measure the sum of the three phase currents.

6.3.11.1   Implementation Considerations

The ground-trip relay must be set to pick up at ≤40% of maximum ground-fault current. A low-ratio CT (25:5 or 50:5; ampere-turns ratio) is expected to be used. Using a high-ratio (≥350:5) CT allows better sensitivity with a voltage-sensitive relay (about 1.5-V pick-up) rather than a current-sensitive relay for a tripping device. Alternatively, manufacturers use a rectifying CT output and DC voltage-sensitive relay. These schemes work with a low-cost, low-burden CT. Because the CT is drastically underused, the CT secondary current cannot be predicted by knowing the turns ratio of the CT. Pick-up values must be determined by testing. Normally open (NO) contacts of the relay usually parallel the undervoltage relay (UVR) of the associated circuit breaker. When the relay pick-up value is exceeded, the relay pick-up's associated contacts close and short out the UVR coil. This technique is adopted to eliminate nuisance tripping due to bounce and vibration which hampers circuits that have contacts in series. The problem with paralleled contacts is that the ground-fault protection is lost if the relay is removed from its socket. To prevent this, UVR power can be supplied through a jumper in the ground-fault relay case. Without the relay, the circuit breaker cannot be closed, except in some small molded-case units (50-A units).

6.3.12   Negative-Sequence Overcurrent Relay

Negative-sequence overcurrent relays are excellent for detecting high-resistance ground faults. Negative-sequence currents arise when a system unbalance is present (e.g., faults, nontransposed lines, load unbalance). Similar to zero-sequence overcurrent protection, negative-sequence current monitors the three phases of current and compares them. Negative-sequence current elements can sense faults at the remote ends of long lines, thus providing better fault coverage than zero-sequence current elements.

6.3.13   Residual Relaying

Residual relaying is used in conjunction with CTs placed about phase conductors. Residual relaying is primarily used on high-voltage distribution circuits that require CTs and inverse-time relays for overcurrent protection. The CTs and phase overcurrent relays are both connected in a wye configuration. The ground-fault or residual relay is connected between the neutral points of the CTs and relays. Current flowing through the residual relay is in proportion to the sum of line current.

The principle of operation of the residual method is similar to that of zero-sequence relaying; however, residually connected relays are often subjected to nuisance tripping and cannot have sensitive or low pick-up settings. Residual relaying arrangement will not always provide consistent repetitive tripping at required tripping levels.

Residual current relays detect earth-fault currents and interrupt the current supply if an earth current flows. The primary application of residual current relays is to prevent electrocution of workers; however, protection equipment, especially from fire, is also used. Residual current relays can be opened and closed manually to switch normal load currents, and they open automatically when an earth-fault current flows that is ≥50% of the rated trip current.

6.3.14 Broken-Delta Relaying

Broken-delta ground-fault protection is similar to the residual relaying method, except that three CTs are wired in series. The resulting output voltages from the transformers form a closed delta when the load is balanced. An unbalanced condition (line-to-neutral fault) will cause formation of an open delta. The resulting voltage causes current to flow through the relay operating coil. Broken delta is sensitive to any unbalance.

6.3.15 Ground-Check Monitoring

The effectiveness of all ground-relaying methods depends on the integrity of the grounding system. A ground-check monitor device is used to continuously monitor grounding connections to verify continuity. If conductivity is inadequate, the function of the monitor is to trip the circuit breaker that feeds power to the system experiencing the defective grounding. A grounding conductor, although not always essential for machine operation, is *imperative for personnel safety.*

The ground-check monitor makes sure via ground connections that equipment frames are at near-neutral potential. The maximum allowable frame potential to earth is 40 V for low- and medium-voltage systems and 100 V for high-voltage systems. Ways of monitoring ground continuity include pilot monitors and pilotless monitors.

6.3.15.1 Pilot Monitor Relays

Pilot monitor relays use a pilot or ground-check conductor. General relay configurations include series loop, transmitter loop, and bridge. A pilot monitor provides selective, high-speed clearing of all faults on the protected line. The pilot monitor uses a pilot wire circuit that compares line currents at all terminals within the line. The pilot wire system circulates a current that is a function of the line current, similar to current differential relaying. The relay is capable of simultaneously clearing all terminals, which minimizes damage and permits high-speed reclosing.

6.3.15.2 Series Loop

In a series loop circuit, a power supply, relay operating coils (instantaneous contacts or a minimal time delay), pilot conductors, and grounding conductors are connected. Relay contacts will reset if the pilot or grounding conductor breaks the loop or if the power supply fails. The circuit can be AC using 60-Hz line frequency or DC. The advantage of a series loop circuit is minimal design. When using a DC source, a series loop is immune to stray AC; however, it is not immune to stray DC. AC monitoring can be subject to nuisance tripping by stray DC current that offsets the signal current. When the relay operating coil is isolated by a blocking capacitor, immunity to stray DC is gained. The relay coil must have a very low impedance to be sensitive to grounding-conductor impedance. Parallel paths and grounded pilot conductors easily negate series loop operation.

6.3.15.3 Transmitter Loop

The transmitter loop is basically the same as the series loop, except the voltage source is installed in the machine. The source must receive its power from the machine. The relay cannot pick up until the circuit is energized. The monitor must be temporarily bypassed to close the circuit breaker.

6.3.15.4 Bridge-Type Monitors

Bridge-type monitors use a series combination of pilot and grounding conductors as one leg of a Wheatstone bridge. The bridge output is sometimes amplified. The relay resets if the preset impedance level is exceeded. The bridge input can be 60-Hz AC, DC, or audiofrequency (5000, 2500, 900 Hz). Simple bridge monitors have the same problems as series loop models and are sensitive to changes in grounding-conductor and pilot-conductor impedance. More elaborate designs can be made immune to AC and DC stray currents. Bridge-type monitors sometimes cannot distinguish between a sound grounding conductor and an illegal parallel path.

6.3.15.5 Interlocking

Pilot conductor ground-check monitors can serve the important function of the safety interlocking feature required on many portions of the power system. When sophisticated interlocking is required, pilot-type monitors must be used.

6.3.16 Pilotless Monitors

Pilotless monitors do not use a pilot conductor. Instead, an audio signal is placed on the phase conductors through a filter. The audio signal is removed at the machine through filters and completes its path back to the source in the grounding conductor. Instead of filters, some models use coils similar to CTs to send and receive the audio signal.

A saturable reactor between the grounding conductor and power-center frame shows high impedance to the monitoring frequency. The purpose of this impedance is to restrict the monitoring signal to its intended path. For coupler grounding, the coupler metallic shell is commonly grounded to a grounding conductor and physically connected through the shell's receptacle to the power-center frame.

The grounding conductor must be isolated from shell ground so the reactor will not be bypassed. Pilotless monitors are superior to pilot designs in some applications. Because the ground-check conductor is not needed, all associated problems are removed. The most elaborate models can distinguish parallel paths and are immune to stray currents.

6.4 PARALLEL GROUND PATHS, STRAY CURRENT, INDUCTION, AND TRANSIENTS

Parallel ground paths may be established by incorrect or faulty wiring, contact to earth through conductive flooring (such as concrete), or grounding conductors on other machines. Alternate ground paths may have a resistance as low as the grounding conductor. These alternate, low-resistance ground paths may be temporary; however, these paths have the potential to do significant harm to employees and electrical system components. Stray AC and DC, and induced AC, are ever-present problems. Stray current is the flow of electricity through conductive parts (i.e., wires, building structures, and equipment) because of imbalances in the electrical supply or because of faulty wiring. Stray current can also result from a high-resistance ground. Alternating current can be induced in grounding conductors if the system current is unsymmetrical. This problem results from cable deterioration, such as may happen in a splice or weak spot in a wire.

Transients are an instantaneous change resulting in a burst of current through a conductor. Power-system transients may occur from lightning, facility load switches, on/off disconnects, device switching, arcing, capacitor bank switching, and tap changing on transformers. High wind has also been known to cause poor connections between points on a power distribution network, resulting in transients. Transients are difficult to detect and may do significant harm to equipment, resulting in shortened life of the equipment or burn out.

6.5 DISCONNECTS

The disconnect for each service, feeder, branch circuit, appliance, and motor is required to be labeled. Labels must include the purpose of the disconnect. Visible disconnect switches should be placed on incoming high-voltage (distribution) circuits within all power equipment. Often the first disconnect is part of a switch yard.

6.6 SINGLE-PHASE PROTECTION

Single-phasing protection is best accomplished at a secondary bus using relays to monitor relay contacts. These relays can activate either (1) undervoltage release via a device that triggers the breaker to open when the voltage falls below a specific threshold, or (2) shunt trip of the main breaker on the secondary. If a shunt trip occurs and the incoming current is too high, the excess power charges an electromagnet in the circuit breaker, which activates a switch to shut off power. If the main breaker is not used, contacts can be connected directly at the output of the control transformer secondary winding. A control transformer can be used when constant voltage or current with a low power or volt-amp rating is required. Control transformers have filtering devices (e.g., capacitors) that minimize variations in the output. In combination with relays, a control transformer maximizes inrush capability and output voltage regulation when the devices are initially energized. If the relay is actuated, all outgoing circuit breakers will trip through their undervoltage releases (Morley, 1990).

6.7　RELAY CONNECTIONS

Basic relay connections must be attached to the power system to sense a malfunction and supply tripping energy to the appropriate circuit breaker. The relay coil receives input information. The coil contacts then pick up or reset, thus affecting control power to the circuit breaker. Basic relay connections are used for protective relaying:

- *For AC systems*—Direct, potential, and differential
- *For DC work*—Direct and potential (differential relaying is also available for DC, but the circuitry is not considered basic)

Most relays have adjustments or tap settings to adapt the relays to as wide an operating range as possible.

6.7.1　SINGLE-POLE CONNECTORS

Single-pole connectors for individual conductors of a circuit used at terminal points must be designed so that all plugs must be completely inserted before the control circuit of the machine can be energized. Conductors must be securely attached to the electrodes in a plug or receptacle and the connections must be totally enclosed. Molded-elastomer connectors are acceptable, provided that

- Any free space within the plug or receptacle is isolated from the exterior of the plug.
- Joints between the elastomer and metal parts are ≥1 inch wide and the elastomer is either bonded to or fits tightly with metal parts.
- The contacts of all line-side connectors are shielded or recessed adequately.

6.8　VOLTAGE AND PHASE PROTECTION

6.8.1　LOW VOLTAGE

Direct relay connections to a monitored circuit are often restricted to low-voltage, low-power circuits, because most relay current or voltage coils are designed to operate in the vicinity of 5 A or 120 V. If power system values exceed these levels, some interface is necessary between the monitored circuit and the relays. Instrument transformers for AC and resistors for DC are used.

6.8.2 High-Voltage, Three-Phase Alternating Current

High-voltage, three-phase AC systems with relaying external from the circuit breakers have the following parameters:

- Line-to-line voltages for undervoltage
- Line overcurrent for overload
- Three-phase or line-to-line faults for short circuit
- Faults causing zero-sequence current for ground overcurrent
- Grounding conductor resistance for ground continuity

6.8.3 Phase Protection by Protective Relaying

Depending on the relays used, phase protection by protective relaying can be overload, short circuit, or both. (*Note:* Molded-case circuit breakers afford the same flexibility depending on the internal tripping element used.) Time-delay relays are employed for overload with instantaneous units for short circuit.

6.9 PREDICTIVE TESTING AND INSPECTION SUSTAINABILITY AND STEWARDSHIP— ELECTRICAL RELAYS

Electrical relays are defined as electromechanical switching devices that open and close electrical contacts to effect the operation of other devices in the same or another electrical circuit. Power relays can be either solid-state or microprocessor-based relays that are designed to protect the following types of systems: transmission lines, generators, transformers, motors, feeders, bus systems, and distribution systems. Their main components include electromagnets, contacts, armatures, and springs. Many times these relay systems are part of a larger package that includes protection, metering, monitoring, and control. Electric relays can also be described as contact-protection circuits, temperature-tolerant relays, contactors, latching relays, and power relays. In this section, electric relays are not considered as being equivalent to printed circuit-board relays, solid-state relays, or specialty (e.g., time-delay) relays.

6.9.1 Required Equipment Information

- Electrical relay type (NEMA enclosure type)
- Electrical relay specifications
- Voltage configuration
- Time-over-current curves (time-delay curves)
- Phase and ground operating curves (shapes)
- Dimensions, weight
- UL certification and electromagnetic interference (EMI) levels, if applicable
- Number and types of output relays
- Current loading

6.9.2   REQUIRED TESTING RESULTS

Required testing results for electrical relays include the following:

- Contact resistance test results
- Insulation resistance test results

6.9.3   ELECTRICAL ACCEPTANCE TESTS

- Perform electrical continuity tests on current, potential, and control circuits.
- Perform a current test on the remainder of the secondary circuit to detect any open or short-circuit connections.

6.10   PREDICTIVE TESTING AND INSPECTION SUSTAINABILITY AND STEWARDSHIP— ELECTRICAL SWITCHES (GENERAL)

6.10.1   REQUIRED EQUIPMENT INFORMATION

- Switch type
- Switch specifications

6.10.2   REQUIRED TESTING RESULTS

Required testing results for the various types of switches are listed below.

6.10.2.1   Electrical Switches, Cutouts

- Contact resistance test results
- Insulation resistance test results
- Airborne ultrasonic test results (optional)
- Power factor test results (optional)
- High-voltage test results (optional)
- Infrared thermography (IRT) test results (optional)

6.10.2.2   Electrical Switches, Low-Voltage Air

- Contact resistance test results
- Insulation resistance test results
- Airborne ultrasonic test results (optional)
- Infrared thermography (IRT) test results (optional)

6.10.2.3   Electrical Switches, Medium- and High-Voltage Air (Open)

- Contact resistance test results
- Insulation resistance test results
- Airborne ultrasonic test results (optional)
- Power factor test results (optional)
- High-voltage test results (optional)
- Infrared thermography (IRT) test results (optional)

6.10.2.4 Electrical Switches, Medium-Voltage Air (Metal Enclosed)

- Contact resistance test results
- Insulation resistance test results
- Airborne ultrasonic test results (optional)
- Power factor test results (optional)
- High-voltage test results (optional)
- Infrared thermography (IRT) test results (optional)

6.10.2.5 Electrical Switches, Medium-Voltage Oil

- Contact resistance test results
- Insulation resistance test results
- Airborne ultrasonic test results (optional)
- Power factor test results (optional)
- High-voltage test results (optional)
- Insulation oil test results
- Infrared thermography (IRT) test results (optional)

6.10.2.6 Electrical Switches, Medium-Voltage SF_6

- Contact resistance test results
- Insulation resistance test results
- Vacuum bottle integrity test results
- SF_6 gas test results
- Airborne ultrasonic test results (optional)
- Power factor test results (optional)
- High-voltage test results (optional)
- Infrared thermography (IRT) test results (optional)

6.10.2.7 Electrical Switches, Medium-Voltage Vacuum

- Contact resistance test results
- Insulation resistance test results
- Vacuum bottle integrity test results
- Insulation oil test results
- Airborne ultrasonic test results (optional)
- Power factor test results (optional)
- High-voltage test results (optional)
- Infrared thermography (IRT) test results (optional)

6.11 PREDICTIVE TESTING AND INSPECTION SUSTAINABILITY AND STEWARDSHIP— ELECTRICAL SWITCHES (MEDIUM-VOLTAGE OIL)

6.11.1 REQUIRED EQUIPMENT INFORMATION

- Switch type
- Switch specifications
- Insulation resistance test

6.11.2 ACCEPTANCE TECHNOLOGIES

6.11.2.1 Insulation Resistance Test

Perform insulation resistance tests on all control wiring with respect to ground. Applied potential should be 500 volts DC for 300-V-rated cable and 1000 volts DC for 600-V-rated cable. Test duration should be 1 minute. For units with solid-state components or control devices that cannot tolerate the applied voltage, follow the manufacturer's recommendation. Perform insulation resistance tests on each pole, phase-to-phase and phase-to-ground, with switch closed and across each open pole for 1 minute. Test voltage should be in accordance with the manufacturer's published data. Use both the dielectric absorption index and the polarization index. The insulation resistance test set should have all of the following minimum requirements:

- Test voltage increments of 500 V, 1000 V, 2500 V, and 5000 V DC
- Resistance range of 0.0 to 500,000 megohms at 500,000 V DC
- A short-circuit terminal current of at least 2.5 milliamps
- Test voltage stability of ±0.1%
- Resistance accuracy of ±5% at 1 megohm

The minimum acceptance criterion is that the results should be better than or equal to the manufacturer's specifications.

6.11.2.2 Insulation Oil Test

Remove a sample of the insulating liquid. The sample should be tested in accordance with the referenced standard:

- Dielectric breakdown voltage (ASTM D877)
- Color (ANSI/ASTM D1500)
- Visual condition (ASTM D1524)

6.12 PREDICTIVE TESTING AND INSPECTION SUSTAINABILITY AND STEWARDSHIP— ELECTRICAL SWITCHES (MEDIUM-VOLTAGE SF$_6$)

6.12.1 REQUIRED EQUIPMENT INFORMATION

- Switch type
- Switch specifications

6.12.2 ACCEPTANCE TECHNOLOGIES

6.12.2.1 Insulation Resistance Test

Perform insulation resistance tests on all control wiring with respect to ground. Applied potential should be 500 V DC for 300-V-rated cable and 1000 volts DC for 600-V-rated cable. Test duration should be 1 minute. For units with solid-state components, follow the manufacturer's recommendation.

TABLE 6.1
SF$_6$ Gas Test

Test	Method	Serviceability Limits
Moisture	Hygrometer	Per manufacturer or 200 ppm
SF$_6$ decomposition byproducts	ASTM D2685	500 ppm
Air	ASTM D2685	5000 ppm
Dielectric breakdown, hemispherical contacts	0.10-inch gap at atmospheric pressure	11.5–13.5 kV

Source: NASA, *Reliability Centered Building and Equipment Acceptance Guide,* National Aeronautics and Space Administration, Washington, DC, 2004.

6.12.2.2 Vacuum Bottle Integrity Test

Perform the vacuum bottle integrity (overpotential) test across each vacuum bottle with the contacts in the open position in strict accordance with the manufacturer's published data. Do not exceed the maximum voltage stipulated for this test. Provide adequate barriers and protection against x-radiation during this test. Do not perform this test unless the contact separation of each interrupter is within the manufacturer's tolerance. (Be aware that some DC high-potential test sets are half-wave rectified and may produce peak voltages in excess of the manufacturer's recommended maximum.)

6.12.2.3 SF$_6$ Gas Test

Remove a sample of SF$_6$ gas if provisions are made for sampling and test in accordance with Table 6.1.

6.13 PREDICTIVE TESTING AND INSPECTION SUSTAINABILITY AND STEWARDSHIP— ELECTRICAL SWITCHES (MEDIUM-VOLTAGE VACUUM)

6.13.1 REQUIRED EQUIPMENT INFORMATION

- Switch type
- Switch specifications

6.13.2 ACCEPTANCE TECHNOLOGIES

6.13.2.1 Insulation Resistance Test

Perform insulation resistance tests on all control wiring with respect to ground. Applied potential should be 500 volts DC for 300-V-rated cable and 1000 volts DC for 600-V-rated cable. Test duration should be 1 minute. For units with solid-state components, follow the manufacturer's recommendation.

6.13.2.2 Vacuum Bottle Integrity Test

Perform the vacuum bottle integrity (overpotential) test across each vacuum bottle with the contacts in the open position in strict accordance with the manufacturer's published data. Do not exceed the maximum voltage stipulated for this test. Provide adequate barriers and protection against x-radiation during this test. Do not perform this test unless the contact separation of each interrupter is within the manufacturer's tolerance. (Be aware that some DC high-potential test sets are half-wave rectified and may produce peak voltages in excess of the manufacturer's recommended maximum.)

6.13.2.3 Insulation Oil Test

Remove a sample of insulating liquid. The sample should be tested in accordance with the referenced standard:

1. Dielectric breakdown voltage (ASTM D877)
2. Color (ANSI/ASTM D1500)
3. Visual condition (ASTM D1524)

6.14 IMPLICATIONS FOR SAFETY AUDITS AND INSPECTIONS

The signals from safeguarding components are typically monitored using safety relays, safety controllers, or safety programmable logic controllers. A good electrical safety inspection determines which relays are safety critical. Points to consider include the following:

1. *Are relays used in the context of worker protection?*
2. *What is the recommended test cycle for relays? Is this inspection occurring and documented?*
3. *If the relay's decision logic is flawed or the relay fails, is a back-up system available to guarantee safety?*

REFERENCES

Morley, L.A. (1990). *Mine Power Systems*, Information Circular 9258, NTIS No. PB 91-241729. Pittsburgh, PA: U.S. Department of the Interior, Bureau of Mines.

NASA. (2004). *Reliability Centered Building and Equipment Acceptance Guide*. Washington, DC: National Aeronautics and Space Administration.

NFPA. (2005). *NFPA 70: National Electrical Code (NEC) Handbook*. Quincy, MA: National Fire Protection Association.

OSHA. (2007). Occupational Safety and Health Standards, 29 CFR, Part 1910, Subpart S, Electrical. Washington, DC: Occupational Safety and Health Administration.

7 Fuses

Gayle Nicoll and Martha J. Boss

CONTENTS

Fuses consist of an internal conductive element that has a particular current threshold. If the current threshold through the element is exceeded, the element heats up and burns through, breaking the circuit. This process is often referred to as "blowing" a fuse. When placed in series with a resistor across a set of leads, a fuse may also be used as a voltage limiter. Fuses are the simplest and oldest device for interrupting an electrical circuit under short-circuit or excessive-overload current. A fuse acts as both a sensing device and an interrupting device. Fuses are installed in series with the protected circuit. Fuses have an inverse-time characteristic; that is, the greater the current, the shorter the time to circuit opening. Fuses may be used in alternating current (AC) or direct current (DC) circuits. Fuses with variations in time–current characteristics are suitable for many special purposes.

7.1 TYPES OF FUSES

Most fuses function in identical ways; however, their construction can dictate their response to particular situations, such as short circuits and long-term overcurrent. Most fuse packaging comes in two forms: one-shot and renewable. One-shot fuses are just as the name suggests—a single-use device. No provision is made to open the housing of a one-shot fuse, and the entire cartridge is intended to be replaced after the fuse has blown. In contrast, a renewable fuse is designed to be opened, and the element within the fuse can be replaced. Four common types of fuses are in use today: non-time-delay, time-delay, dual-element, and self-resetting. Care must be taken when sizing a fuse for a circuit. If the voltage across the circuit is sufficient to exceed the dielectric insulation provided by the fuse, an arc flash may occur when the fuse's element burns through.

7.1.1 NON-TIME-DELAY FUSES

Non-time-delay fuses have no intentional built-in delay. Their two end terminals are joined together by a copper or zinc fusible element link. The link is current (heat) sensitive to melting. These fuses respond to instant circuit conditions and will burn through their element as soon as a threshold current has been reached. This makes non-time-delay fuses useful for protection against sudden surges; however, in cases where a circuit is subject to non-damaging short-term overcurrents, non-time-delay fuses have limited effectiveness. The lack of intentional time delay and limited inter-rupting rating (≤10,000 A) have substantially reduced the popularity of these fuses.

7.1.2 TIME-DELAY FUSES

The metal alloy used in time-delay fusible links is sensitive to the specific current existing for a specified time period. This arrangement permits harmless high-magnitude, short-duration currents to exist. Time-delay fuses are often needed for proper system operation, as in motor starting.

7.1.3 DUAL-ELEMENT FUSES

Dual-element fuses are designed primarily for motor-circuit protection and combine the features of non-time-delay and time-delay units. This type of fuse may contain a metal strip that melts instantly at high current and an additional low-melting element that responds to long-term overload of low current values. This allows the same fuse to provide short-circuit protection as well as long-term overcurrent protection. The elements within the fuse must be properly matched to protect the load from surge current and must be sized to protect the wiring and other equipment more sensitive to heating induced by long-term overcurrent.

7.1.4 SELF-RESETTING FUSES

Self-resetting fuses are technically not fuses. Self-resetting fuses are constructed from a thermoplastic conductor element, called a *polymeric positive temperature coefficient thermistor* (PPTCT). These elements are variable resistors that change resistance as the resistors heat up. For self-resetting fuses, the thermistor starts at a very low resistance value. As the heat generated by the current flowing through the thermistor increases, the resistance of the thermistor increases. At critical currents, the heat dissipation causes the resistance of the thermistor to become nearly infinite, effectively breaking the circuit. The resistance of the thermistor will lower as the thermistor cools, allowing the current to flow when the thermistor has cooled sufficiently.

7.1.5 FUSES OVER 600 VOLTS, NOMINAL, PROTECTION AND OSHA

According to Occupational Safety and Health Administration (OSHA) regulations, fuses must be located or shielded such that employees will not be burned or otherwise injured by their operation. In addition, where fuses are used to protect conductors and equipment, a fuse must be placed in series with each ungrounded conductor. Two fuses may be used in parallel to protect the same load, if a sufficiently rated fuse is unavailable. In this case, the fuses must have identical ratings, and they must have a common mounting that ensures that current is shared equally across the two fuses. Because fuses are simply low-ohm resistors, fuses can generate heat during normal operation as current passes through them. Power fuses that use venting to cool the fuse element may not be used indoors or underground. Vented fuses may be used in metal enclosures; however, those enclosures must be properly marked and rated.

7.2 HIGH-VOLTAGE FUSES

High-voltage fuses provide usable protection for 2.3- to 161-kV systems and have two general categories: power fuses and fuse cutouts.

7.2.1 POWER FUSES AND ARCS

As with any physical circuit interruption technology, the risk of an electrical arc being generated is present when a fuse operates. Some fuses are designed specifically to prevent the generation of an arc, while others have arc suppression mechanisms built in. The likelihood and magnitude of an arc depend on the point of the waveform where initiation of the fault occurs, as well as on the size and design of the fuse element.

7.2.1.1 Silver-Sand Fuses

A silver-sand fuse uses a silver element for current interruption; the remainder of the fuse casing is filled with quartz sand. Silver vaporizes quickly at the appropriate current, allowing the quartz sand, which has a high dielectric constant, to fill the small gap created. This electrical insulation prevents an arc from forming. These fuses can be used indoors or in small-size enclosures on surface or underground, with no noise from operation, gas, or flame discharge.

7.2.1.2 Expulsion Fuses

High-voltage expulsion fuses are usually plastic tubes reinforced with glass or fiber-glass and filled with a substance that turns to gas (a "gas-evolving" substance) when subjected to a large amount of heat. High-voltage expulsion fuses start the current-interruption process by melting a fusible link, much as other fuses; however, as the fuse melts high-voltage arcs may form. The heat and energy from these arcs are absorbed by the gas-evolving substance, which is turned into gas. The limited volume of the tube and the heat from the arc create a high pressure in the tube, which is then expelled through the ends of the tube. The expansion and expulsion process can be very noisy. This process quenches the arc and may expel potentially hazardous material into the air. For this reason, expulsion fuses are not for use indoors. Expulsion fuses are typically filled with boric acid or other solid material.

7.2.2 Fuse Cutouts

A fuse cutout is a combination of a fuse and a switch, designed primarily to protect transformers from current surges. The fuse cutout operates like a standard fuse; however, a switch that allows the fuse to be blown manually by service personnel is also incorporated. Manual operation of the switch does not actually blow the fuse's element; instead, the fuse's holder is ejected from the circuit when the switch is actuated and the holder is typically physically tethered to prevent loss. An arc flash risk does occur during the ejection process, and many holders are held in place by spring-loaded contacts because of this risk. As the spring is released, the distance between the contacts is rapidly widened, reducing (but not eliminating) the potential for an arc flash. Distribution fuse cutouts are designed for overhead distribution circuits for the protection of residential distribution transformers. Use of distribution fuses in utility-type systems is extensive.

7.2.2.1 Fuse Cutouts and OSHA

Fuse cutouts installed in buildings or transformer vaults must be of a type identified for the purpose. Distribution cutouts may not be used indoors, underground, or in metal enclosures. All cutouts must be readily accessible for fuse replacement. Where fuse cutouts are not suitable to interrupt the circuit manually while carrying a full load, an approved means must be installed to interrupt the entire load. Unless the fuse cutouts are interlocked with the switch to prevent opening of the cutouts under load, a conspicuous sign must be placed at such cutouts reading:

WARNING—DO NOT OPERATE UNDER LOAD

Suitable barriers or enclosures must be provided to prevent contact with unshielded cables or energized parts of oil-filled cutouts.

7.3 TYPICAL USES

The uses for fuses are many and varied, and the most common uses of fuses in power distribution systems are listed below.

7.3.1 EQUIPMENT ISOLATION

Fuses are used to completely isolate electrical equipment for inspection and repair. Almost all fuses used for this purpose are fuse cutouts; however, other configurations are possible. Fuse cutouts are preferred because a visible means is provided to indicate that the equipment to be worked on is isolated from the equipment's power source.

7.3.2 OVERCURRENT PROTECTION

Fuses are also commonly used for overcurrent protection. Although fuses are typically used in ungrounded conductors, fuses may generally be placed in any location. Fuses are commonly used in feeder circuits, transformer inputs, and many other places. Fuses should be selected based on their timely reactions to short-circuit currents so as to prevent conductor or conductor insulation damage and to protect circuit loads.

7.3.2.1 OSHA, Fuses, and Feeder Circuits

When selecting a fuse for a feeder circuit, OSHA requires the following criteria to be considered in the selection of a fuse:

- The continuous ampere rating of a fuse may not exceed three times the ampacity of the conductors.
- The long-time trip element setting of a breaker or the minimum trip setting of an electronically actuated fuse may not exceed six times the ampacity of the conductor.
- For fire pumps, conductors may be protected for short circuit only.
- Conductors tapped to a feeder may be protected by the feeder overcurrent device where that overcurrent device also protects the tap conductor.

Feeders and branch circuits over 600 volts, nominal, must have overcurrent protection in each ungrounded conductor located at the point where the conductor receives its supply or in a location in the circuit determined under engineering supervision.

7.3.3 SHORT-CIRCUIT PROTECTION

Short-circuit protection requires that a fuse limit energy delivered by a short circuit to a faulted component. The energy any interrupting device lets through under fault conditions cannot exceed the protected component's withstand rating.

7.3.4 Transformer Primaries

Fuses are recommended to protect transformer primaries for control-power circuits when inadvertent deactivation of load centers is a concern. Control-power fuses can be mounted in insulated dead-front holders or in a spring-clip arrangement that is only accessible through a bolted cover.

7.3.5 Cable Limiters

Cable limiters are used in multicable circuits (paralleled cables) and are placed in series with each cable in parallel. These fuses provide short-circuit protection to each cable by removing the cable from power in case of failure. Fuses are rated according to cable size.

7.3.6 Semiconductor Protection

Semiconductor-protection fuses and semiconductor-isolation fuses are used in series with the application being protected. Protection fuses are used where solid-state devices are to be protected rather than isolated after a failure. Protection fuses have lower let-through characteristics than other current-limiting fuses. An example application is protecting a rectifier or thyristor in case of an overload. Isolation fuses are high-speed fuses used to isolate a defective solid-state device in case of the device's failure. Isolation fuses are mandatory fuses for individual power diodes paralleled in large rectifier banks.

7.3.7 Capacitor Protection

Capacitor fuses are applied in series with power-factor correction (or other type) capacitors. These fuses are used to isolate a failed component by clearing the short-circuit current before excessive gas is generated in the capacitor.

7.3.8 Welding Fuses

Welding fuses are current-limiting fuses for use in welder circuits only. Due to time–current characteristics, welding fuses allow a longer intermittent overload than general-purpose fuses, yet still provide short-circuit protection.

7.4 TESTING AND CHANGING OF FUSES

Testing and changing control-circuit fuses are frequent electrical maintenance tasks. Typical spring-clip fuse holders have uninsulated exposed metal clips. If workers remove the fuses from these holders for testing without deenergizing the circuits, their hands will be in close proximity to energized clips. For this reason, dead-front fuse mounting is recommended for all energized components enclosed in an insulated housing. Fuses can be removed and replaced without exposing workers to metallic clips and fuse ends. Incorrect removal of fuses or removal of fuses from an energized circuit can put the proximate workers and those around them at risk (electrocution, arc flash). Installation of dead-front fuse mountings and proper training are a more sustainable approach than using unguarded fuse holders (Morley, 1990).

7.5 IMPLICATIONS FOR SAFETY AUDITS AND INSPECTIONS

Some things to consider when inspecting fuses include the following:

1. *Are the fuses of the appropriate type and suitable for the load?* Determine the type of fuses being used and whether these fuses are rated for the service implementation.
2. *Are personnel appropriately trained?* Those expected to work on the fuses should have adequate training, and those who are not expected to work on the fuses should be aware of their responsibilities as well.
3. *Are fuse boxes appropriately labeled?* Fuse boxes should be identified by type with hazard labeling.

REFERENCES

Morley, L.A. (1990). *Mine Power Systems*, Information Circular 9258, NTIS No. PB 91-241729. Pittsburgh, PA: U.S. Department of the Interior, Bureau of Mines.

NFPA. (2005). *NFPA 70: National Electrical Code (NEC) Handbook*. Quincy, MA: National Fire Protection Association.

OSHA. (2007). Occupational Safety and Health Standards, 29 CFR, Part 1910, Subpart S, Electrical. Washington, DC: Occupational Safety and Health Administration.

8 Substations and Switchyards

Gayle Nicoll and Martha J. Boss

CONTENTS

Generator outputs are sent through transmission lines from the generator to a local electrical substation. This substation then converts the generated power into differing voltages and currents for transmission to additional substations that form the core of the electrical distribution system. Transmission voltages vary but are usually in the hundreds of thousands of volts. Substations form the backbone of the electrical distribution grid. Several different types of substations are available, although a single substation may combine the properties from the several types:

- *Distribution substation*—This substation takes power from the power transmission system and directs this power to the distribution system for a geographic area.
- *Transmission substation*—This substation connects two or more transmission lines and may convert between the voltages carried on each line. These substations may also have power factor correction devices (capacitors, reactors, or volt-ampere reactive [VAR] compensators) to address shifts in phase during transmission.
- *Converter substation*—This substation is used to connect two or more transmission systems that operate at significantly different voltages or frequencies.
- *Collector substation*—This substation is used to collect generated electrical inputs from many sources and then inject the combined generated power into the transmission system.
- *Switching substation (switchyard)*—This substation functions only at a single voltage level and does no transformation of voltages. Switching substations are used to direct power flow from generation points to distribution points along a power grid and are used to provide backup routes for power in the event of transmission system failure.

Substations can be located in private facilities or in large, industrial sites shared by many utility companies and owned by a third party. For purposes of safety auditing and regulation, each utility company's systems are considered separable elements to be audited separately. Regardless of ownership, the Occupational Safety and Health Administration (OSHA) requires that a power system's equipment be marked to indicate the hazard, including arc flash. If these OSHA-required markings are not provided by the utility company, the entire site may be out of compliance if and when their personnel enter the substation yard area or even touch the fence (OSHA, 2007). Although substations have differing functions, the same electrical components are often used. These components are discussed in Chapters 1 through 6.

8.1 SUBSTATION SAFETY CONSIDERATIONS

Electrical substations provide a wide range of hazards for personnel onsite. A number of factors must be considered when constructing, operating, and auditing an electrical substation.

8.1.1 SUBSTATION GROUNDING

The substation ground bed components include surge arresters, grounding conductors, static conductors, metallic frames, and substation fencing. These station bed components located within a fenced interior are considered the *substation area*. If the utility controls the substation/switchgear yard, then grounding continuity should be routinely evaluated by the utility company and this information provided to the industrial site. If the industrial site controls the yard, the grounding continuity should be routinely checked by the industrial site and this information should be provided, as agreed upon, to the utility company.

8.1.1.1 Station/System Ground Bed and Substation Ground Mat

The station/system ground bed is usually a substation ground mat or mesh located underneath the substation area. The ground bed is tied to the substation area components (e.g., surge arresters, grounding conductors, static conductors, metallic frames, substation fencing) within the substation's fenced interior. Lightning discharges and other transformer primary surging conditions must be directed to the station ground bed system. The grounding wire from the fence to the ground bed should be visible when examining the fencing; however, the continuity of the ground bed system from all substation area components to the grounding bed may not be visually obvious. Continuity testing as part of an assured grounding program should be conducted and documented at regular intervals. The system ground bed holds the voltage of the substation floor at ground potential. Electric-shock protection must be provided for personnel in and around the substation during lightning, short circuits, equipment failures, and incidents caused by human error. The substation must be enclosed completely by a continuous buried ground conductor that should extend approximately 3 feet beyond the outer substation fence. All substation equipment must be bonded to this substation ground mat to ensure continuity.

The substation ground mat covers a relatively large area with conductors arranged to reduce high-voltage gradients. It provides grounding of all accessible surfaces to limit step and touch potentials to safe levels and to limit insulation stress by conducting surges to earth. It also limits voltage rise during lightning, thus reducing stress on power-system insulation. The substation ground mat must be buried beneath the soil surface. The substation ground mat is usually a series of interconnected conductors and earthing ground rods. Within the perimeter, the conductors form a regular grid or mesh of wires (the ground mat) spaced approximately 5 to 15 feet apart to cover the entire substation area. Conductors are bonded together at all intersecting points using either welded connections or heavy clamps designed for grounding. Substation fence posts and gates must also be bonded to the grounding grid.

Maximum acceptable grid spacing depends on the available fault current resistivity of earth, clearing time of the protective devices, and overall bed conformation. Wires should be uniformly spaced and located along rows of equipment to facilitate grounding connections. Extra conductors should be added to the grid at the corners of the substation and for substation work areas.

To limit the magnitude of the voltage produced across a bed during current flow, low resistance (≤5 ohms) is mandatory for safety ground beds. To ensure that step and touch potentials are not hazardous in the grounding location, potential gradients during surging conditions must be restricted. In order for step and touch potentials to remain non-hazardous, the grounds must be intact, continuous, and at a low resistance. These factors should be routinely checked in an assured grounding program.

The neutral of substation transformers secondaries must be connected to the safety ground bed through a neutral grounding resistor. Safety grounds must be kept separate so current flow intended for one safety ground will not enter another safety ground.

8.1.2 Substation Ground-Fault Protection

For substations, the ground-fault protection must be

- *Neutral (direct) relaying* between the grounding resistor and the neutral bushing of the power transformer or grounding transformer
- *Zero-sequence relaying* for each outgoing circuit, which measures the zero-sequence current using directional overcurrent relays
- *Potential relaying* about the grounding resistor, which measures the change in voltage between two points

Protective relaying to detect grounding-transformer failures may include use of fuses or relays. In some instances, fuses are not sufficiently sensitive to detect the fault and relays must be used. Protective relaying requires ground-check monitoring of the circuits feeding portable and mobile equipment and other loads. Frames of all direct loads for the main substation must be connected via grounding conductors to a safety ground bed with ground-check monitoring of the connecting conductors, or part of an established low-resistance ground bed on the unit substation end (Morley, 1990).

8.1.3 Outgoing Circuit Protective Relaying

Outgoing circuit protective relaying establishes a primary protection zone from the initial receiving switchyards to the first switchyard downstream. Outgoing circuit protective relaying acts to protect against

- Ground faults
- Overloads
- Short circuits

Ground-fault protection must be coordinated with the protection available in down-stream switch yards. Backup ground-fault relays may be provided and will include the following:

- Neutral-current sensing between the transformer busing and the top of the grounding resistor to ensure backup if the resistor shorts
- Potential relaying about the grounding resistor to provide protection against open-resistor hazards

Line overcurrent relays must coordinate with the power generator's primary trans-former requirements and with downstream protection.

8.1.4 Surge Protection

Surge arresters should be used throughout the substation to protect against a wide variety of surge conditions. Surge arresters should be inspected regularly, as deterio-ration over time may occur due to use in their operating environment. To determine the size of a surge arrester to use, a worst-case discharge current of 20,000 A can be used for most substations. The discharge current can be used to calculate the discharge voltage across the arrester. The margin of protection for a surge arrester should be at least 20% greater than a worst-case discharge scenario. The basic insu-lation level (BIL) for substation components (transformers) may be reduced if proper coordination is maintained with deployed surge arresters. The surge arrester should clamp on voltages well below the BIL of the transformer.

8.1.4.1 Current-Limiting Fuses

Whenever current-limiting fuses are applied with surge arresters, the arc voltage produced by fuse operation could result in arrester sparkover and damage. Station-class and intermediate-class arresters are more susceptible to damage because of lower protective characteristics. The arrester can be placed on either the source (line) side or the load side of a current-limiting fuse. If the arrester is placed on the line side, the surges will be diverted by the arrester; however, the fuse cannot limit the energy to the failed arrester, and the fuse arc voltage may spark over the arrester. If the arrester is placed on the load side, the arc voltage does not appear across the arrester; however, the fuse is subject to surge transients. The possibility of arc volt-age production resulting from fuse placement should be evaluated during arc flash hazard analysis studies of the site.

8.1.4.2 Breakers and Disconnect Switches

Breakers and disconnect switches are designed so that the impulse withstand level over open switches is greater than that to ground. This reduces the possibility of flashover. Consequently, a surge on these devices will flash to ground and will not usually cause permanent damage to equipment. The necessity to flash a surge to ground is one reason why the grounding mat under the substation/switchgear yard must be intact and functional.

8.1.4.3 Lightning

Lightning strikes cause transient overvoltages. Surface installations must be protected using overhead static wires and shielding masts. Surge arresters are mandatory to limit transient overvoltages.

8.1.4.4 Incoming and Outgoing Lines

Surge arresters on incoming lines should be located as close as possible to the transformer terminals. Station-class valve arresters should be used. Arresters should be selected based on the primary (input) voltage of the transformer, effectiveness of the local grounding mat, and insulation coordination between the transformer's basic insulation level and the arrester's discharge rating. Surge arresters on outgoing lines should be located as close as possible to the point where outgoing lines leave the substation. These surge arresters can be station-class arresters; however, intermediate (distribution) class arresters are commonly used.

8.2 TRANSFORMER DISCONNECTING SWITCHES

The disconnect switch is a mechanically operated air-type switch used to provide a quick means of disconnecting the primary or power-center transformer. Spring-loaded or torsion-bar mechanisms provide quick-make and quick-break operation. Load-break disconnecting switches or disconnecting means are used to interrupt currents that are not in excess of the continuous-current rating. These switches are rated for a maximum interrupting capacity and designed for one-time interruption only. The only means of activating a load-break switch is by manually operating a handle accessible from outside the load-center enclosure. *Note:* The operating mechanism should not be tied into the power center protective circuitry.

8.3 LOAD INTERRUPTER SWITCHES AND OSHA

Load interrupter switches are used only if suitable fuses or circuits are used in conjunction with these devices to interrupt fault currents. When used in combination, these devices must be coordinated electrically to safely withstand the effects of closing, carrying, or interrupting all possible currents up to the assigned maximum short-circuit rating. Where more than one switch is installed with interconnected load terminals to provide for alternative connections to different supply conductors, each switch must be provided with a conspicuous sign reading:

WARNING—SWITCH MAY BE ENERGIZED BY BACKFEED

8.4 ASSOCIATED EQUIPMENT SUSTAINABILITY AND STEWARDSHIP

8.4.1 Switchgear and Motor Control Centers

Bus insulation for both switchgear and motor control centers should be measured. Readings below the minimum values (based on design parameters) are indicative of improperly installed or wet insulation or loose bus connections and should be resolved prior to continuing testing and commissioning. In addition to the insulation resistance tests, the integrity of the connected cable installations associated with switchgear and motor control centers can also be tested by impressing a high voltage, commonly called a *withstand test voltage*.

8.4.1.1 Required Equipment Information
- Switchgear type
- Switchgear specification data (voltage rating)
- Switchgear assemblies
- Equipment foundation drawings and data

8.4.1.2 Required Testing Results
Required testing results for motor control centers include the following:

- Insulation resistance test results
- Airborne ultrasonic test results
- Infrared thermography (IRT) test results

Required testing results for switchgear include the following:

- Contact resistance test results
- Insulation resistance test results
- Airborne ultrasonic test results
- Power factor test results
- High-voltage test results
- Infrared thermography (IRT) test results
- Weatherproof test results

8.4.1.3 Power Factor
For switchgear rated 5 kV and above the bus should be power factored to the values shown in Table 8.1.

8.4.1.4 Electrical Acceptance Tests
- Perform the high-voltage test to verify the insulation in a new switchgear and to ensure that no excessive leakage current is present. This is a potentially destructive test. The minimum acceptance criterion is that the results should be better than or equal to the manufacturer's specifications.

TABLE 8.1
Switchgear Power Factor Values

Voltage Rating (Volts)	Test Voltage (Volts)	Maximum Reading
5000	5000	2%
7000	5000	2%
15,000	10,000	2%
35,000	10,000	2%

- Perform electrical continuity tests on current, potential, and control circuits.
- Perform ratio and polarity tests on current and potential transformers.
- Perform a current test on the remainder of the secondary circuit to detect any open or short-circuit connections.

8.4.1.5   Other Tests
- Use airborne ultrasonic testing to verify the non-existence of electrical arcing and other high-frequency events.
- Determine that the power factor does not exceed the manufacturer's data.

8.4.2   ELECTRICAL POWER CENTERS

Electrical power centers combine electrical distribution equipment (distribution panels) and building management controls (control panels) into a single factory-assembled and factory-wired integrated system. This approach replaces the traditional method of independently mounting each panelboard, lighting control, and building management system. These switchboard enclosure units:

- Contain the same components as electrical distribution and control panels
- May also contain lighting control, power and control cable wiring, contactors, and terminal blocks
- May be customized further with the addition of third-party components (e.g., building management systems, automatic transfer switches, power conditioners)

Power centers also describe certain kinds of quality power filtering systems, system protection enclosures, and power distribution systems.

8.4.2.1   Required Equipment Information
- Electrical power center type (NEMA enclosure type)
- Electrical power center specifications
- Voltage configuration (120/240 V AC, 12/24 V DC)
- Amperage (panel main bus maximum)
- Dimensions, weight
- UL certification and electromagnetic interference (EMI) levels, if applicable
- Number of circuit breaker positions (outputs)
- Electrical power centers impedance

8.4.2.2   Infrared Thermography Survey
Perform a thermographic survey to detect uneven heating indicative of loose or dirty terminal connections, other internal electrical connections to the individual components, or uneven heating patterns in the oil and windings of transformers and oil breakers that may be integral to these panels. This survey may also indicate internal corrosion or other flaws. Localized heating may be indicative of flaws in the windings or insufficient ventilation of the surrounding area. A bank of same-type

electrical distribution panels with significantly different temperature readings may indicate unbalanced loading or a defective electrical distribution panel. Electrical panel loading should be greater than 50% for performing this test. Minimum acceptable criteria for this test include the use of predefined relative difference limits, hot and cold spots, and deviations from normal or expected temperature ranges consistent with the manufacturer's design data. Minimum acceptable limits for this test are as follows:

- Qualitative verification of good terminal connections, internal connections to the individual components, and relative equivalent temperatures of the adjacent components during operation
  - Acceptable deviations in temperature which will vary with the size and type of the electrical panel but do not exceed ±5°C
  - Terminal connections or internal connections to the individual components displaying temperatures that are ≤11.3°C ambient

8.4.2.3   Electrical Acceptance Tests
- Perform electrical continuity tests on current, potential, and control circuits.
- Perform ratio and polarity tests on current and potential transformers (if applicable).
- Perform a current test on the remainder of the secondary circuit to detect any open or short-circuit connections.

8.4.2.4   Other Tests
- Use airborne ultrasonic testing to verify the non-existence of electrical arcing and other high-frequency events.
- Determine that the power factor does not exceed the manufacturer's data.

8.4.3   ELECTRICAL CONTROL PANEL

The electrical control panel provides control and/or monitoring of a variety of electrical equipment. The panel is primarily defined to be the NEMA-type enclosure unit that contains components such as fuses, contact relays, power supply transformers, push button switches, contactors, starters, programmable logic controllers (PLCs), and monitoring/control interface equipment. The control panel includes the enclosure and all of the above items, as well as the terminal strips or connection points that connect each of these items. Other common and alternative names for a control panel include operator interface systems, alarm panels, remote panels, pump panels, and controllers.

8.4.3.1   Required Equipment Information
- Electrical control panel type (NEMA enclosure type)
- Electrical control panel specifications
- Voltage configuration (120/240 V AC, 12/24 V DC)
- Amperage
- Dimensions, weight
- UL certification and electromagnetic interference (EMI) levels, if applicable

8.4.3.2 Required Testing Results

Required testing results for the electrical control panel include the following:

- Contact resistance test results
- Insulation resistance test results
- Airborne ultrasonic test results
- Power factor test results
- Infrared thermography (IRT) test results

8.4.3.3 Infrared Thermography Survey

Perform a thermographic survey to detect uneven heating indicative of loose or dirty terminal connections and other internal electrical connections to the individual components. This survey may also indicate internal corrosion or other flaws. Localized heating may be indicative of flaws in the windings or insufficient ventilation of the surrounding area. A bank of same-type electrical distribution panels with significantly different temperature readings may indicate unbalanced loading or a defective electrical distribution panel. Electrical panel loading should be greater than 50% for performing this test.

Minimum acceptable criteria for this test include the use of predefined relative difference limits, hot and cold spots, and deviations from normal or expected temperature ranges consistent with the manufacturer's design data. Minimum acceptable limits for this test are as follows:

- Qualitative verification of good terminal connections, internal connections to the individual components, and relative equivalent temperatures of the adjacent components during operation
  - Acceptable deviations in temperature which will vary with the size and type of the electrical panel but do not exceed ±5°C
  - Terminal connections or internal connections to the individual components displaying temperatures that are ≤11.3°C ambient

8.4.3.4 Electrical Acceptance Tests

- Perform electrical continuity tests on current, potential, and control circuits.
- Perform a current test on the remainder of the secondary circuit to detect any open or short-circuit connections.

8.4.3.5 Other Tests

- Use airborne ultrasonic testing to verify the non-existence of electrical arcing and other high-frequency events.
- Determine that the power factor does not exceed manufacturer's data.

8.4.4 ELECTRIC BUS

Required testing results for the electric bus include the following:

- Contact resistance test results
- Insulation resistance test results
- Overpotential test results
- Airborne ultrasonic test results (optional)
- Infrared thermography (IRT) test results (optional)

8.4.4.1 Overpotential Test

Perform an overpotential test on each busway, phase-to-ground with phases not under test grounded, in accordance with the manufacturer's published data. Where no direct current (DC) test value is available, an alternating current (AC) value should be used. The test voltage should be applied for 1 minute. The minimum acceptance criterion is that the results should be better than or equal to the manufacturer's specifications. DC withstand tests are recommended for flexible bus to avoid the loss of insulation life that may result from the dielectric heating that occurs with rated frequency withstand testing. Because of the variable voltage distribution encountered when making DC withstand tests and variances in leakage currents associated with various insulation systems, the manufacturer should be consulted for recommendations before applying DC withstand tests to this equipment.

8.4.5 ELECTRICAL DISTRIBUTION PANEL

An electrical distribution panel provides power at a designated voltage, current, and frequency to a variety of electrical equipment. This panel is primarily defined to be the NEMA type enclosure that contains components (e.g., busbars, circuit breakers, step-down transformers, automatic/manual transfer switches, optional monitoring/control equipment). The electrical distribution panel includes the enclosure and all of the above items, the terminal strips or connection points that connect each of these items, and the foundation, enclosure, enclosure access doors, and terminal connection elements.

8.4.5.1 Required Equipment Information

- Electrical distribution panel type (NEMA enclosure type)
- Electrical distribution panel specifications
- Voltage configuration (120/240 V AC, 12/24 V DC)
- Amperage (panel main bus maximum)
- Dimensions, weight
- UL certification and electromagnetic interference (EMI) levels, if applicable
- Number of circuit breaker positions (outputs)
- Electrical distribution panel impedance

8.4.5.2 Required Testing Results

Required testing results for the electrical distribution panel include the following:

- Contact resistance test results
- Insulation resistance test results

- Airborne ultrasonic test results
- Power factor test results
- High-voltage test results
- Infrared thermography (IRT) test results
- Equipment foundation data (if applicable)

8.4.5.3   Infrared Thermography Survey

Perform a thermographic survey to detect uneven heating indicative of loose or dirty terminal connections and other internal electrical connections to the individual components. This survey may also indicate internal corrosion or other flaws. Localized heating may be indicative of flaws in the windings or insufficient ventilation of the surrounding area. A bank of same-type electrical distribution panels with significantly different temperature readings may indicate unbalanced loading or a defective electrical distribution panel. Electrical panel loading should be greater than 50% for performing this test.

Minimum acceptable limits for this test include qualitatively verifying good terminal connections, internal connections to the individual components, and relative equivalent temperatures of the adjacent components during operation. Acceptable deviations in temperature will vary with the size and type of the electrical panel but should not exceed ±5°C.

8.4.5.4   Electrical Acceptance Tests

- Perform electrical continuity tests on current, potential, and control circuits.
- Perform a current test on the remainder of the secondary circuit to detect any open or short-circuit connections.

8.4.5.5   Other Tests

- Use airborne ultrasonic testing to verify the non-existence of electrical arcing and other high-frequency events.
- Determine that the power factor does not exceed the manufacturer's data (NASA, 2004).

REFERENCES

Morley, L.A. (1990). *Mine Power Systems*, Information Circular 9258, NTIS No. PB 91-241729. Pittsburgh, PA: U.S. Department of the Interior, Bureau of Mines.

NASA. (2004). *Reliability Centered Building and Equipment Acceptance Guide*. Washington, DC: National Aeronautics and Space Administration.

NFPA. (2005). *NFPA 70: National Electrical Code (NEC) Handbook*. Quincy, MA: National Fire Protection Association.

OSHA. (2007). Occupational Safety and Health Standards, 29 CFR, Part 1910, Subpart S, Electrical. Baltimore, MD: Occupational Safety and Health Administration.

9 Direct Current Utilization

Gayle Nicoll and Martha J. Boss

CONTENTS

Direct current (DC) is sometimes needed and has special requirements.

9.1 RECTIFIER

If the equipment does not have direct access to a feeder line, a rectifier must be used to convert three-phase alternating current (AC) voltage to DC voltage. The rectifier can be (1) a separate piece of equipment housed in its own enclosure and powered by means of a feeder cable from the AC power center, or (2) incorporated into a single enclosure within the AC power center, with the total unit being referred to as the *AC–DC combination power center.*

9.1.1 Full-Wave Bridge Rectifier

A full-wave bridge is the most popular rectifier configuration used in combination power centers or section rectifiers. A DC current is much more difficult to interrupt than an AC current. The amount of available short-circuit current must be limited to less than that of the AC equipment with similar capacity.

9.1.2 Rectifier Configuration and Transformers

The output of the transformer tertiary winding feeds the rectifier via a main circuit breaker. The rectifier configuration is a three-phase full-wave bridge using six diodes (one connected to each end of each of the transformer secondary windings).

9.1.3 Power Center Three-Winding Transformers

Three-winding transformers commonly used in a combination power center consist of primary, secondary, and tertiary windings. The primary winding is delta connected, with the secondary and tertiary windings wye connected. The transformer may be designed so the primary winding can be connected in either a delta or a wye configuration applied at two different distribution voltages. If a resistance-grounded system is used on a delta-connected winding, a zig-zag or grounding transformer is used to provide a neutral.

9.1.4 Fault Current

A separate transformer to limit the maximum fault current to a level that can be safely interrupted is sometimes required. A magnitude that will not damage the rectifier bridge main circuit breaker can be provided to protect the transformer winding that supplies the DC circuit. The thermal-trip setting of this breaker is usually based on 125% of the rated current of the windings.

9.2 CURRENT RATING

The current rating of the rectifier bridge (and thus diodes) is selected based on the expected bolted-fault current (faulted DC output) with the power center input connected to an infinite bus. Current should also incorporate a multiplier to account for a worst-case offset. The rectifier should be capable of handling this current for the time required for the circuit breakers to interrupt the current flow.

9.3 PARALLEL DIODES

To achieve adequate current-carrying capacity, diodes are paralleled in each rectifier leg to share the current. The number of parallel diodes is determined by the individual diode current rating (all must be rated equally). The maximum available fault current of each diode must also be rated in terms of the peak inverse voltage (PIV), which is the maximum voltage that can be applied across the diode in a reverse-biased mode without causing breakdown. Matched diode character-istics should allow sufficient means for sharing current when used in a parallel configuration.

9.4   CIRCUIT PROTECTION AND FUSES

Replacement due to the failure of any diode would easily upset the balance. Although rectifiers are supplied with matched diodes, most use current-balancing reactors to force uniform conduction of all diodes in parallel. Each diode is accompanied by a circuit-protection fuse. The purpose of the fuse is not to protect the diode (which fails in shorted mode); rather, the fuse is intended to prevent catastrophic failure to a bridge leg. The fuse current rating must not initiate interruption for a bolted fault at the DC bus until the clearing time for the downstream circuit-interrupting device has elapsed. The time–current characteristics must be coordinated. Each fuse should have a match-ing light to indicate when the fuse has blown and the diode needs to be replaced.

9.5   SURGE SUPPRESSION DEVICES

Surge suppression devices (also known as surge arresters or surge protective devices) should always be used to protect the rectifier from transient overvoltages occurring from either the AC or DC side. "Clipping off" surge arresters operate by diverting the highest level of voltage away from the load. Solid-state devices use metal oxide varistors (MOVs) or silicon avalanche suppression diodes (SASDs) that turn on at a predetermined voltage and conduct the overvoltage away. This process is sometimes called *voltage clamping*. Transient voltage surge suppressors (TVSSs) are a series of MOVs that dissipate a spike. Selenium voltage suppressors are also popular surge suppressors, competing with MOVs.

9.6   DIRECT CURRENT GROUND-FAULT PROTECTION SYSTEMS

Direct current ground-fault protection systems include the following:

• Grounded diode
• Basic grounding-conductor
• Relayed grounding conductors
• Neutral shift
• Differential current

Of these, only the differential current system provides a sensitive and selective technique for ground-fault protection. In one type of differential current system, the grounding resistor is placed between the transformer neutral and the load-cen-ter frame. Both positive and negative conductors pass through a saturable reactor. Differential current created by a ground fault causes magnetic saturation of the reac-tor core, which in turn allows the ground-fault relay to pick up. Each outgoing DC circuit thus has practical ground-fault protection.

9.7   DIRECT CURRENT INTERRUPTING DEVICES

Interruption of direct current requires an interrupting device to force the current to zero with an arc voltage greater than the line voltage. The device must be capable of withstanding energy dissipated during the time that the arc exists across the contacts of the device. Arc energy is a function of the circuit inductance and the current magnitude. The two basic interrupters used to protect DC machine circuits are the DC contactor and the molded-case circuit breaker. For both, the actual interrupting capability varies with the inductance of the circuit being interrupted.

The contactors most commonly used as interrupting devices for DC machine circuits have a continuous-current rating of 250 or 500 A. Their interrupting capability is approximately 10 times their continuous-current rating, which is significantly less than that given for molded-case circuit breakers. DC contactors can be applied at 600 V. Sometimes molded-case circuit breakers are used as backup protection for DC contactors. Molded-case circuit breakers that are rated at 600 volts alternating current (VAC) are also normally rated at 300 volts direct current (VDC). Molded-case devices cannot adequately interrupt DC faults, whereas low-voltage power circuit breakers appear to work satisfactorily (Morley, 1990).

9.8   ASSOCIATED EQUIPMENT SUSTAINABILITY AND STEWARDSHIP

Electrical rectifiers are defined as electrical circuits that convert an AC signal into a DC signal. Rectifier systems may contain such components as high-power thyristors, transformers, cooling units, overvoltage protection resistors, pulse amplifiers, heat-sink panels, and monitoring/control interface equipment. Sometimes referred to as rectifier diodes, rectifiers are electronic devices that allow current to flow in one direction only. Similarly, silicon controlled rectifiers (SCRs) are thyristors for forward bias, unidirectional power switching, and control. These rectifier systems can also be described as thyristors, converters, bridge rectifiers, and diode rectifiers.

9.8.1   Required Equipment Information

- Electrical rectifier type (enclosure type)
- Electrical rectifier specifications
  - DC voltage range (and DC current supply, kA)
  - Thyristor configurations (bridge, double-star, parallel)
  - Pulse number per unit
  - Dimensions, weight
  - UL certification
- Equipment foundation data (if applicable)

9.8.2   Required Testing Results

Required testing results for electrical rectifiers include the following:

- Contact resistance test results
- Insulation resistance test results
- Airborne ultrasonic test results
- Power factor test results
- High-voltage test results
- Infrared thermography (IRT) test results
- Turns ratio test results

9.8.3   Infrared Thermography Survey

Perform a thermographic survey to detect uneven heating that is indicative of loose or dirty terminal connections and other internal electrical connections to the individual components. The thermographic survey may also indicate internal corrosion or other flaws. Localized heating may be indicative of flaws in the windings or insufficient ventilation of the surrounding area. A bank of same-type electrical rectifiers with significantly different temperature readings may indicate unbalanced loading or a defective electrical rectifier. Electrical panel loading should be greater than 50% for performing this test. Minimum acceptable limits for this test are as follows:

- Qualitative verification of good terminal connections, internal connections to the individual components, and relative equivalent temperatures of the adjacent components during operation
  - Acceptable deviations in temperature which vary with the size and type of the electrical panel but do not exceed ±5°C
  - Terminal connections or internal connections to the individual components displaying temperatures that are ≤11.3°C than ambient

9.8.4   Electrical Acceptance Tests

- Perform electrical continuity tests on current, potential, and control circuits.
- Perform ratio and polarity tests on current and potential transformers (if applicable).
- Perform a current test on the remainder of the secondary circuit to detect any open or short-circuit connections.

9.8.5   Other Tests

- Use ultrasonics to verify the non-existence of electrical arcing and other high-frequency events.
- Determine that the power factor does not exceed manufacturer's data.
- Verify acceptable turns ratio (minimum acceptance criterion, ±2% design specifications) (NASA, 2004).

9.9   IMPLICATIONS FOR SAFETY AUDITS AND INSPECTIONS

Some people erroneously believe that direct current is a less hazardous form of electricity than alternating current. In truth, however, electricity is potentially harmful, regardless of whether the energy comes from DC or AC. Thus, verifying the electrical safety of DC systems is just as important as for AC systems. Some points to consider include the following:

1. *Are circuit protection means (i.e., fuses) adequate for the implementation?* Verify that circuit protection devices are being used and that the circuit protection is appropriate for the design specifications of the circuit.
2. *Are surge suppression devices being used and are the surge suppression devices adequate for the implementation?* Determine whether surge suppression devices are being used on all DC systems and that the devices are adequate to protect not only the electrical equipment itself but also human beings in the vicinity.
3. *Are ground-fault protection devices adequate for the implementation?*

REFERENCES

Morley, L.A. (1990). *Mine Power Systems*, Information Circular 9258, NTIS No. PB 91-241729. Pittsburgh, PA: U.S. Department of the Interior, Bureau of Mines.
NASA. (2004). *Reliability Centered Building and Equipment Acceptance Guide*. Washington, DC: National Aeronautics and Space Administration.
NFPA. (2005). *NFPA 70: National Electrical Code (NEC) Handbook*. Quincy, MA: National Fire Protection Association.
OSHA. (2007). Occupational Safety and Health Standards, 29 CFR, Part 1910, Subpart S, Electrical. Baltimore, MD: Occupational Safety and Health Administration.

10 Circuit Disconnects, On/Off Switches, and OSHA

Martha J. Boss and Dennis Day

CONTENTS

The term "switch" is somewhat confusing, as the Occupational Safety and Health Administration (OSHA) uses this term to denote both disconnecting (from power source) means and resistance insertion to delimit current flow (to run a motor). In this chapter, switches intended as disconnects are discussed as disconnecting switches or disconnects, whereas switches intended to prevent sufficient current flow to run a motor (without completely disconnecting from the power source) are discussed as on/off switches. *Note:* Snap switches are considered to be on/off switches.

Motors, motor-control apparatus, and motor branch-circuit conductors must be protected against overheating due to motor overloads or failure to start and against short circuits or ground faults. These provisions do not require overload protection that will stop a motor where a shutdown is likely to introduce additional or increased hazards. Examples include a fire pump, where continued operation of a motor is necessary for a safe equipment/process shutdown, and motor overload sensing devices connected to a supervised alarm.

10.1   TOTAL IMPEDANCE AND COMPONENT SHORT-CIRCUIT CURRENT RATINGS

The overcurrent protective devices, the total impedance, and the component short-circuit current ratings must be selected and coordinated to allow the circuit protective devices to clear a fault and to do so without the occurrence of extensive damage to the electrical components of the circuit. This fault may be assumed to be either between two or more of the circuit conductors or between any circuit conductor and the grounding conductor or enclosing metal raceway.

10.2   DISCONNECTING MEANS ON THE SUPPLY SIDE

Disconnecting means on the supply side are provided to independently disconnect a circuit of any voltage if the circuit includes readily accessible cartridge fuses or if the circuit is over 150 volts to ground and contains any type of fuse. Current-limiting devices without a disconnecting means are permitted by OSHA on the supply side of the service disconnecting means. Single disconnecting means are permitted by OSHA under the following circumstances:

- On the supply side of more than one set of fuses
- By the exception in 29 CFR 1910.305(j)(4)(vi) for group operation of motors
- For fixed electric space-heating equipment

10.3   DISCONNECTING MEANS FOR SERVICE, FEEDER, AND BRANCH CIRCUITS

Each service, feeder, and branch circuit, at its disconnecting means or overcurrent device, must be legibly marked to indicate its purpose, unless located and arranged so the purpose is evident.

10.4   UNGROUNDED CONDUCTORS AND SERVICE-ENTRANCE CONDUCTORS

Means must be provided to simultaneously disconnect all ungrounded conductors from the service-entrance conductors. The service disconnecting means must

- Plainly indicate whether the disconnect is in the open or closed position.
- Be installed at a readily accessible location nearest the point of entrance of the service-entrance conductors.
- Be suitable for prevailing conditions.

10.4.1   For Services over 600 Volts, Nominal

Service-entrance conductors installed as open wires must be guarded to make them accessible only to qualified persons. Signs warning of high voltage must be posted where unqualified employees might come in contact with live parts.

10.5   FEEDERS AND BRANCH CIRCUITS OVER 600 VOLTS, NOMINAL

Feeders and branch circuits over 600 volts, nominal, must have overcurrent protection in each ungrounded conductor located at (1) the point where the conductor receives its supply, or (2) a location in the circuit determined under engineering supervision.

10.6   THREE-PHASE CIRCUITS

Circuit breakers used for overcurrent protection of three-phase circuits must have a minimum of three overcurrent relays operated from three current transformers. On three-phase, three-wire circuits, an overcurrent relay in the residual circuit of the current transformers may replace one of the phase relays. An overcurrent relay that is operated from a current transformer that links all phases of a three-phase, three-wire circuit may replace the residual relay and one other phase conductor current transformer. Where the neutral is not grounded on the load side of the circuit, the current transformer may link all three phase conductors and the grounded circuit conductor (neutral).

10.7 OVERCURRENT PROTECTION, 600 VOLTS, NOMINAL, OR LESS

Conductors and equipment must be protected from overcurrent. Overcurrent devices should not interrupt the continuity of the grounded conductor unless all conductors of the circuit are opened simultaneously (the exception being motor overload protection). These devices must be

- Readily accessible to each employee
- Protected from exposure to physical damage
- Located away from easily ignitable material

10.8 DISCONNECTING MEANS FOR MOTOR CONTROL CENTERS AND CIRCUITS

Each disconnecting means required for motors and appliances must be legibly marked to indicate its purpose (i.e., what is being disconnected), unless located and arranged so the purpose is evident. Another term for motor disconnects is *motor control centers* (MCCs). MCCs can supply circuits that provide power to more than one motor at time, or they can be dedicated to only one motor supply on a particular piece of equipment.

The disconnecting means must disconnect the motor and the controller from all ungrounded supply conductors and must be so designed such that no pole can be operated independently. The disconnecting means must be capable of being locked in the open position and must plainly indicate the open (off) and closed (on) positions.

The disconnecting means must be readily accessible. If more than one disconnect is provided for the same piece of equipment, only one need be readily accessible. If one piece of equipment must be within sight of another piece of equipment, the pieces of equipment must be visible and ≤15.24 meters (50.0 feet) from each other.

At a minimum, an individual disconnecting means must be provided for each motor controller. A single disconnecting means may be used for a group of motors if

- A number of motors drive special parts of a single machine or piece of apparatus (e.g., metal or woodworking machine, crane, hoist).
- A group of motors is under the protection of one set of branch-circuit protective devices.
- A group of motors is in a single room within sight of the disconnecting means.

The disconnecting means must be located within sight of the controller location; however, a single disconnecting means may be located adjacent to a group of coordinated controllers mounted adjacent to each other (on a multiple-motor continuous process machine). The controller disconnecting means for motor branch circuits over 600 volts, nominal, may be out of sight of the controller if the controller is marked with a warning label giving the location and identification of the disconnecting means that is to be locked in the open position. Motor-circuit disconnect switches are rated in horsepower and can interrupt the motor's maximum operating overload current if the motor has the same horsepower rating.

10.9 CRANES, MONORAIL HOISTS, AND ALL RUNWAYS PER OSHA

A disconnecting means must be provided between the runway contact conductors and the power supply. Disconnecting means for cranes and monorail hoists must

- Include a motor-circuit switch, molded-case switch, or circuit breaker provided in the leads from the runway contact conductors or other power supply on all cranes and monorail hoists.
- Be capable of being locked in the open position.

Means must be provided at the operating station to open the power circuit to all motors of the crane or monorail hoist where the disconnecting means is not readily accessible from the crane or monorail hoist operating station. The disconnecting means may be omitted where a monorail hoist or hand-propelled crane bridge installation meets all of the following conditions:

- The unit is controlled from the ground or floor level.
- The unit is within view of the power supply disconnecting means.
- No fixed work platform has been provided for servicing the unit.

A limit switch or other device must be provided to prevent the load block from passing the safe upper limit of travel of any hoisting mechanism.

10.10 DISCONNECT SWITCH

The prime function of this disconnect switch is isolating outgoing circuits from the power source and consequently preventing ingoing circuits from delivering power to a load.

10.10.1 KNIFE DISCONNECTING SWITCHES

Single-throw knife disconnecting switches, molded-case disconnecting switches, disconnecting switches with butt contacts, and circuit breakers (used as switches) must be connected such that the terminals supplying the load are deenergized when the disconnecting means is in the open position. Where the disconnecting switch is connected to circuits or equipment inherently capable of providing a backfeed source of power,

- Blades and terminals supplying the load of a disconnecting switch may be energized when the disconnecting switch is in the open position.

• A permanent sign must be installed on the disconnecting switch enclosure or immediately adjacent to open disconnecting switches that reads as follows:

WARNING—LOAD SIDE TERMINALS
MAY BE ENERGIZED BY BACKFEED

Knife disconnecting switches may be either single or double throw, with various contacts. Each has specific requirements.

10.10.1.1   Single-Throw Knife Switches

Single-throw knife switches must be placed such that gravity will not tend to close them. If approved for use in the inverted position, they must be provided with a locking device that will ensure that the blades remain in the open position when so set. Both single-throw knife switches and switches with butt contacts must be connected so the blades are deenergized when the switch is in the open position.

10.10.1.2   Double-Throw Knife Switches

Double-throw knife switches may be mounted in such a way that the throw will be either vertical or horizontal; however, if the throw is vertical, a locking device must be provided to ensure that the blades remain in the open position when so set.

10.10.2   Snap (On/Off) Switches, Including Dimmer Switches

Snap switches must be effectively grounded and provide a means to ground a metal faceplate, whether or not a metal faceplate is installed based on 29 CFR 1910, Subpart S. A snap switch without a grounding connection is permitted for replacement purposes only in legacy installations that do not have grounding. These replacement snap switches must be provided with a faceplate of non-conducting, non-combustible material if located within reach of conducting floors or other conducting surfaces. Snap switches mounted in boxes must have faceplates installed to completely cover the opening and seat against the finished surface.

10.10.3   Disconnecting Switchboards and Panelboards

Panelboards must be mounted in cabinets, cutout boxes, or enclosures designed for the purpose and be dead front (unless accessible only to qualified persons). Exposed blades of knife disconnect switches mounted in switchboards or panelboards must be dead when open. Switchboards that have any exposed live parts must be located in permanently dry locations and accessible only to qualified persons (OSHA, 2007).

10.10.4   Isolating Switches

The isolating disconnect (switch) can be operated only after the circuit has been opened by some other means. Isolating disconnect switches do not have an interrupting rating. A trapped-key interlock system may be needed to ensure proper sequence of operation.

10.11   INTERRUPTER SWITCH

An interrupter switch is a combination of an air disconnect switch and a circuit interrupter. It has a continuous current interrupting rating less than or equal to the continuous rating of the switch at rated voltage. A full-load interrupter switch is an interrupter switch having a current interrupt rating equal to the continuous-current rating of the switch at the rated voltage. The interrupter can terminate currents that do not exceed the continuous-current rating. Interrupters have a quick-make/quick-break mechanism that provides a fast switch-operation speed (independent of the handle speed). Interrupters can be motor driven with remote or automatic operation.

Load-break switches require a close-and-latch rating. Where interlocks are not employed, the rating indicates the margin of safety when the switch is closed into a faulted circuit. Load-break switches and interlocks cause interruption of the source power prior to contact separation. This operation is usually performed through ground-monitoring circuitry, and load-break switches may be used in conjunction with fuses employed as interrupters. Equipment intended to interrupt current at fault levels must have an interrupting rating sufficient for the nominal circuit voltage and the current that is available at the line terminals of the equipment. Equipment intended to interrupt current at other than fault levels must have an interrupting rating at nominal circuit voltage sufficient for the current that must be interrupted (OSHA, 2007).

10.12   MECHANICALLY INTERLOCKED CONNECTORS

If a mechanical interlock is provided, the design must provide that the plug cannot be withdrawn before the circuit has been interrupted and the circuit cannot be established with the plug partially withdrawn.

10.13   ELECTRICALLY DRIVEN OR CONTROLLED IRRIGATION MACHINES

If an irrigation machine has a stationary point, a grounding electrode system must be connected to the machine at the stationary point for lightning protection. The main disconnecting means for a center pivot irrigation machine must be (1) located at the point of connection of electrical power to the machine or (2) visible and ≤15.2 meters (50 feet) from the machine. A disconnecting means must be provided for each motor and controller. The disconnecting means must be readily accessible and capable of being locked in the open position.

10.14   SWIMMING POOLS, FOUNTAINS, AND SIMILAR INSTALLATIONS

Equipment in or adjacent to all swimming, wading, therapeutic, and decorative pools and fountains; hydromassage bathtubs; and metallic auxiliary equipment (such as pumps, filters, and similar equipment) must be grounded. *Note:* Therapeutic pools in healthcare facilities are exempt from these provisions.

10.14.1 Receptacles

A single receptacle of the locking and grounding type that provides power for a permanently installed swimming pool recirculating pump motor must be located not less than 1.52 meters (5 feet) from the inside walls of a pool. All other receptacles on the property must be located at least 3.05 meters (10 feet) from the inside walls of a pool. Receptacles that are located within 4.57 meters (15 feet)—or 6.08 meters (20 feet) if the installation was built after August 13, 2007—of the inside walls of the pool must be protected by ground-fault circuit interrupters.

Where a pool is installed permanently at a dwelling unit, at least one 125-volt, 15- or 20-ampere receptacle on a general-purpose branch circuit must be located a minimum of 3.05 meters (10 feet) and ≤6.08 meters (20 feet) from the inside wall of the pool. This receptacle must be located ≤1.98 meters (6.5 feet) above the floor, platform, or grade level serving the pool. *Note:* When determining these dimensions, the distance to be measured is the shortest path the supply cord of an appliance connected to the receptacle would follow without piercing a floor, wall, or ceiling of a building or other effective permanent barrier.

10.14.2 Lighting Fixtures, Lighting Outlets, and Ceiling-Suspended (Paddle) Fans

Lighting in outdoor pool areas may not be installed over the pool or over the area extending 1.52 meters (5 feet) horizontally from the inside walls of a pool—unless no part of the lighting fixture of a ceiling-suspended (paddle) fan is less than 3.66 meters (12 feet) above the maximum water level. Lighting fixtures and lighting outlets installed in the area extending between 1.52 meters (5 feet) and 3.05 meters (10 feet) horizontally from the inside walls of a pool must be protected by a ground-fault circuit interrupter unless they are installed 1.52 meters (5 feet) above the maximum water level and are rigidly attached to the structure adjacent to or enclosing the pool.

Flexible cords used with the following equipment may not exceed 0.9 meters (3 feet) in length and must have a copper equipment grounding conductor with a grounding-type attachment plug: (1) cord- and plug-connected lighting fixtures installed within 4.88 meters (16 feet) of the water surface of permanently installed pools, and (2) other cord- and plug-connected fixed or stationary equipment used with permanently installed pools.

10.14.3 Underwater Equipment

A ground-fault circuit interrupter must be installed in the branch circuit supplying underwater fixtures operating at more than 15 volts. Equipment installed underwater must be identified for the purpose. No underwater lighting fixtures may be installed for operation at over 150 volts between conductors. A lighting fixture facing upward must have the lens adequately guarded to prevent contact by any person.

10.14.4 FOUNTAINS

All electric equipment, including power supply cords, operating at more than 15 volts and used with fountains must be protected by ground-fault circuit interrupters. Fountains are defined by OSHA as ornamental pools, display pools, and reflection pools. This definition does not include drinking fountains. The requirements for fountains in 29 CFR 1910.306(j)(5) do not apply to drinking fountains and water coolers. *Note:* The National Electric Code (NEC) does not intend to apply the requirements on fountains to drinking fountains. NEC states that the definition of "fountains" does not include drinking fountains.

10.15 SERIES COMBINATION RATINGS

Where circuit breakers or fuses are applied in compliance with the series combination ratings marked on the equipment by the manufacturer:

- Equipment enclosures must be legibly marked in the field to indicate that the equipment has been applied with a series combination rating.
- The marking must be readily visible and read as follows:

CAUTION—SERIES COMBINATION SYSTEM RATED [INSERT NUMBER] AMPERES. IDENTIFIED REPLACEMENT COMPONENT REQUIRED.

10.16 ENCLOSURES FOR DAMP OR WET LOCATIONS

Cabinets, cutout boxes, fittings, boxes, and panelboard enclosures in damp or wet locations must be

- Weatherproof
- Installed to prevent moisture or water from entering and accumulating within the enclosures
- Mounted with 6.35-mm (0.25-in.) airspace between the enclosure and the wall or other supporting surface, although non-metallic enclosures may be installed without the airspace on a concrete, masonry, tile, or similar surface

Switches, circuit breakers, and switchboards installed in wet locations must be enclosed in weatherproof enclosures (OSHA, 2007).

10.17 DISCONNECTING MEANS FOR ELEVATORS, ESCALATORS, AND LIFTS

Disconnects must have a single means for disconnecting all ungrounded main power supply conductors for each unit. The disconnecting means must be

- An enclosed externally operable fused motor circuit switch or circuit breaker capable of being locked in the open position
- A listed device (by a nationally recognized testing laboratory, or NRTL)

- Capable of being restored (to power) only by manual means, with no automatic restoration allowed
- Located where readily accessible to qualified persons

Opening or closing this disconnecting means from any other part of the premises must not be allowed; however, if sprinklers are installed in hoistways, machine rooms, or machinery spaces, the disconnecting means may automatically open the power supply to the affected elevators prior to the application of water. Control panels not located in the same space as the drive machine must be located in cabinets with doors or panels capable of being locked closed.

10.17.1   ELEVATORS WITHOUT GENERATOR FIELD CONTROL

On elevators without generator field control, the disconnecting means must be located within sight of the motor controller. Driving machines (or motion and operation controllers) not within sight of the disconnecting means must be provided with a manually operated switch installed in the control circuit adjacent to the equipment in order to prevent starting. Where the driving machine is located in a remote machinery space, a single disconnecting means for disconnecting all ungrounded main power supply conductors must be provided and be capable of being locked in the open position.

10.17.2   ELEVATORS WITH GENERATOR FIELD CONTROL

On elevators with generator field control, the disconnecting means must be located within sight of the motor controller for the driving motor of the motor–generator set. Driving machines, motor–generator sets, or motion and operation controllers not within sight of the disconnecting means must be provided with a manually operated switch installed in the control circuit to prevent starting. The manually operated switch should be installed adjacent to this equipment. Where the driving machine or the motor–generator set is located in a remote machinery space, a single means for disconnecting all ungrounded main power supply conductors must be provided and be capable of being locked in the open position. On escalators and moving walks, the disconnecting means must be installed in the space where the controller is located. On wheelchair lifts and stairway chair lifts, the disconnecting means must be located within sight of the motor controller.

10.17.3   IDENTIFICATION AND SIGNS

The disconnecting means must be provided with a sign to identify the location of the supply-side overcurrent protective device. For installations with more than one driving machine in a machine room, the disconnecting means must be numbered to correspond to the identifying number of the driving machine controlled. Warning sign for multiple disconnecting means must be clearly legible and read as follows:

WARNING—PARTS OF THE CONTROLLER
ARE NOT DEENERGIZED BY THIS SWITCH

Where multiple disconnecting means are used and parts of the controllers remain energized from a source (other than the one disconnected), the warning sign should be mounted on or next to the disconnecting means. Where interconnections between controllers are necessary for the operation of the system on multicar installations that remain energized from a source (other than the one disconnected), the warning sign should be mounted on or next to the disconnecting means.

10.17.4 SINGLE-CAR AND MULTICAR INSTALLATIONS

On single-car and multicar installations, equipment receiving electrical power from more than one source must be provided with a disconnecting means for each source of electrical power and within sight of the equipment served.

10.17.5 MOTOR CONTROLLERS

Motor controllers may be located outside the spaces where equipment is immediately located provided the motor controllers are in enclosures with doors or removable panels and are capable of being locked closed. The disconnecting means must be located adjacent to or be an integral part of the motor controller. Motor controller enclosures for escalators or moving walks may be located in the balustrade on the side located away from the moving steps or moving treadway. Disconnects that are an integral part of the motor controller must be operable without opening the enclosure.

10.18 DISCONNECTING MEANS FOR ELECTRIC WELDERS

10.18.1 ARC WELDERS

For each arc welder not equipped with an integral (mounted as part of the welder) disconnect, a disconnecting means must be provided in the supply circuit. The disconnecting means must be a switch or circuit breaker, and it must be rated at no less than necessary to accommodate overcurrent protection.

10.18.2 RESISTANCE WELDERS

A switch or circuit breaker must be provided by which each resistance welder (and control equipment) can be disconnected from the supply circuit. The ampere rating of this disconnecting means may not be less than the supply conductor ampacity. Where the circuit supplies only one welder, the supply circuit switch may be used as the welder disconnecting means.

10.19 DISCONNECTING MEANS FOR INFORMATION TECHNOLOGY EQUIPMENT

An approved means must be provided to disconnect power to all electronic equipment in an information technology equipment room, as well as a similar means to disconnect the power to all dedicated heating, ventilating, and air-conditioning

(HVAC) systems serving the room and to cause all required fire/smoke damp-ers to close. *Note:* A single means to control both the electronic equipment and HVAC system is permitted. The control for these disconnecting means will be grouped and identified and will be readily accessible at the principal exit doors. *Note:* Integrated electrical systems used for fire prevention equipment need not have disconnecting means.

10.20   DISCONNECTING MEANS FOR X-RAY EQUIPMENT

A disconnecting means must be provided in the supply circuit. The disconnect-ing means must be operable from a location readily accessible from the x-ray con-trol. For equipment connected to a 120-volt branch circuit of 30 amperes or less, a grounding-type attachment plug cap and receptacle of proper rating may serve as a disconnecting means. If more than one piece of equipment is operated from the same high-voltage circuit, then each piece or each group of equipment as a unit must be provided with a high-voltage switch or equivalent disconnecting means. Also, the disconnecting means must be constructed, enclosed, or located so as to avoid contact by employees with the live parts of the disconnecting means.

10.21   INDUSTRIAL AND COMMERCIAL LABORATORY EQUIPMENT

Radiographic and fluoroscopic-type equipment must be effectively enclosed or have interlocks that deenergize the equipment automatically to prevent ready access to live current-carrying parts. Diffraction- and irradiation-type equipment must have a pilot light, readable meter deflection, or equivalent means to indicate when the equipment is energized, unless the equipment or installation is effec-tively enclosed or is provided with interlocks to prevent access to live current-carrying parts during operation.

10.22   INDUCTION AND DIELECTRIC HEATING EQUIPMENT FOR INDUSTRIAL AND SCIENTIFIC APPLICATIONS

(*Note:* The OSHA provisions provided here do not apply to medical or dental appli-cations or to appliances.) The converting apparatus (including the DC line) and high-frequency electric circuits (excluding the output circuits and remote-control circuits) must be completely contained within enclosures of non-combustible material. All panel controls must be of dead-front construction.

Doors or detachable panels must be employed for internal access. Where doors are used to give access to voltages from 500 to 1000 volts AC or DC, either door locks must be provided or interlocks must be installed. Where doors are used to give access to voltages of over 1000 volts AC or DC, either (1) mechanical lockouts with a disconnecting means to prevent access until circuit parts within the cubicle are deen-ergized or (2) both door interlocking and mechanical door locks must be provided.

Detachable panels not normally used for access to such parts must be fastened in a manner that will make them difficult to remove (e.g., requiring the use of tools). Warning labels or signs must be posted that read as follows:

DANGER—HIGH VOLTAGE—KEEP OUT

These labels must be attached to the equipment and be plainly visible where persons might contact energized parts when doors are opened or closed or when panels are removed from compartments containing over 250 volts AC or DC.

Induction and dielectric heating equipment must be protected by

- Protective cages or adequate shielding used to guard work applicators other than induction heating coils
- Induction heating coils with insulation or refractory materials or both
- Interlock switches on all hinged access doors, sliding panels, or other such means of access to the applicator, unless the applicator is an induction heating coil at DC ground potential or operating at less than 150 volts AC
- Interlock switches connected in such a manner as to remove all power from the applicator when any one of the access doors or panels is open

A readily accessible disconnecting means must be provided so each piece of heating equipment can be isolated from its supply circuit. The ampere rating of this disconnecting means may not be less than the nameplate current rating of the equipment. *Note:* The supply circuit disconnecting means is permitted as a heating equipment disconnecting means where the circuit supplies only one piece of equipment.

If remote controls are used for applying power, a selector switch must be provided and interlocked to provide power from only one control point at a time. Switches operated by foot pressure must be provided with a shield over the contact button to avoid accidental closing of the switch (OSHA, 2007).

10.23 IMPLICATIONS FOR SAFETY AUDITS AND INSPECTIONS

When inspecting electrical switches of any kind, consider the following points:

1. *Is the switch appropriate for the application?*
2. *Is the switch rated for the environment?* For example, if the switch is located in a wet environment, the switch box should be rated for operation in a wet environment.
3. *Is the approach to the switch unobstructed?* For emergency shut-off purposes, the approach to the switch should not be obstructed, and it should not be mounted so high that a person could not be expected to reach the switch without additional aid.
4. *Is the switch labeled as to intended purpose?* The switch should be labeled clearly to identify the purpose of the switch.

5. *Have personnel that would be expected to operate the switch been trained on safe operation of the switch?* Verify training records.
6. *Are personnel that would be expected to operate the switch familiar with the proper procedure for operating the switch?* Verify by interviewing personnel onsite.

REFERENCES

NFPA. (2005). *NFPA 70: National Electrical Code (NEC) Handbook.* Quincy, MA: National Fire Protection Association.
OSHA. (2007). Occupational Safety and Health Standards, 29 CFR, Part 1910, Subpart S, Electrical. Baltimore, MD: Occupational Safety and Health Administration.

11 Grounding

Martha J. Boss and Dennis Day

CONTENTS

11.1 ISOLATION

Grounding isolates offending sections of the electrical system by selective relaying of ground faults. Sensitivity and time delays of the protective circuitry are adjusted. Protective circuitry is adjusted so that a fault in an area will cause a local breaker to sense a malfunction and quickly remove power from only that affected section. If relative tripping levels and speeds are not established correctly, then nearby breakers may not trip at the correct time. The relaying system must be arranged so that even at the lowest level of the power-distribution chain sufficient fault current can flow to enable the protective circuitry to sense the fault and take remedial action.

11.2 GROUNDING

Grounding works by limiting potential gradients, limiting energy, or controlling overvoltages.

11.2.1 LIMIT POTENTIAL GRADIENTS

Grounding limits the potential gradients between conducting materials in a given area. For example, during a ground fault, a phase conductor comes into contact with a machine frame. Current flows through equipment and the potential of the equipment becomes elevated above the ground potential by an amount equal to the voltage on the conductor. If a person touches the machine while simultaneously in contact with the ground (earth), the potential of the person's body is elevated. The maximum potential when touching a machine frame is equal to the voltage drop along the grounding conductors. The grounding system must provide a low-resistance path for fault current to return to the source. Ground conductors should have a low resistance to allow the conductors to carry the maximum expected fault current without excessive voltage drop.

11.2.2 LIMIT ENERGY

Grounding limits the energy that leads to premature failures, reduced component life, and mysterious nuisance trips by providing a path between the transformer neutral and ground. Most of the sources of transient overvoltages should be reduced or possibly eliminated by grounding. During a surge event, the grounding conductor becomes ionized, making the conductor capable of carrying tremendous amounts of current. High-energy faults can vaporize breakers, switchgear, and phase conductors. Protective enclosures may be blown apart with explosive force. Controlling the maximum and allowable current significantly reduces the danger of fire and holds equipment damage to a minimum.

11.2.3 CONTROL OVERVOLTAGES

An overvoltage condition may occur by accidental contact of equipment with a higher voltage system or from transient phenomena due to lightning strikes, intermittent ground faults, autotransformer connections, and switching surges. The maximum ratings may be temporarily exceeded in cable insulation, transformer windings, and relay contactors. Component parts of the electrical system are successively overstressed and weakened by repeated exposure. Overvoltages lead to premature failures and reduced component life. Mysterious nuisance trips can occur without apparent reason. By providing a path between the transformer neutral and ground, most of the sources of transient overvoltages can be reduced or possibly eliminated.

11.3 GROUNDING SYSTEMS AS REQUIRED BY OSHA

The Occupational Safety and Health Administration (OSHA) provides requirements for grounding that assume grounding electrodes can be directly inserted at locations proximate to power usage; however, in some applications, cables carrying grounding conductors to exterior grounding beds must be used.

11.3.1 THREE-WIRE DC SYSTEMS

All three-wire direct current (DC) systems must have their neutral conductor grounded. Two-wire DC systems operating at >50 V through 300 V between conductors must be grounded unless these systems (1) supply only industrial equipment in limited areas and are equipped with a ground detector, (2) are rectifier-derived from an alternating current (AC) system, or (3) are fire-alarm circuits having a maximum current of 0.030 amperes.

11.3.2 AC CIRCUITS <50 VOLTS

Alternating current circuits of <50 volts must be grounded if installed as overhead conductors outside of buildings or supplied by transformers with a primary supply system that is ungrounded or exceeds 150 volts to ground.

11.3.3 AC CIRCUITS ≥50 VOLTS TO 1000 VOLTS

Alternating current systems of ≥50 volts to 1000 volts must be grounded if

- The system can be so grounded that the maximum voltage to ground on the ungrounded conductors is not >150 volts.
- The system is nominally rated three-phase, four-wire wye connected, in which the neutral is used as a circuit conductor.
- The system is nominally rated three-phase, four-wire delta connected, in which the midpoint of one phase is used as a circuit conductor.
- The service conductor is uninsulated.

AC systems ranging from ≥50 volts to 1000 volts are not required to be grounded if the electric system is

- Used exclusively to supply industrial electric furnaces for melting, refining, and tempering
- Separately derived and is used exclusively for rectifiers supplying only adjustable-speed industrial drives
- Used exclusively for control circuits, is separately derived, is supplied by a transformer that has a primary voltage rating <1000 volts, and all of the following conditions are met:
  - The system is used exclusively for control circuits.
  - Conditions of maintenance and supervision ensure that only qualified persons will service the installation.
  - Continuity of control power is required.
  - Ground detectors are installed on the control system.
- An isolated power system that supplies circuits in healthcare facilities
- A high-impedance grounded neutral system in which a grounding impedance (usually a resistor) limits the ground-fault current to a low value for three-phase AC systems of 480 to 1000 volts and all of the following conditions are met:
  - Conditions of maintenance and supervision ensure that only qualified persons will service the installation.
  - Continuity of power is required.
  - Ground detectors are installed on the system.
  - Line-to-neutral loads are *not* served.

11.3.4   PREMISES WIRING

Systems supplying premises wiring must be grounded. All three-wire DC systems must have the neutral conductor grounded. Two-wire DC systems operating at >50 volts through 300 volts between conductors must be grounded unless the systems:

- Supply only industrial equipment in limited areas and are equipped with a ground detector.
- Are rectifier-derived from an AC system.
- Are fire-alarm circuits having a maximum current of 0.030 amperes.

The conductor to be grounded for AC premises wiring systems must be

- One conductor of a single-phase, two-wire system
- A neutral conductor of a single-phase, three-wire system
- A common conductor of a multiphase system having one wire common to all phases
- A one-phase conductor of a multiphase system where one phase is grounded
- A neutral conductor of a multiphase system in which one phase is used as a neutral conductor

11.4 GROUNDING SYSTEMS

The basic resistance-grounded system consists of a resistor inserted between the power-system neutral point and the ground. Concerns when selecting the grounding resistor are resistance, time rating, insulation, and available connection to the power-system neutral. Ground current can be limited at a level less than the restricted maximum for high-resistance grounding. The smallest value chosen has two concerns: ground-fault relaying and charging current. For maximum safety ground, protective circuitry should sense ground current at a fraction of the current limit.

Reliable relay operation with electromechanical devices can be a problem if the maximum current is <15 to 20 A. The limitation is that the ground-fault current should always be greater than the system-charging current (the current required to charge system capacitance) when the system is energized. When very low ground-relay settings are used, charging current may itself cause tripping.

11.4.1 GROUNDING RESISTOR

Another concern in selecting a grounding resistor is the time rating (ability to dissipate heat). The grounding resistor carries only a very small current under normal system operation. When a ground fault occurs, current may approach full value. This high current exists until the circuit breaker removes power from the faulted circuit. Power removal may take from a fraction of a second to several seconds. Protection devices have been known to malfunction and ground current may continue to flow until power is removed manually. The resistor must be able to dissipate power produced from full ground current for an extended time when portable or mobile equipment is involved. If the power is not dissipated, the resistor can burn open and unground the system.

To provide a safety margin, the transformer-neutral side of the resistor (i.e., the hot side) must be insulated from ground at a level to withstand line-to-line system voltage. Both resistor ends are at ground potential with normal operation. Under a ground fault, the transformer end can approach line-to-neutral potential. To afford good insulation, the resistor frame is placed on porcelain insulators. For wye-connected secondaries, the transformer-neutral bushing must be insulated to at least line-to-neutral voltage. The grounding resistor is installed between the transformer neutral and the safety ground bed.

In substations, using insulated conductors is important because bare conductors can easily compromise the required separation between the system and safety ground beds. The grounding conductors must extend from the ground-bed side of the resistor. To minimize resistor conductor lengths, the resistor must be located on the power-source end distribution as close as possible to the source power transformer. Distances greater than 100 feet are usually too long.

11.4.2 GROUNDING TRANSFORMERS

Delta–wye, wye–delta, and delta–delta power transformers offer very high impedance to zero-sequence currents. A ground fault existing on the secondary will only raise the primary line current. If the transformer has a delta secondary, no neutral point exists to which a grounding system can be connected. A separate grounding transformer is needed to obtain an artificial neutral. Two types of grounding transformers are in general use: zigzag and wye–delta.

11.5   GROUNDING SYSTEM TYPES

Several different types of AC grounded systems are available:

- *Ungrounded neutral*—Usually used on delta systems; however, it can also be used on a floating wye. Because the system is ungrounded, a detection system must be used.
- *Solidly grounded*—Usually used in wye systems. A solidly grounded system uses a grounding electrode to ground the entire system to earth potential. This is usually a safe system with good lightning protection.
- *Resistance-grounded*—These systems can be either low-resistance or high-resistance grounded. Low resistance uses a low-resistance electrode to ground. This method limits the fault current and minimizes the damage at the fault. Resistance grounding is usually used with multiple sources (e.g., transformers, generators). High resistance uses a higher resistance electrode to ground. High resistance allows the system to operate with one phase grounded. Current is allowed to flow in the grounding wires, which allows for detection and monitoring within the network. However, the current is extremely low, so no damage to the system occurs.
- *Capacitive ground*—A capacitor is placed between the system and the grounding electrode.
- *Reactance-grounded*—In this case, a reactor is connected between the system neutral and the grounding electrode. This reduces the ground-fault current between the phase and the ground. Reactance grounding is mostly used on systems above 5000 volts. Reactance-grounded systems are not normally used in industrial power systems.
- *Ground-fault neutralizer*—This is actually a subclass of reactance grounding in which the reactor has a high reactance value. As a specialty method, ground-fault neutralizers are mostly used on systems above 15,000 volts phase to ground.

The selection of which system to use should depend on an evaluation of the following factors:

- Safety
- Ground-fault detection
- Signal reference
- Available fault currents
- Operation stability
- System response

11.5.1 Ungrounded Neutral

The ungrounded neutral system was the first grounding method. No intentional ground connections are used; however, a perfect ungrounded system cannot exist. Any current-carrying conductor may be coupled to ground through numerous paths (e.g., distributed capacitance of conductors, wiring, motor windings). A ground fault has very little effect on the equipment within the system, as fault current cannot find a complete circuit back to the source. When selecting this method, keep in mind that the magnitude of the hazard must be very small or nil (very low fault current, no flash hazard, no equipment damage). During the hazard, the circuit operation continues normally (no interruption of power). The first fault is often difficult to locate because the fault's effects are negligible; therefore, no repair effort is made until a second fault occurs with concomitant hazards (arcing, heavy current flow, equipment damage). The entire system is *floating* with regard to the control of transients. Overvoltages are a problem if accidental contact occurs with a higher voltage system. All other overvoltage sources mentioned previously are enhanced because of distributed capacitance to ground.

11.5.2 Solidly Grounded Neutral

The first ground fault produces a substantial neutral current flow that may be quickly sensed by protective circuitry, shutting down the faulting section's overvoltage. The controlled system has its neutral solidly referenced to ground hazards due to the magnitude of the fault current. The detection equipment must be sensitive enough to detect low-level fault currents and fast enough to disconnect faulting circuits before heavy faults can disrupt system integrity. Large fault currents, typically several thousand amperes, can explode protective enclosures, destroy equipment, and start fires. This grounding technique must not be used in explosive atmospheres.

11.5.3 Low-Resistance Grounded Neutral

The low-resistance grounded neutral is established by inserting a resistor between the system neutral and a ground resistance. The ground-fault currents are limited from 50 to 600 amperes, commonly about 400 amperes. Transients are controlled by a ground connection. Ample fault current is available for actuating protective relays. Flash hazard is not as serious as in solidly grounded neutral systems; however, current flow of 400 amperes can still do considerable damage. To limit damage, the least sensitive ground relay should respond to 10% of maximum ground-fault current.

11.5.4 High-Resistance Grounded Neutral/Safety Ground System

The high-resistance grounded neutral is established by inserting a high-rating resistor between the system neutral and the ground resistance. The safety ground system neutral grounding resistor is sized according to the system voltage level to limit ground-fault current to ≤50 amperes. Where line-to-neutral potential is ≤1000 volts, the grounding resistor must limit fault current to ≤25 amperes. For >1000 volts, the voltage drop in the grounding circuit external to the resistor must be ≤100 volts under fault conditions. With this system, sensitive relaying must detect faults on the order of a few amperes to provide fault isolation and facilitate quick location of the trouble spot. The level of fault current must be low enough to practically eliminate arcing and flashover dangers. The ground connection will also serve to limit the amplitude of overvoltages. Loads cannot be connected line-to-neutral and the grounding conductor must not carry any load current.

11.5.5   CAPACITIVE GROUND

The electrical conductors and the windings of all components are capacitively connected to the ground although not physically connected to ground. A small current will flow to ground from each phase. Current does not occur at any particular location and instead is distributed throughout the system. Capacitance to ground is distributed throughout the system.

11.6   CORD SETS AND RECEPTACLES: TESTING AND OSHA

Tests must be performed on all cord sets and receptacles required to be grounded:

- All equipment grounding conductors must be tested for continuity and must be electrically continuous.
- Each receptacle and attachment cap or plug must be tested for correct attachment of the equipment grounding conductor.
  - The equipment grounding conductor must be connected to its proper terminal.
  - All required tests must be performed before first use, before equipment is returned to service following repairs, before equipment is used after an incident (which can be reasonably suspected to have caused damage), and at intervals not to exceed 3 months, with the exception that cord sets and receptacles that are fixed and not exposed to damage must be tested at intervals not exceeding 6 months.

Equipment that has not met the test requirements must not be used. Tests performed must be documented. This test record must identify each receptacle, cord set, and cord- and plug-connected equipment that passed the test and must indicate the last date on which this equipment was tested or the test interval. This record must be kept by means of logs, color coding, or other effective means and must be maintained until replaced by a more current record. The record must be made available on the jobsite.

11.7   GROUND CONDUCTOR IDENTIFICATION

A grounding conductor must be identifiable and distinguishable from all other conductors. An equipment grounding conductor must also be identifiable and distinguishable from all other conductors.

11.7.1 MULTIWIRE BRANCH CIRCUITS

The National Electrical Code (NEC) defines a multiwire branch circuit as

> A branch circuit consisting of two or more ungrounded conductors having a potential difference between them, and a grounded circuit conductor having equal potential difference between it and each ungrounded conductor of the circuit and which is connected to the neutral (grounded) conductor of the system.

The rationale for using this type of circuit may be that loads are intermittent (machines run for short periods of time) and, therefore, the possibility of faulting out the system with a total combined ground fault from all circuit branches is not likely. However, in some situations the neutral conductor can carry the short-circuit/ground-fault load from both (multiple) circuits and be overwhelmed by this situation (e.g., arc or fire ensues).

Section 210.4(B) of the National Electric Code states that in the panelboard where the branch circuit originates all ungrounded conductors must be provided with a means to disconnect them simultaneously. Section 210.4(D) also requires that the ungrounded and grounded circuit be grouped by cable ties or similar means in at least one location within the panelboard.

Multiwire branch circuits should be protected by a double-pole common-internal trip circuit breaker, including the physical "trip tie" that bonds the two circuit breaker switches together. This is a safety measure that helps ensure that the circuit is wired properly at the panel. OSHA requires that, where more than one nominal voltage system exists in a building containing multiwire branch circuits, each ungrounded conductor of a multiwire branch circuit, where accessible, must be identified by phase and system. The means of identification must be permanently posted at each branch-circuit panelboard. Multiwire branch circuits must be provided with a means to disconnect simultaneously all ungrounded conductors at the power outlet or panelboard where the branch circuit originated. Circuit breakers with their handles connected by approved handle ties are considered a single disconnecting means for the purpose of this requirement.

11.7.2 GROUNDING FOR CONDUCTORS

Conductors to be grounded for AC wiring systems that are required to be grounded include the following:

- One conductor of a single-phase, two-wire system
- The neutral conductor of a single-phase, three-wire system
- The common conductor of a multiphase system having one wire common to all phases
- One phase conductor of a multiphase system where one phase is grounded
- The neutral conductor of a multiphase system in which one phase is used as a neutral conductor

11.7.3 GROUNDING CONNECTIONS

For a grounded system, a grounding electrode conductor must be used to connect both the equipment grounding conductor and the grounded circuit conductor to the grounding electrode. Both the equipment grounding conductor and the grounding electrode conductor must be connected to the grounded circuit conductor on the supply side of the service disconnecting means or on the supply side of the system disconnecting means or overcurrent devices if the system is separately derived.

For an ungrounded service-supplied system, the equipment grounding conductor must be connected to the grounding electrode conductor at the service equipment. For an ungrounded separately derived system, the equipment grounding conductor must be connected to the grounding electrode conductor at, or ahead of, the system disconnecting means or overcurrent devices.

On extensions of existing branch circuits that do not have an equipment grounding conductor, grounding-type receptacles may be grounded to a grounded cold-water pipe near the equipment if the extension was installed before August 13, 2007. When any element of this branch circuit is replaced, the entire branch circuit will use an equipment grounding conductor.

11.7.4 CIRCUIT GROUNDS FOR RECEPTACLES AND CORD CONNECTORS

Receptacles installed on 15- and 20-ampere branch circuits must be of the grounding type. Grounding-type receptacles must be installed only on circuits of the voltage class and current as rated, except as provided in Tables 11.1 and 11.2. Receptacles and cord connectors having grounding contacts must be effectively grounded; that is, the grounding contacts must be grounded by connection to the equipment grounding conductor of the supply circuit (branch circuit wiring method). Exceptions to this rule may include receptacles mounted on portable and vehicle-mounted generators and replacement receptacles.

TABLE 11.1

Maximum Cord- and Plug-Connected Load to Receptacle

Circuit Rating (Amperes)	Receptacle Rating (Amperes)	Maximum Load (Amperes)
15–20	15	12
20	20	16
30	30	24

Source: OSHA, Occupational Safety and Health Standards, 29 CFR, Part 1910, Subpart S, Electrical, Occupational Safety and Health Administration, Baltimore, MD, 2007.

TABLE 11.2

Receptacle Ratings for Various Size Circuits

Circuit Rating (Amperes)	Receptacle Rating (Amperes)
15	Not over 15
20	15 or 20
30	30
40	40 or 50
50	50

Source: OSHA, Occupational Safety and Health Standards, 29 CFR, Part 1910, Subpart S, Electrical, Occupational Safety and Health Administration, Baltimore, MD, 2007.

11.7.5 GROUNDING PATH

The path to ground from circuits, equipment, and enclosures must be permanent, continuous, and effective. Supports, enclosures, and equipment must be grounded.

11.7.6 SERVICE-SUPPLIED SYSTEM AND GROUNDING

For a grounded system, a grounding electrode conductor must be used to connect both the equipment grounding conductor and the grounded circuit conductor to the grounding electrode (to earth). Both the equipment grounding conductor and the grounding electrode conductor must be connected to the grounded circuit conductor on the supply side of the service disconnecting means or on the supply side of the system disconnecting means or overcurrent devices if the system is separately derived.

For an ungrounded service-supplied system, the equipment grounding conductor must be connected to the grounding electrode conductor at the service equipment. For an ungrounded separately derived system, the equipment grounding conductor must be connected to the grounding electrode conductor at, or ahead of, the system disconnecting means or overcurrent devices. The path to ground from circuits, equipment, and enclosures must be permanent, continuous, and effective.

11.7.7 GROUNDING ELECTRICAL EQUIPMENT AND FRAMES

Electrical equipment is considered to be effectively grounded if the equipment is secured to, and in electrical contact with, a structure that is grounded. For circuitry grounding, the grounding prong is provided to ground the equipment through a cord back to the electrical socket, which is in turn connected to either a building ground or an electrical supply ground. For electrical equipment frames (not normally electrified), grounding can be achieved using a metal rack or structure provided for the equipment's support that is grounded. The metal rack or structure must be grounded by an equipment grounding conductor that ultimately is continuous to an earthing ground. For DC circuits only, the equipment grounding conductor may be run separately from the circuit conductors.

11.7.7.1   Metal Cable Trays, Metal Raceways, and Metal Enclosures

Metal raceways, cable trays, cable armor, cable sheath, enclosures, frames, fittings, and other metal non-current-carrying parts that are to serve as grounding conductors, with or without the use of supplementary equipment grounding conductors, must be effectively bonded to ensure electrical continuity and the capacity to safely conduct fault current. Any nonconductive paint, enamel, or similar coating must be removed at threads, contact points, and contact surfaces or be connected by means of fittings designed so as to make such removal unnecessary. Where necessary for the reduction of electrical noise (electromagnetic interference) of the grounding circuit, an equipment enclosure supplied by a branch circuit may be isolated from a raceway containing circuits supplying only that equipment. Isolation may be achieved by using one or more listed non-metallic raceway fittings located at the point of attachment of the raceway to the equipment enclosure. The metal raceway must be supplemented by an internal insulated equipment grounding conductor installed to ground the equipment enclosure.

Note: The following items do not require grounding per OSHA:

- Metal enclosures (e.g., sleeves that are used to protect cable assemblies from physical damage)
- Metal enclosures for conductors added to existing installations of open wire, knob-and-tube wiring, and non-metallic-sheathed cable if *all* of the following conditions are met:
  - Runs are less than 7.62 meters (25.0 feet).
  - Enclosures are free from probable contact with ground, grounded metal, metal laths, or other conductive materials.
  - Enclosures are guarded against employee contact.

11.7.7.2   Service Equipment Metal Enclosures

Metal enclosures for service equipment must be grounded. Exposed non-current-carrying metal parts of fixed equipment that may become energized must be grounded if these parts are

- Within 2.44 meters (8 feet) vertically or 1.52 meters (5 feet) horizontally of ground or grounded metal objects and subject to employee contact
- Located in a wet or damp location and not isolated
- In electrical contact with metal
- In a hazardous (classified) location
- Supplied by a metal-clad, metal-sheathed, or grounded metal raceway wiring method
- Operating with any terminal at over 150 volts to ground.

11.7.7.3   Grounding Fixed Equipment

Non-current-carrying metal parts of fixed equipment must be grounded by an equipment grounding conductor that is contained within the same raceway, cable, or cord or runs with or encloses the circuit conductors. For installations made before

April 16, 1981, electric equipment is also considered to be effectively grounded if the metal frame of the equipment is secured to, and in metallic contact with, the grounded structural metal frame of a building. *Note:* When any element of this branch circuit is replaced, the entire branch circuit must use an equipment grounding conductor.

For DC circuits only, the equipment grounding conductor may be run separately from the circuit conductors. Electric equipment is considered to be effectively grounded if the metal frame of the equipment is secured to, and in electrical contact with, a supporting metal rack or structure. The metal rack or structure must be grounded by the method specified for the non-current-carrying metal parts of fixed equipment. Effective grounding may also be achieved when metal frames are secured to metal car frames supported by metal hoisting cables attached to or running over metal sheaves or drums of grounded elevator machines.

Ideally, all points within an electrical system should be at the same relative voltage. In order to protect employees, the ground should be of sufficiently low resistance to discharge a ground fault and yet correctly chosen so that overcurrrent protection devices will activate. In order to comply with regulations and to protect workers, the building's low-resistance ground should be checked on a routine basis as part of a holistic electrical safety program.

When referring to electrical grounding, the terms "ground" and "earth" can have several different meanings. "Ground" may be the reference point in an electrical circuit from which other voltages are measured, a common return path for electric current, or a direct physical connection to the Earth. Also, the terms "grounding" and "earthing" are frequently confused with "neutral," as the neutral conductor on many wires is connected to a ground; however, these terms are not interchangeable. Current carried on a grounding conductor can result in objectionable or dangerous voltages appearing on equipment enclosures, so the installation of grounding conductors and neutral conductors is carefully defined in electrical regulations.

Ground consists of grounded or grounding conductors extending from ground beds to the equipment. A power conductor is tied to the grounding system. The grounding conductor is separate from the power conductors and is used only to ground the exposed metallic parts of the power system. The ground bed/mesh/electrode is a complex of conductors placed in earth to provide a low-resistance connection to *infinite* earth.

Ground in an AC system is a conductor providing a low-impedance path to earth to prevent hazardous voltages. Normally a grounding conductor does not carry current. Neutral is a circuit conductor that carries current in normal operation, which is connected to ground, usually at the service panel with the main disconnecting switch or breaker. Different industries use different systems to minimize the voltage difference between neutral and local earth ground.

11.7.7.4 Fixed Equipment for Which No Ground Is Required Per OSHA

Exposed non-current-carrying metal parts of the following types of fixed equipment need not be grounded:

- Enclosures for switches or circuit breakers used for other than service equipment and accessible to qualified persons only
- Electrically heated appliances that are permanently and effectively insulated from ground
- Distribution apparatus (e.g., transformer, capacitor cases) mounted on wooden poles and at a height exceeding 2.44 meters (8.0 feet) above ground or grade level
- Listed equipment protected by a system of double insulation, or its equivalent, and distinctively marked as such

11.7.8 NON-CURRENT-CARRYING METAL PARTS OF CORD- AND PLUG-CONNECTED EQUIPMENT

Exposed non-current-carrying metal parts of cord- and plug-connected equipment that may become energized must be grounded under any of the following conditions:

- In hazardous (classified) locations
- When operated at over 150 volts to ground, except for guarded motors and metal frames of electrically heated appliances if the appliance frames are permanently and effectively insulated from ground
- When equipment is of the following types:
  - Sump pumps
  - Refrigerators, freezers, and air conditioners
  - Clothes-washing, clothes-drying, and dishwashing machines; sump pumps; and electric aquarium equipment
  - Hand-held motor-operated tools, stationary and fixed motor-operated tools, and light industrial motor-operated tools
  - Motor-operated appliances of the following types: hedge clippers, lawn mowers, snow blowers, and wet scrubbers
  - Cord- and plug-connected appliances used in damp or wet locations, or by employees standing on the ground or on metal floors or working inside of metal tanks or boilers
  - Portable and mobile x-ray and associated equipment
  - Tools likely to be used in wet and conductive locations
  - Portable hand lamps

11.7.9 MOBILE EQUIPMENT REQUIRING GROUNDING

Mobile machines must be provided with a metallic enclosure for enclosing the terminals of the power cable. The enclosure must include provisions for a solid connection for the grounding terminal to effectively ground the machine frame. The method of cable termination used must prevent any strain or pull on the cable from stressing the electrical connections. The enclosure must have provision for locking so only authorized qualified persons may open the enclosure and must be marked with a sign warning of the presence of energized parts.

All energized switching and control parts must be enclosed in effectively grounded metal cabinets or enclosures. Locked circuit breaker enclosures and protective equipment must have the operating means projecting through the metal cabinet or enclosure so that these units can be reset without the locked doors being opened. Enclosures and metal cabinets must have provisions for locking so that only authorized qualified persons have access and must be marked with a sign warning of the presence of energized parts. Collector ring assemblies on revolving-type machines (shovels, draglines) must be guarded.

11.7.10 MOBILE EQUIPMENT NOT REQUIRING GROUNDING

The following equipment need not be grounded:

- Tools likely to be used in wet and conductive locations if supplied through an isolating transformer with an ungrounded secondary of less than 50 volts
- Listed or labeled portable tools and appliances if protected by an approved system of double insulation, or its equivalent, and distinctively marked

11.7.11 NON-ELECTRICAL EQUIPMENT

The metal parts of non-electrical equipment must be grounded, including the following:

- Frames and tracks of electrically operated cranes and hoists
- Frames of non-electrically driven elevator cars to which electric conductors are attached
- Hand-operated metal shifting ropes or cables of electric elevators
- Metal partitions, grill work, and similar metal enclosures around equipment of over 750 volts between conductors

11.8   GROUND-FAULT CIRCUIT INTERRUPTORS AND OSHA

All 125-volt, single-phase, 15-, 20-, and 30-ampere receptacle outlets must have ground-fault circuit interruptor (GFCI) protection. A cord connector on an extension cord set is a *receptacle outlet* if the cord set is used for temporary electric power. Cord sets and devices with GFCIs connected to the receptacle closest to the source of power are acceptable forms of protection. Where such GFCI protection is not available for receptacles other than 125-volt, single-phase, 15-, 20-, and 30-amphere, an assured equipment grounding conductor program must be in place. An assured equipment grounding conductor program includes the following elements:

- A written description of the program, including the specific procedures adopted by the employer that is available at the jobsite for inspection and copying by the Assistant Secretary of Labor and any affected employee
- One or more competent persons designated to implement the program

Each cord set and receptacle must be visually inspected before each day's use for external defects (e.g., deformed or missing pins or insulation damage) and for internal damage. Equipment found damaged or defective must not be used until repaired.

11.8.1 GROUND-FAULT CIRCUIT INTERRUPTORS IN BATHROOMS AND ROOFTOPS

All 125-volt, single-phase, 15- and 20-ampere receptacles installed in bathrooms or on rooftops must have GFCI protection for personnel.

11.8.2 REPLACEMENT

When receptacle outlets now required to be GFCI protected are replaced, GFCI-protected receptacles must be installed. If a grounding means does not exist in the receptacle enclosure, the new receptacle will be considered a non-grounding type receptacle and may be replaced with (1) another non-grounding type of receptacle; (2) a GFCI-type receptacle marked "NO EQUIPMENT GROUND" (note that an equipment grounding conductor may not be connected from the GFCI receptacle to any outlet supplied from the GFCI receptacle); or (3) a grounding-type receptacle and supplied through a GFCI. This type of receptacle must be marked "GFCI PROTECTED" and "NO EQUIPMENT GROUND." *Note:* An equipment grounding conductor may not be connected to this grounding-type receptacle.

11.8.3 SYSTEMS GREATER THAN 1000 VOLTS (HIGH-VOLTAGE)

If high-voltage systems greater than 1000 V are grounded, these systems must comply with the above requirements and additional requirements. For systems supplying portable or mobile high-voltage equipment (other than surface substations installed on a temporary basis),

- The system's neutral must be grounded through an impedance. If a delta-connected high-voltage system is used to supply the equipment, a system neutral must be derived.
- Exposed non-current-carrying metal parts of portable and mobile equipment must be connected by an equipment grounding conductor to the point at which the system neutral impedance is grounded.
- Ground-fault detection and relaying must be provided to automatically de-energize any high-voltage system component that has developed a ground fault. The continuity of the equipment grounding conductor must be continuously monitored so as to automatically de-energize the high-voltage feeder to the portable equipment upon loss of continuity of the equipment grounding conductor.

The grounding electrode to which the portable equipment system neutral impedance is connected must be isolated from and separated in the ground by at least 6.1 meters (20.0 feet) from any other system or equipment grounding electrode. No direct connection may exist between the grounding electrodes.

All non-current-carrying metal parts of portable equipment and fixed equipment (e.g., associated fences, housings, enclosures, supporting structures) must be grounded; however, equipment that is guarded by location and isolated from ground need not be grounded. Additionally, pole-mounted distribution apparatus at a height exceeding 2.44 meters (8.0 feet) above ground or grade level need not be grounded.

11.9   SURGE ARRESTERS

Surge arresters are installed near the ends of long conductors and used to divert any transients (e.g., from a lightning strike, from the switch faulting of a high-voltage system) to ground without damaging any electrical equipment. Also called a *surge protection device* (SPD) or *transient voltage surge suppressor* (TVSS), a surge arrester is designed to protect against electrical transients only, not direct electrical surges to the conductors themselves.

Although transients are much smaller than the original electrical surge, the transients still carry enough energy to cause arcing between different circuit pathways, especially those containing microprocessors. Because transients usually are initiated at some point between the two ends of a conductor, most applications install a surge arrester at each end of the conductor. Each conductor must provide a pathway to earth to safely divert the transient away from the protected component. The one exception is in high-voltage distribution systems. In general, the induced voltage is not sufficient to do damage at the electric generation end of the lines, so no surge arrester is installed; however, installation at the service entrance to a building is key to protecting downstream products that are not as robust.

The effectiveness of a surge arrester is based on its ability to limit the rate-of-rise of transient overvoltages and to reduce system characteristic impedance, which is accomplished by

- Discharging energy associated with a transient overvoltage
- Limiting and interrupting 60-Hz current that follows transient current through the arrester
- Returning to an insulating state without interrupting the supply of power to the load

The main power transformer is protected by a surge arrester. Protection is obtained through coordination between the transformer insulation withstand or the basic insulation level (BIL) and characteristics that the transient voltage arrester lets through. Many types of arresters exist, including the following:

- *Low-voltage surge arrester*—Used in low-voltage distribution systems and low-voltage distribution transformer windings.
- *Distribution arrester*—Used in 3-, 6-, and 10-kV power distribution systems to protect distribution transformers, cables, and power station equipment.
- *Valve arrester*—Used to protect 3- to ~220-kV transformer station equipment. Valve-surge arresters of low-sparkover distribution class are used in power centers.

- *Magnetic blow valve station arrester*—Used to protect 35- to ~500-kV transformers.
- *Neutral protection arrester*—Used for the transformer's neutral protection.

11.9.1 DRY TRANSFORMERS AND ARRESTERS

Dry transformers are used almost exclusively in power centers. Their insulation strength does not increase significantly above their BIL as duration of applied pulse decreases. The margin of protection at any given time is the difference between the transformer BIL and the arrester discharge characteristic. This value should be ≥20% if a 5000-ampere surge discharge current is assumed.

11.9.2 DISTANCES

To maximize arrester performance, the arrester leads must be as short and straight as possible and made of No. 6 American wire gauge (AWG) solid copper or larger. Critical distances are conductor lengths to the line conductor and power-center frame ground, and the distance between the arrester and the transformer. The surge arrester and arrester connections to line conductors should be as close to the transformer primary terminals as possible. Surge arrester leads should be connected at the primary terminals.

11.10 PORTABLE AND VEHICLE-MOUNTED GENERATORS

The frame of a portable generator need not be grounded and may serve as the grounding electrode for a system supplied by the generator if

- The generator supplies only equipment mounted on the generator or cord- and plug-connected equipment through receptacles mounted on the generator, or both.
- The non-current-carrying metal parts of equipment and equipment grounding conductor terminals of the receptacles are bonded to the generator frame.

The frame of a vehicle need not be grounded and may serve as the grounding electrode for a system supplied by a generator located on the vehicle if

- The frame of the generator is bonded to the vehicle frame.
- The generator supplies only equipment located on the vehicle and cord- and plug-connected equipment through receptacles mounted on the vehicle.
- The non-current-carrying metal parts of equipment and equipment grounding conductor terminals of the receptacles are bonded to the generator frame.
- The system complies with all other provisions of 29 CFR 1910.304(g).

A system conductor that is required to be grounded must be bonded to the generator frame where the generator is a component of a separately derived system.

11.11   ELECTROLYTIC CELLS

This section is applicable to the installation of the electrical components and accessory equipment of electrolytic cells, electrolytic cell lines, and process power supply for the production of aluminum, cadmium, chlorine, copper, fluorine, hydrogen peroxide, magnesium, sodium, sodium chlorate, and zinc. *Note:* Cells used as a source of electric energy and for electroplating processes and cells used for production of hydrogen are not included in this section. Some portions of electrolytic installations are excepted from some provisions of 29 CFR 1910.304, and OSHA should be directly consulted to determine the compliance status of

- Electrolytic cell DC process power circuit overcurrent protection
- Equipment located or used within the cell line working zone or associated with the cell line DC power circuits

Note: If more than one DC cell line process power supply serves the same cell line then a disconnecting means must be provided. This disconnection means must be on the cell line circuit side of each power supply to disconnect from the cell line circuit. Removable links or removable conductors may be used as the disconnecting means.

- Electrolytic cells, cell line conductors, cell line attachments, and the wiring of auxiliary equipment and devices within the cell line working zone

The frames and enclosures of some portable electric equipment used within the cell line working zone do not require grounding if the cell line circuit voltage does not exceed 200 volts DC. Ungrounded portable electric equipment must be distinctively marked and employ plugs and receptacles of a configuration that prevents

- Connection of this equipment to grounding receptacles
- Inadvertent interchange of ungrounded and grounded portable electric equipment

Circuits supplying power to ungrounded receptacles for hand-held and cord- and plug-connected equipment must meet the following requirements:

- The circuits must be electrically isolated from any distribution system supplying areas other than the cell line working zone and will be ungrounded.
- The circuits must be supplied through isolating transformers with primaries operating at ≤600 volts between conductors and protected with proper overcurrent protection.
- The secondary voltage of the isolating transformers may not exceed 300 volts between conductors.
- All circuits supplied from the secondaries must be ungrounded and have an approved overcurrent device of proper rating in each conductor.

Receptacles and their mating plugs for ungrounded equipment may not have provision for a grounding conductor and must be of a configuration that prevents their use for equipment required to be grounded. Receptacles on circuits supplied by an isolating transformer with an ungrounded secondary must

- Have a distinctive configuration.
- Be distinctively marked.
- Not be used in any other location in the facility.

Note: Auxiliary equipment (auxiliaries) includes motors, transducers, sensors, control devices, and alarms mounted on an electrolytic cell or other energized surfaces connected to the premises wiring systems by any of the following means:

- Multiconductor hard usage or extra hard usage flexible cord
- Wire or cable in suitable non-metallic raceways or cable trays
- Wire or cable in suitable metal raceways or metal cable trays installed with insulating breaks that will not cause a potentially hazardous electrical condition

Auxiliary non-electrical connections (e.g., air hoses, water hoses) may have non-contiguous conductive reinforcing wire, armor, or braids, or the like. These hoses are constructed of a nonconductive material.

11.11.1   Fixed and Portable Electric Equipment

The following need not be grounded: (1) AC systems supplying fixed and portable electric equipment within the cell line working zone, and (2) exposed conductive surfaces, such as electric equipment housings, cabinets, boxes, motors, raceways, and the like, that are within the cell line working zone.

11.11.2   Fixed Electric Equipment

Fixed electric equipment may be bonded to the energized conductive surfaces of the cell line, its attachments, or auxiliaries. If fixed electric equipment is mounted on an energized conductive surface, bonding to that surface is required.

11.11.3   Cranes and Hoists

The conductive surfaces of cranes and hoists that enter the cell line working zone need not be grounded. The portion of an overhead crane or hoist that contacts an energized electrolytic cell or energized attachments must, however, be insulated from ground. Remote crane or hoist controls that may introduce hazardous electrical conditions into the cell line working zone must have the following (OSHA, 2007):

- Isolated and ungrounded control circuits
- Non-conductive rope operators
- Pendant pushbuttons with non-conductive supporting means and with non-conductive surfaces or ungrounded exposed conductive surfaces
- Radios

11.12   WELDING

Arc welding involves open-circuit (when not welding) voltages that are typically from as low as 20 volts to as high as 100 volts. The primary (supply/service/circuit) voltage inside welding equipment is from 120 to 575 volts or more. The welder secondary (or welding) voltage is much lower. A shock from the primary (input) voltage can occur if contact is made. Such contact could happen by touching a lead or other electrically hot component inside the welder or by simultaneously touching the welder case or other grounded metal with the power to the welder turned on. To turn the power inside the welder off, the input power cord must be unplugged or the power disconnect switch turned off. The case must be (circuit) grounded so that if a problem develops inside the welder a fuse will blow, disconnecting the power.

11.12.1   Grounding Practices

Grounding of electrical circuits is a safety practice that is documented in various codes and standards. A typical arc welding setup may consist of several electrical circuits. Applying and maintaining proper grounding methods within the welding area is important to promote electrical safety in the workplace. Associated processes such as plasma cutting will also benefit from proper grounding. (See Figure 11.1.)

11.12.2   Supply Circuitry Ground

Welding machines that use a flexible cord- and plug-connected arrangement or those that are permanently wired into an electrical supply system contain a grounding conductor. The grounding conductor connects the metal enclosure of the welding

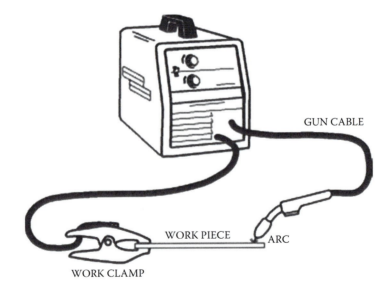

FIGURE 11.1   Welding machine ground. (Courtesy of Lincoln Electric, Euclid, OH.)

machine to ground. This circuitry grounding wire goes back through the electrical power distribution system and is connected to earth, usually through a metal rod driven into the earth. The purpose of connecting the equipment enclosure to ground is to ensure that the metal enclosure of the welding machine and ground are at the same potential. When at the same potential, a person will not experience an electrical shock when touching the two points. Grounding the enclosure also limits the voltage on the enclosure in the event that insulation should fail within the equipment.

The current-carrying capability of the grounding conductor is coordinated with the overcurrent device of the electrical supply system. The coordination of ampacity allows the grounding conductor to remain intact even if an electrical fault occurs within the welding machine. Some welding machines may have a double insulated design. In this case, a grounding conductor connection is not required. This type of welding machine relies on extra insulation to protect the user from shock. When double insulation is present, a "box within a box" symbol is present on the rating plate.

For small welding machines that utilize a plug on the end of a power cord, the grounding conductor connection is made automatically when the welding machine is plugged into the receptacle. The grounding pin of the plug makes a connection within the receptacle. Receptacle circuit testers will easily check the continuity of the grounding conductor. Consult with a qualified electrician to test circuits greater than 120 volts.

11.12.3   Arc Welding Circuit

The welding circuit consists of all conductive material through which the welding current is intended to flow. Welding current flows through the welding machine terminals, welding cables, workpiece connection, gun, torch, electrode holder, and workpiece. The welding circuit is not connected to ground within the welding machine and in fact is isolated from ground. The work cable (sometimes erroneously called the *welding ground cable*) is part of the arc welding circuit and only carries the welding current. The work cable does not ground the welder case. The workpiece is connected to a welding cable using a spring-loaded clamp or screw clamp. Unfortunately, this workpiece connection is often incorrectly called a *ground clamp*, and the workpiece lead is incorrectly called a *ground lead*. However the welding cable does not bring a ground connection to the workpiece. The ground connection is separate from the workpiece connection.

11.12.4   Earthing Ground from the Workpiece or Welding Table

The workpiece or the metal table that the workpiece rests upon must be separately grounded to earth. The building metal building frame may be a suitable pathway if a continuous low-resistance (less than 25 ohm) pathway can be provided. Alternatively, a conductor connected directly to a similarly proved earthing ground can be used. The grounded workpiece is at the same potential as other grounded objects in the area. In the event of insulation failure in the arc welding machine or other equipment, the voltage between the workpiece and ground will be limited (ANSI, 2005).

11.12.5   HIGH-FREQUENCY GROUND

Some welding machines utilize starting and stabilizing circuits that contain a high-frequency voltage. This method is common on tungsten inert gas (TIG) welding machines. The high-frequency voltage may have frequency components that extend into the megahertz region. In contrast, the welding voltage may be as low as 60 Hz. High-frequency signals have a tendency to radiate away from the welding area. These signals may cause interference with nearby radio and television reception or other electrical equipment. One method to minimize the radiation of high-frequency signals is to ground the welding circuit. The welding machine instruction manual will have specific instructions on how to ground the welding circuit and components in the surrounding area to minimize the radiation effect.

11.12.6   PORTABLE AND VEHICLE-MOUNTED WELDING GENERATOR GROUNDING

Portable and vehicle-mounted arc welding generators often have the capability to supply 120- and 240-volt auxiliary power. These generators are used in remote locations away from an electrical power distribution system. A convenient earth ground is not usually available for connection. The rules for grounding depend on the specific use and design of the auxiliary power generator. Most applications fall into one of the two categories summarized below:

1. If all of the following requirements are met, then an earthing ground to the generator frame is not required: (a) The generator is mounted to truck or trailer; (b) the auxiliary power is taken from receptacles on the generator using a cord and plug arrangement; (c) the receptacles have a grounding pin; and (d) the frame of the generator is bonded, or electrically connected, to the truck or trailer frame.
2. If either of the following conditions is met, then an earthing ground to the generator frame is required: (a) The generator is connected to a premises wiring system (i.e., to supply power for remote site work), or (b) the auxiliary power is hard-wired into the generator without the use of cords and plugs.

11.12.7   EXTENSION CORD GROUNDING

Extension cords should be periodically tested for ground continuity. The use of a receptacle circuit tester will confirm that all of the connections are intact within the cord, plug, and receptacle.

11.12.8   PRECAUTIONS

Precautions must be taken to insulate the welder from the welding circuit. Use dry insulating gloves and other insulating means. Also, maintain insulation on weld cables, electrode holders, guns, and torches to provide protection (Lincoln Electric, 2014).

11.13 PREDICTIVE TESTING AND INSPECTION SUSTAINABILITY AND STEWARDSHIP— ELECTRIC GROUNDING GRID

11.13.1 REQUIRED EQUIPMENT INFORMATION

• Grid identification (type)

11.13.2 REQUIRED TESTING RESULTS

Required testing results for the electric grounding grid include the following:

• Layout drawings
• Test point locations
• Fall-of-potential test results
• Point-to-point test results

11.13.3 FALL-OF-POTENTIAL TEST

The fall-of-potential test is used to test the grounding resistance of a single grounding electrode or grounding electrode system:

1. The method involves passing a known current through the electrode to be tested (X) and another test electrode (Z) placed a specific distance away.
2. A second test electrode (Y) is placed at a specified distance between the X electrode and the Z electrode.
3. The voltage drop between electrodes X and Y is measured by a voltmeter, and the current flow between electrodes X and Z is measured with an ammeter.
4. The required resistance is calculated by dividing the measured voltage by the known current. Actual Y and Z distances are obtained from standardized tables.

The minimum acceptance criterion is that the resistance between the main grounding electrode and ground should be no greater than 5 ohms for large commercial or industrial systems and 1.0 ohm or less for generating or transmission station grounds, unless otherwise specified by the owner. Other acceptance criteria may also be acceptable on a site-specific basis.

11.13.4 POINT-TO-POINT TEST

Point-to-point tests are used to determine the resistance between the main grounding system and all major electrical equipment frames, system neutral, and/or derived neutral points. Investigate point-to-point resistance values that exceed 0.5 ohm. Establish criteria for grounding electrodes on a site-specific basis.

11.14   PREDICTIVE TESTING AND INSPECTION SUSTAINABILITY AND STEWARDSHIP— ELECTRICAL LIGHTNING PROTECTION

Electrical lightning protection provides a designated path for the lightning current to travel. The system neither attracts nor repels a lightning strike but simply intercepts and guides the current harmlessly to ground. A lightning protection system is made up of several components:

- Air terminals (also known as lightning rods)
- Main conductors (cables that interconnect the air terminals to grounds)
- Ground terminations (typically copper or copper-clad rods driven into the earth a minimum of 10 feet deep)
- Bonding connections (made to equalize the potential between grounded metal objects)
- Lightning arresters (to protect wiring from lightning-induced damage)
- Surge suppressors (may be added to further protect valuable electronic equipment)

11.14.1   REQUIRED EQUIPMENT INFORMATION

- Electrical lightning protection for Class I or II building structures
  - Class I type structures are buildings below 75 feet in height.
  - Class II type structures are buildings at or above 75 feet in height.
- Class I and Class II type structures where the structural steel will be used in lieu of download or vertical cable
- Electrical lightning protection specifications
- UL certification (master label rating by an Underwriters Laboratories inspector)
- Installation configuration

11.14.2   REQUIRED TESTING RESULTS

Required testing results for the electric lightning protection include the following:

- Continuity test results
- Insulation resistance test results

11.14.3   REQUIRED ACCEPTANCE DOCUMENTATION

- Parts list and recommended spare parts list
- Acceptance documentation (with dates and signatures)

11.14.4 CONTINUITY TEST

Perform a continuity test of the circuits, specifically the electrical lightning protection circuit and current, potential, and control circuits. Perform a current test on the remainder of the secondary circuit to detect any open or short-circuit connections. The minimum acceptance criterion is no discontinuity.

11.14.5 INSULATION RESISTANCE TEST

Perform the insulation resistance test to determine insulation resistance to ground. Use both the dielectric absorption index and the polarization index. The insulation resistance test set will have

- Test voltage increments of 500, 1000, 2500, and 5000 VDC
- A resistance range of 0.0 to 500,000 megohms at 500,000 VDC
- A short-circuit terminal current of at least 2.5 milliamps
- Test voltage stability of ±0.1%
- Resistance accuracy of ±5% at 1 megohm

The minimum acceptance criteria are that the complete ground system will have no more than 10 ohms of resistance as measured per the fall of potential method and certification that surge protection is installed on the electrical service with a minimum of 160 kA per phase surge capacity (NASA, 2004).

11.15 IMPLICATIONS FOR SAFETY AUDITS AND INSPECTIONS

Grounding is an important part of any electrical safety inspection, but one that may be overlooked. When conducting an electrical safety audit or inspection, consider the following points:

1. *Has the facility's ground to earth been periodically verified?* Grounding rods can degrade over time and need to be put on a routine inspection schedule.
2. *Has the facility's structure undergone any renovation?* When additions are put on the building, often the new electrical system is simply "piggy-backed" onto the existing system; however, the old system may not have been designed to handle such additions and changes to the building. In such scenarios, the existing building ground may be insufficient to effectively ground the entire structure.
3. *Are the transformers on a separate ground from the building ground?* Transformers and other high-voltage systems should have their own, dedicated ground to earth.

4. *Is an alternative path to ground available?* Depending on the type of facility, a secondary path to ground is necessary or warranted. In the case of a facility with large amounts of electrical equipment, sensitive equipment, or expensive equipment, a secondary path to ground is desirable. This secondary path to ground serves as a back-up in case the primary path to ground fails (e.g., due to corrosion).

5. *Is all electrical equipment connected to ground?* A path to ground other than through the electrical socket may be necessary for large or high-voltage equipment. This provides an assured path to ground should the ground associated with the electrical receptacle fail. The ground should be physical and visible for inspection.

6. *Is the path to an earthing ground from the frames or metallic containers provided, adequate, and continuous where metal enclosures not normally electrified are present, such as those associated with cranes, electric hoists on cranes, equipment used in wet locations, cable enclosures (racks), welding equipment frames, battery charger frames, and flammable-containing environments?*

REFERENCES

American National Standards Institute. (2005). *ANSI Z49.1:2005. Safety in Welding, Cutting and Allied Processes.* Miami, FL: American Welding Society (http://www.aws.org/technical/facts/Z49.1-2005-all.pdf).

Lincoln Electric. (2014). *Grounding and Arc Welding Safety.* Euclid, OH: Lincoln Electric (http://www.lincolnelectric.com/en-us/support/process-and-theory/Pages/grounding-safety-detail.aspx).

NASA. (2004). *Reliability Centered Building and Equipment Acceptance Guide.* Washington, DC: National Aeronautics and Space Administration.

NFPA. (2005). *NFPA 70: National Electrical Code (NEC) Handbook.* Quincy, MA: National Fire Protection Association.

OSHA. (2007). Occupational Safety and Health Standards, 29 CFR, Part 1910, Subpart S, Electrical. Baltimore, MD: Occupational Safety and Health Administration.

12 Electrical Safety and OSHA

Martha J. Boss and Dennis Day

CONTENTS

12.1 EXAMINATION, INSPECTION, AND USE

The safety of equipment must be determined by considering its suitability for the installation and its use in conformity with the following provisions:

- Mechanical strength and durability, including, for parts designed to enclose and protect other equipment, the adequacy of the protection thus provided
- Wire-bending and connection space
- Electrical insulation
- Short circuits and grounds (completed wiring installations must be free from short circuits and grounds other than those required or permitted)
- Heating effects under all conditions of use
- Arcing effects
- Classification by type, size, voltage, current capacity, and specific use

Listed or labeled equipment must be installed and used in accordance with any instructions included in the listing or labeling.

12.2   INSTALLATION

Electric equipment must be (1) firmly secured to the mounting surface and (2) installed in a neat and workmanlike manner. Unused openings in boxes, raceways, auxiliary gutters, cabinets, equipment cases, or housings must be effectively closed to afford protection substantially equivalent to the wall of the equipment. Conductors must be racked to provide ready and safe access in underground and subsurface enclosures that persons enter for installation and maintenance. Internal parts of electrical equipment (including busbars, wiring terminals, and insulators) may not be damaged or contaminated by foreign materials (e.g., paint, plaster, cleaners, abrasives, corrosive residues). Damaged parts that may adversely affect safe operation or mechanical strength (e.g., parts that are broken, bent, cut, or deteriorated by corrosion, chemical action, or overheating) must be repaired or replaced.

12.2.1   Pipes or Ducts

Pipes and ducts may not be located in the vicinity of electrical equipment if

- The pipes and ducts are foreign to the electrical installation and require periodic maintenance. (Piping and other facilities are not considered foreign if provided for the electrical equipment's fire protection.)
- Malfunctions would endanger the electrical system operation.

Protection must be provided where necessary to avoid damage from condensation, leaks, and breaks in such foreign systems.

12.2.2   Deteriorating Agents

No conductors or equipment should be located in damp or wet locations; where exposed to gases, fumes, vapors, liquids, or other agents that have a deteriorating effect on the conductors or equipment; or where exposed to excessive temperatures.

12.2.3   Air Flow

Electric equipment that depends on the natural circulation of air and convection principles for cooling of exposed surfaces must be installed so room air flow over such surfaces is not prevented by walls or by adjacent installed equipment. For equipment designed for floor mounting, clearance between the top surfaces and adjacent surfaces must be provided to dissipate rising warm air. Electric equipment with ventilating openings must be installed so walls or other obstructions do not prevent the free circulation of air through the equipment.

12.2.4   Electrical Connections

Due to the dissimilarity of metals, verification of the connection material's suitability is required in devices such as pressure terminal or pressure splicing connectors and soldering lugs. Conductors of generally dissimilar metals (e.g., copper and aluminum, copper and copper-clad aluminum, aluminum and copper-clad alumi-num) may not be intermixed in a terminal or splicing connector. Exceptions must be documented as to their material and use requirements. Materials such as solder, fluxes, inhibitors, and compounds, where employed, must be suitable for such use and not adversely affect the conductors, installation, or equipment.

12.2.5 TERMINALS (CONDUCTORS TO TERMINAL PARTS)

Terminals must ensure a good connection without damaging the conductors. The terminals must be made by using pressure connectors (set-screw type), solder lugs, or splices to flexible leads. No. 10 or smaller conductors must be connected using wire binding screws or studs and nuts having upturned lugs. Terminals for more than one conductor and terminals used to connect aluminum must be identified.

12.2.6 SPLICES

Conductors must be spliced with splicing devices identified for the use or by braz-ing, welding, or soldering with a fusible metal or alloy. Soldered splices must first be spliced or joined to be mechanically and electrically secure without solder and then soldered. All splices and joints and the free ends of conductors must be covered with an insulation equivalent to that of the conductors or with an insulating device. Wire connectors or splicing means installed on conductors for direct burial must be listed (approved by Underwriters Laboratories or Factory Mutual Global).

12.2.7 MARKING

The marking must be of sufficient durability to withstand the environment involved. Required markings that are to be affixed before use include the manufacturer's name and trademark, as well as the voltage, current, and wattage.

12.3 ILLUMINATION

Illumination must be provided for all working spaces about service equipment, switchboards, panelboards, and motor control centers installed indoors. Workers must not enter spaces containing exposed energized parts, unless illumination is provided that enables the workers to perform the work safely. Where lack of illumi-nation or an obstruction precludes observation of the work to be performed, work-ers may not perform tasks near exposed energized parts. Workers must not reach blindly into areas that may contain energized parts. Additional lighting fixtures are not required where the working space is illuminated by an adjacent light source. In electric equipment rooms, the illumination must not be controlled by automatic means only. Lighting outlets must be arranged so that workers changing lamps or making repairs on the lighting system will not be endangered by live parts or other equipment. Points of control must be located to prevent contact with any live part or equipment moving part while turning on lights.

12.4   INFORMATION TECHNOLOGY (COMMUNICATIONS)

The relevant electrical safety requirements for information technology equipment are discussed in 29 CFR 1910.306. A disconnecting means must be provided to disconnect power to all electronic equipment in an information technology equipment room. Similar disconnecting means must be provided to all dedicated heating, ventilating, and air-conditioning (HVAC) systems serving the room and to cause all required fire/smoke dampers to close. The control for these disconnecting means must be grouped and identified and must be readily accessible at the principal exit doors. A single means to control both the electronic equipment and HVAC system is permitted. The dimension of the working space in the direction of access to live parts that may require examination, adjustment, servicing, or maintenance while alive must be a minimum of 762 mm (2.5 ft). Where controls are enclosed in cabinets, the doors must either open at least 90° or be removable (OSHA, 2007).

12.5   CONDUCTORS FOR GENERAL WIRING

(*Note:* The following requirements do not apply to conductors that are an integral part of factory-assembled equipment.) All conductors used for general wiring must be insulated. The conductor insulation must be of a type that is approved for the voltage, operating temperature, and location of use. Insulated conductors must be distinguishable by appropriate color or other suitable means as being grounded conductors, ungrounded conductors, or equipment grounding conductors.

12.6   IDENTIFICATION

A conductor of a flexible cord or cable that is used as a grounded conductor or an equipment grounding conductor must be distinguishable from other conductors. The following types must be durably marked on the surface: S, SC, SCE, SCT, SE, SEO, SEOO, SJ, SJE, SJEO, SJEOO, SJO, SJT, SJTO, SJTOO, SO, SOO, ST, STO, and STOO flexible cords, as well as G, G–GC, PPE, and W flexible cables. The conductors must be marked at intervals not exceeding 610 mm (24 in.) and indicate the type designation, size, and number of conductors. No wiring systems of any type should be installed in ducts used to transport dust, loose stock, or flammable vapors. No wiring system of any type should be installed in any duct used for vapor removal or for ventilation of commercial-type cooking equipment, or in any shaft containing only such ducts.

12.7   FEEDERS

Feeders must originate in an approved distribution center. Conductors must be run as multiconductor cord or cable assemblies; however, if accessible only to qualified persons, feeders may be run as single insulated conductors and should be evaluated on a case-by-case basis to determine National Electrical Code (NEC) compliance.

12.8   BRANCH CIRCUITS

Branch circuits must originate in an approved power outlet or panelboard. Conductors must be multiconductor cord or cable assemblies or open conductors. If run as open conductors, the conductors should be fastened at ceiling height every 3.05 meters (10.0 feet). No branch circuit conductor should be laid on the floor. Each branch circuit that supplies receptacles or fixed equipment must contain a separate equipment grounding conductor if run as open conductors.

12.9   RECEPTACLES

Receptacles must be of the grounding type. Unless installed in a continuous grounded metallic raceway or metallic covered cable, each branch circuit must contain a separate equipment grounding conductor, and all receptacles must be electrically connected to the grounding conductor.

12.10   CONDUCTOR PROTECTION

Open wiring on insulators is only permitted on systems of 600 volts, nominal, or less for industrial or agricultural establishments, indoors or outdoors, in wet or dry locations, where subject to corrosive vapors, and for services. Conductors smaller than No. 8 must be rigidly supported on non-combustible, non-absorbent insulating materials and may not contact any other objects. Supports must be installed:

- Within 152 mm (6 in.) from a tap or splice
- Within 305 mm (12 in.) of a dead-end connection to a lampholder or receptacle
- At intervals not exceeding 1.37 m (4.5 ft), and at closer intervals sufficient to provide adequate support where likely to be disturbed

In dry locations, where not exposed to severe physical damage, conductors may be separately enclosed in flexible non-metallic tubing. The tubing must be in continuous lengths not exceeding 4.57 m (15.0 ft) and secured to the surface by straps at intervals not exceeding 1.37 m (4.5 ft).

Open conductors passing through walls, floors, wood cross members, or partitions must be separated from contact with these structures. Contact is prevented through the use of tubes or bushings of non-combustible, non-absorbent insulating material that surrounds the interior bore of the opening. If the bushing is shorter than the hole, (1) a waterproof sleeve of non-conductive material must be inserted in the hole and (2) an insulating bushing slipped into the sleeve at each end in such a manner as to keep the conductors absolutely out of contact with the sleeve. Each conductor must be carried through a separate tube or sleeve. Protection is required where open conductors cross ceiling joints and wall studs and are exposed to physical damage—for example, when located within 2.13 meters (7.0 feet) of the floor.

12.11   CABINETS, BOXES, AND FITTINGS

Conductors entering cabinets, cutout boxes, or fittings must be protected from abrasion, and openings through which conductors enter must be effectively closed. Unused openings in cabinets, boxes, and fittings must be effectively closed. Where cable is used, each cable must be secured to the cabinet, cutout box, or meter socket enclosure; however, where cable with an entirely non-metallic sheath enters the top of a surface-mounted enclosure through one or more nonflexible raceways not less than 457 mm (18 in.) or more than 3.05 m (10.0 ft) in length, the cable need not be secured to the cabinet, box, or enclosure provided

- Each cable is fastened within 305 mm (12 in.) of the outer end of the raceway, measured along the sheath.
- The raceway extends directly above the enclosure and does not penetrate a structural ceiling.
- A fitting is provided on each end of the raceway to protect the cable from abrasion, and the fittings remain accessible after installation.
- The raceway is sealed or plugged at the outer end using approved means so as to prevent access to the enclosure through the raceway.
- The cable sheath is continuous through the raceway and extends into the enclosure not less than 6.35 mm (0.25 in.) beyond the fitting.
- The raceway is fastened at its outer end and at other points as necessary.
- Where installed as conduit or tubing, the allowable cable fill does not exceed that permitted for complete conduit or tubing systems.

12.12   COVERS AND CANOPIES

All pull boxes, junction boxes, and fittings must be provided with covers approved for the purpose. If metal covers are used, they must be grounded. In completed installations, each outlet box must have a cover, faceplate, or fixture canopy. Covers of outlet boxes having holes through which flexible cord pendants pass must be provided with bushings designed for the purpose or must have smooth, well-rounded surfaces on which the cords may bear. Where a fixture canopy or pan is used, any combustible wall or ceiling finish exposed between the edge of the canopy or pan and the outlet box must be covered with non-combustible material.

With regard to pull and junction boxes for systems over 600 volts, nominal, the following conditions (in addition to other requirements in this section) apply:

- Boxes must provide a complete enclosure for the contained conductors or cables.
- Boxes must be closed by suitable covers securely fastened in place. Underground box covers weighing over 45.4 kg (100 lb) meet this requirement.
- Covers for boxes must be permanently marked "HIGH VOLTAGE." The marking must be on the outside of the box cover and must be readily visible and legible.

12.13   POLARITY OF CONNECTIONS

No grounded conductor may be attached to any terminal or lead so as to reverse the designated polarity.

12.14   GROUNDING TERMINALS AND DEVICES

A grounding terminal or grounding-type device on a receptacle, cord connector, or attachment plug may not be used for purposes other than grounding.

12.15   OUTLET DEVICES

Outlet devices must have an ampere rating not less than the load to be served.

12.16   LIGHTING FIXTURES, LAMPHOLDERS, AND HANDLAMPS

All lamps for general illumination must be protected from accidental contact or breakage by a suitable fixture or lampholder with a guard. Brass-shell, paper-lined sockets or other metal-cased sockets may not be used unless the shell is grounded. Where connected to a branch circuit having a rating in excess of 20 amperes, lampholders must be of the heavy-duty type. A heavy-duty lampholder must have a rating of not less than 660 watts if of the admedium type and not less than 750 watts if of any other type. Lighting fixtures, lampholders, lamps, rosettes, and receptacles must not have live parts normally exposed to employee contact; however, rosettes and cleat-type lampholders and receptacles located at least 2.44 meters (8.0 feet) above the floor may have exposed terminals. Handlamps of the portable type supplied through flexible cords must be equipped with a handle of molded composition or other material identified for the purpose, and a substantial guard must be attached to the lampholder or the handle. Metal-shell, paper-lined lampholders may not be used. Lampholders of the screw-shell type must be installed for use as lampholders only. Where supplied by a circuit having a grounded conductor, the grounded conductor must be connected to the screw shell. Lampholders installed in wet or damp locations must be of the weatherproof type.

12.16.1   WET OR DAMP LOCATIONS

Fixtures installed in wet or damp locations must be identified for the purpose and must be so constructed or installed that water cannot enter or accumulate in wireways, lampholders, or other electrical parts.

12.16.2   OUTDOOR LAMPS

Lamps for outdoor lighting must be located below all energized conductors, transformers, or other electric equipment, unless (1) such equipment is controlled by a disconnecting means that can be locked in the open position, or (2) adequate clearances or other safeguards are provided for relamping operations.

12.17   FIXTURE WIRES

Fixture wires must be approved for the voltage, temperature, and location of use. A fixture wire that is used as a grounded conductor must be identified. Fixture wires may be used only for (1) lighting fixtures and in similar equipment where they are enclosed or protected and not subject to bending or twisting in use; or (2) connecting lighting fixtures to the branch circuit conductors supplying the fixtures. Fixture wires may not be used as branch circuit conductors except as permitted for Class 1 power-limited circuits and for fire alarm circuits.

12.18   RECEPTACLES, CORD CONNECTORS, AND ATTACHMENT PLUGS (CAPS)

All 15- and 20-ampere attachment plugs and connectors must be constructed so that no exposed current-carrying parts except the prongs, blades, or pins are present. The cover for wire terminations must be a part that is essential for the operation of an attachment plug or connector (dead-front construction). Attachment plugs must be installed so that their prongs, blades, or pins are not energized unless inserted into an energized receptacle. Receptacles must not be installed so as to require an energized attachment plug as the receptacle's source of supply. Receptacles, cord connectors, and attachment plugs must be constructed so that no receptacle or cord connector may accept an attachment plug with a different voltage or current rating than that for which the device is intended; however, a 20-ampere T-slot receptacle or cord connector may accept a 15-ampere attachment plug of the same voltage rating. Non-grounding-type receptacles and connectors may not be used for grounding-type attachment plugs.

12.18.1   OUTDOOR LOCATIONS

A receptacle installed outdoors in a location protected from the weather must have an enclosure for the receptacle that is weatherproof when the receptacle is covered (attachment plug cap not inserted and receptacle covers closed). *Note:* A receptacle is considered to be in a location protected from the weather when it is located under roofed open porches, canopies, or marquees and where not subjected to a beating rain or water runoff. A receptacle installed in a wet or damp location must be suitable for the location. The following conditions apply to receptacles installed in a wet location and intended to have equipment plugged into the receptacle:

- If the receptacle is not attended while in use (e.g., sprinkler system controllers, landscape lighting, holiday lights), the receptacle must have an enclosure that is weatherproof with the attachment plug cap inserted or removed.
- If the receptacle is attended while in use (e.g., portable tools), then the receptacle must have an enclosure that is weatherproof when the attachment plug cap is removed.

12.18.2 ATTACHMENT PLUGS AND DIFFERING CIRCUIT PARAMETERS

Receptacles connected to circuits having different voltages, frequencies, or types of current (AC or DC) are not interchangeable.

12.19 RECEPTACLE OUTLETS

A single receptacle installed on an individual branch circuit must have an ampere rating of not less than the ampere rating of the branch circuit. Where connected to a branch circuit supplying two or more receptacles or outlets, a receptacle may not supply a total cord- and plug-connected load in excess of the maximum specified in Table 12.1. Where connected to a branch circuit supplying two or more receptacles or outlets, receptacle ratings must conform to the values listed in Table 12.2 or, when larger than 50 amperes, the receptacle rating must equal or exceed the branch-circuit rating. However, receptacles of cord- and plug-connected arc welders may have ampere ratings that equal or exceed the minimum branch circuit conductor ampacity.

TABLE 12.1
Maximum Cord- and Plug-Connected Load to Receptacle

Circuit Rating (Amperes)	Receptacle Rating (Amperes)	Maximum Load (Amperes)
15–20	15	12
20	20	16
30	30	24

Source: OSHA, Occupational Safety and Health Standards, 29 CFR, Part 1910, Subpart S, Electrical, Occupational Safety and Health Administration, Baltimore, MD, 2007.

TABLE 12.2
Receptacle Ratings for Various Size Circuits

Circuit Rating (Amperes)	Receptacle Rating (Amperes)
15	Not over 15
20	15 or 20
30	30
40	40 or 50
50	50

Source: OSHA, Occupational Safety and Health Standards, 29 CFR, Part 1910, Subpart S, Electrical, Occupational Safety and Health Administration, Baltimore, MD, 2007.

12.20   APPLIANCES

Appliances may have no live parts normally exposed to contact other than parts functioning as open-resistance heating elements, such as the heating elements of a toaster, which are necessarily exposed. Each appliance must have a means to disconnect the appliance from all ungrounded conductors. If an appliance is supplied by more than one source, the disconnecting means must be grouped and identified. Each electric appliance must be provided with a nameplate giving the identifying name and the rating in volts and amperes or in volts and watts. Information markings must be provided if (1) the appliance is to be used on a specific frequency or frequencies, or (2) motor overload protection external to the appliance is required. Marking must be located so as to be visible or easily accessible after installation (OSHA, 2007).

12.21   OUTSIDE OPEN CONDUCTORS: 600 VOLTS, NOMINAL, OR LESS

The following requirements apply to branch circuit, feeder, and service conductors rated 600 volts, nominal, or less and run outdoors as open conductors.

12.22   CONDUCTORS ON POLES

Where not placed on racks or brackets, conductors on poles must have a separation greater than or equal to 305 mm (1.0 ft). Conductors supported on poles must provide a horizontal climbing space not less than the values provided below:

- Power conductors below communication conductors—762 mm (30 in.)
- Power conductors alone or above communication conductors:
  - 300 volts or less—610 mm (24 in.)
  - Over 300 volts—762 mm (30 in.)
- Communication conductors below power conductors—Same as power conductors
- Communications conductors alone—No requirement

12.22.1   CLEARANCE FROM GROUND

Open conductors, open multiconductor cables, and service-drop conductors of not over 600 volts, nominal, must conform to the minimum clearances specified in Table 12.3.

12.22.2   CLEARANCE FROM BUILDING OPENINGS

Service conductors installed as open conductors or multiconductor cable without an overall outer jacket must have clearance of not less than 914 mm (3.0 ft) from windows that are designed to be opened, doors, porches, balconies, ladders, stairs, fire escapes, and similar locations. An exception is that conductors running above the top level of a window may be less than 914 mm (3.0 ft) from the window. The vertical clearance of final spans above or within 914 mm (3.0 ft), measured horizontally, of platforms, projections, or other surfaces from which conductors might be reached must (per the conductor distances) conform to the minimum clearances specified in Table 12.3. Overhead service conductors must not be installed (1) beneath openings through which materials may be moved or (2) where the entrance to these building openings would be obstructed by the conductor.

TABLE 12.3
Clearances from Ground

| Distance | Installations Built before August 13, 2007 | | Installations Built on or after August 13, 2007 | |
	Maximum Voltage	Conditions	Voltage to Ground	Conditions
3.05 m (10.0 ft)	<600 V	Above finished grade or sidewalks, or from any platform or projection from which they might be reached (if these areas are accessible to other than pedestrian traffic, then one of the other conditions applies)	<150 V	Above finished grade or sidewalks, or from a platform or projection from which they might be reached (if these areas are accessible to other than pedestrian traffic, then one of the other conditions applies)
3.66 m (12.0 ft)	<600 V	Over areas, other than public streets, alleys, roads, and driveways, subject to vehicular traffic other than truck traffic	<300 V	Over residential property and driveways; over commercial areas subject to pedestrian traffic or to vehicular traffic other than truck traffic[a]
4.57 m (15.0 ft)	<600 V	Over areas, other than public streets, alleys, roads, and driveways, subject to truck traffic	301–600 V	Over residential property and driveways; over commercial areas subject to pedestrian traffic or to vehicular traffic other than truck traffic[a]
5.49 m (18.0 ft)	<600 V	Over public streets, alleys, roads, and driveways	<600 V	Over public streets, alleys, roads, and driveways; over commercial areas subject to truck traffic; other land traversed by vehicles, including land used for cultivating or grazing and forests and orchards

Source: OSHA, Occupational Safety and Health Standards, 29 CFR, Part 1910, Subpart S, Electrical, Occupational Safety and Health Administration, Baltimore, MD, 2007.

[a] This category includes conditions covered under the 3.05-m (10.0-ft) category where the voltage exceeds 150 V.

12.22.3 ABOVE ROOFS

Overhead spans of open conductors and open multiconductor cables must have a vertical clearance of not less than 2.44 m (8.0 ft) above the roof surface. The vertical clearance above the roof level must be maintained for a distance not less than 914 mm (3.0 ft) in all directions from the edge of the roof. The area above a roof surface subject to pedestrian or vehicular traffic must have a vertical clearance from the roof surface that conforms to the minimum clearances specified in Table 12.3. A reduction in clearance to 914 mm (3.0 ft) is permitted where the voltage between conductors does not exceed 300 V and the roof has a slope of 102 mm (4 in.) in 305 mm (12 in.) or greater. A reduction in clearance above only the overhanging portion of the roof to not less than 457 mm (18 in.) is permitted where the voltage between conductors does not exceed 300 if the conductors do not pass above the roof overhang for a distance of more than 1.83 m (6.0 ft) or 1.22 m (4.0 ft) horizontally and are terminated at a through-the-roof raceway or approved support. The requirement for maintaining a vertical clearance of 914 mm (3.0 ft) from the edge of the roof does not apply to the final conductor span where the conductors are attached to the side of a building (OSHA, 2007).

12.23 OVERHEAD LINES, CONDUCTORS, AND CABLES

For voltages normally encountered with overhead power lines, objects that do not have an insulating rating for the voltage involved are considered to be conductive. These objects must not contact overhead power lines. Conductors or cables that are sufficiently insulated as rated may be contacted (touched) by workers; however, if this insulation fails, workers would be exposed to hazardous electrical energy. Conductors, whether in overhead lines or cables, that are not sufficiently insulated are considered to be uninsulated lines. If work is to be performed near uninsulated lines, the lines must be de-energized and grounded, or other protective measures must be provided before work is started. If protective measures (e.g., guarding, isolating, insulating) are provided, these precautions must prevent workers from contacting such lines directly with any part of their body or indirectly through conductive materials, tools, or equipment. (*Note:* The work practices used by qualified or authorized persons installing insulating devices on overhead power transmission or distribution lines are covered by 29 CFR 1910.269, not by 29 CFR 1910.332–335.)

12.24 FLEXIBLE CORDS AND CABLES

Flexible cords and cables must be approved for conditions of use and location. Flexible cords and cables may be used only for

- Pendants
- Wiring of fixtures
- Connection of portable lamps or appliances
- Portable and mobile signs
- Elevator cables
- Wiring of cranes and hoists

- Connection of stationary equipment to facilitate their frequent interchange
- Prevention of the transmission of noise or vibration
- Appliances where the fastening means and mechanical connections are designed to permit removal for maintenance and repair
- Data processing cables approved as a part of the data processing system
- Connection of moving parts
- Temporary wiring

12.24.1 RECEPTACLE OUTLET AND CORD CONNECTIONS

The flexible cord must be equipped with an attachment plug and must be energized from an approved receptacle outlet. A receptacle outlet must be installed wherever flexible cords with attachment plugs are used. Where flexible cords are permitted to be permanently connected, receptacles may be omitted.

12.24.2 SPLICES

Flexible cords will be used only in continuous lengths without splice or tap. Hard-service cord and junior hard-service cord No. 14 and larger may be repaired if the splice retains the insulation, outer sheath properties, and usage characteristics of the cord being spliced.

12.24.3 PROHIBITIONS TO USE

Although some exceptions do apply, flexible cords and cables may not be used

- As a substitute for the fixed wiring of a structure
- Where run through holes in walls, ceilings, or floors
- Where run through doorways, windows, or similar openings
- Where attached to building surfaces
- Where concealed behind building walls, ceilings, or floors
- Where installed in raceways

12.24.4 SHOWROOMS AND SHOWCASES

Flexible cords used in show windows and showcases must be type S, SE, SEO, SEOO, SJ, SJE, SJEO, SJEOO, SJO, SJOO, SJT, SJTO, SJTOO, SO, SOO, ST, STO, or STQO, except for the wiring of chain-supported lighting fixtures and supply cords for portable lamps and other merchandise being displayed or exhibited.

12.24.5 SUPPORT AND STRAIN RELIEF

Flexible cords and cables must be protected from accidental damage, as might be caused by sharp corners, projections, and doorways or other pinch points. The cords and cables must be supported in place at intervals that ensure protection from physical damage. Such support must be in the form of staples, cables ties, or straps installed so as not to cause damage. Flexible cords and cables must be connected to devices and fittings so that strain relief is provided that prevents the pull from being directly transmitted to joints or terminal screws.

12.24.6 CABLE TRAYS

Only the following wiring methods may be installed in cable tray systems: armored cable; electrical metallic tubing; electrical non-metallic tubing; fire alarm cables; flexible metal conduit; flexible metallic tubing; instrumentation tray cable; intermediate metal conduit; liquid-tight flexible metal conduit; liquid-tight flexible non-metallic conduit; metal-clad cable; mineral-insulated, metal-sheathed cable; multiconductor service-entrance cable; multiconductor underground feeder and branch-circuit cable; multipurpose and communications cables; non-metallic-sheathed cable; power and control tray cable; power-limited tray cable; optical fiber cables; and other factory-assembled, multiconductor control, signal, or power cables that are specifically approved for installation in cable trays, rigid metal conduit, and rigid non-metallic conduit. In industrial establishments where conditions of maintenance and supervision ensure that only qualified persons service installed cable tray systems, the following cables may also be installed in ladder, ventilated-trough, or ventilated-channel cable trays:

- *Single-conductor cable.* The cable must be No. 1/0 or larger and must be of a type listed and marked on the surface for use in cable trays. Where Nos. 1/0 through 4/0 single-conductor cables are installed in the ladder cable tray, the maximum allowable rung spacing for the ladder cable tray is 229 mm (9 in.). Where exposed to the direct rays of the sun, cables must be identified as being sunlight resistant.
- *Welding cables installed in dedicated cable trays*
- *Single conductors used as equipment grounding conductors.* These conductors, which may be insulated, covered, or bare, must be No. 4 or larger.
- *Multiconductor cable, type MV.* Where exposed to the direct rays of the sun, the cable must be identified as being sunlight resistant.

Metallic cable trays may be used as equipment grounding conductors only where continuous maintenance and supervision ensure that only qualified persons service the installed cable tray system. Cable trays in hazardous (classified) locations may contain only the cable types permitted in such locations. Cable tray systems may not be used in hoistways or where subjected to severe physical damage.

12.24.7 TEMPORARY WIRING DISCONNECTS

Suitable disconnecting switches or plug connectors must be installed to permit the disconnection of all ungrounded conductors of each temporary circuit. Temporary wiring must be removed immediately upon completion of the project or purpose for which the wiring was installed. Temporary electrical power and lighting installations of 600 volts, nominal, or less may be used as follows:

- During and for remodeling, maintenance, or repair of buildings, structures, or equipment, and similar activities
- For a period not to exceed 90 days for Christmas decorative lighting, carnivals, and similar purposes
- For experimental or development work and during emergencies

Temporary electrical installations of more than 600 volts may be used only during periods of tests, experiments, emergencies, or construction-like activities. *Note:* Temporary wiring includes all cord sets and receptacles that are not a part of the permanent wiring of the building or structure, and cord- and plug-connected equipment (OSHA, 2007).

12.25 PREDICTIVE TESTING AND INSPECTION SUSTAINABILITY AND STEWARDSHIP—CABLES

Perform high-voltage tests to verify the insulation and to ensure that no excessive leakage current is present. *Note:* This is a potentially destructive test. The minimum acceptance criteria are results better than or equal to the manufacturer's specifications and limits in accordance with ANSI/IEEE Standard 400.

12.25.1 Cables (General)

12.25.1.1 Required Testing Results

Required testing results for the various types of cables include the following:

- Dielectric test results
- High-voltage test results
- Radiographic test results

12.25.2 Cables, Low-Voltage (600 Volts Maximum)

12.25.2.1 Required Testing Results

Required testing results for the various types of cables include the following:

- Insulation resistance test results
- Airborne ultrasonic test results (optional)

12.25.2.2 Acceptance Technologies

Use ultrasonics to verify the non-existence of electrical arcing and other high-frequency events. A visual inspection of the cable should reveal no abnormalities or defects. The minimum acceptance criterion is that the power factor must not exceed the manufacturer's data.

12.25.3   Cables, Medium-Voltage (600–33,000 Volts)

12.25.3.1   Required Testing Results

Required testing results for the various types of cables include the following:

- Insulation resistance test results
- Airborne ultrasonic test results (optional)
- Power factor test results (optional)
- High-voltage test results (optional)

12.25.3.2   Acceptance Technologies

Use ultrasonics to verify the non-existence of electrical arcing and other high-frequency events. Use the power factor test to verify acceptable dissipation and power loss from the insulation to ground. Measure with the grounded specimen test (GST), ungrounded specimen test (UST), and GST with guard. The power factor test set should have all of the following minimum requirements:

- Test voltage range of 500 V to 12 kV
- Ability to perform UST, GST, and GST with guard tests
- Readings for power factor, dissipation factor, capacitance, and watts-loss
- Power factor/dissipation factor range of 0 to 200%
- Capacitance measuring range of 0 to 0.20 picofarads

A visual inspection of the cable should reveal no abnormalities or defects. The minimum acceptance criterion is that the power factor must not exceed the manufacturer's data. Testing can be performed by means of direct current (DC), alternating current (AC), partial discharge (pd), or very low frequency (VLF). The selection can be made only after an evaluation of the alternative methods. Test voltages should not exceed 80% of the cable manufacturer's factory test value. During testing,

- Ensure that the input voltage to the test set is regulated.
- Ensure that current-sensing circuits in test equipment, when available, measure only the leakage current associated with the cable under test and do not include internal leakage of the test equipment.
- Record wet- and dry-bulb temperatures or relative humidity and temperature.
- Test each cable section individually.
- Test each conductor individually with all other conductors grounded. Ground all shields.
- Ensure that terminations are adequately corona-suppressed by guard ring, field reduction sphere, or other suitable method, as necessary.
- Ensure that the maximum test voltage does not exceed the limits for terminators specified in ANSI/IEEE 48, IEEE 386, or the manufacturer's specifications.
- Raise the conductor to the specified maximum test voltage and hold for 15 minutes

- If performed by means of DC, reduce the test set potential to zero and measure residual voltage at discrete intervals.
- Apply grounds for a time period adequate to drain all insulation stored charge.

When new cables are spliced into existing cables, the acceptance test should be performed on the new cable prior to splicing. After the splice is completed, an insulation resistance test and a shield-continuity test should be performed on the length of new and existing cable including the splice and utilizing a test voltage that does not exceed 60% of the factory test value.

12.25.4   CABLES, HIGH-VOLTAGE (33,000 VOLTS MINIMUM)

12.25.4.1   Required Testing Results

Required testing results for the various types of cables include the following:

- Insulation resistance test results
- Airborne ultrasonic test results (optional)
- Power factor test results (optional)
- High-voltage test results (optional)

12.25.4.2   Acceptance Technologies

Use ultrasonics to verify the non-existence of electrical arcing and other high-frequency events. Use the power factor test to verify acceptable dissipation and power loss from the insulation to ground. Measure with the grounded specimen test (GST), ungrounded specimen test (UST), and GST with guard. The power factor test set should have all of the following minimum requirements:

- Test voltage range of 500 V to 12 kV
- Ability to perform UST, GST, and GST with guard tests
- Readings for power factor, dissipation factor, capacitance, and watts-loss
- Power factor/dissipation factor range of 0 to 200%
- Capacitance measuring range of 0 to 0.20 picofarads

A visual inspection of the cable should reveal no abnormalities or defects. The minimum acceptance criterion is that the power factor must not exceed the manufacturer's data.

Testing can be performed by means of direct current (DC), alternating current (AC), partial discharge (pd), or very low frequency (VLF). The selection can be made only after an evaluation of the alternative methods. Test voltages should not exceed 80% of the cable manufacturer's factory test value. During testing,

- Ensure that the input voltage to the test set is regulated.
- Ensure that current-sensing circuits in test equipment, when available, measure only the leakage current associated with the cable under test and do not include internal leakage of the test equipment.

- Record wet- and dry-bulb temperatures or relative humidity and temperature.
- Test each cable section individually.
- Test each conductor individually with all other conductors grounded. Ground all shields.
- Ensure that terminations are adequately corona-suppressed by guard ring, field reduction sphere, or other suitable method, as necessary.
- Ensure that the maximum test voltage does not exceed the limits for terminators specified in ANSI/IEEE 48, IEEE 386, or the manufacturer's specifications.
- Raise the conductor to the specified maximum test voltage and hold for 15 minutes.
- If performed by means of DC, reduce the test set potential to zero and measure residual voltage at discrete intervals.
- Apply grounds for a time period adequate to drain all insulation stored charge.

When new cables are spliced into existing cables, the acceptance test should be performed on the new cable prior to splicing. After the splice is completed, an insulation resistance test and a shield-continuity test should be performed on the length of new and existing cable including the splice and utilizing a test voltage that does not exceed 60% of the factory test value.

REFERENCES

NASA. (2004). *Reliability Centered Building and Equipment Acceptance Guide*. Washington, DC: National Aeronautics and Space Administration.

NFPA. (2005). *NFPA 70: National Electrical Code (NEC) Handbook*. Quincy, MA: National Fire Protection Association.

OSHA. (2007). Occupational Safety and Health Standards, 29 CFR, Part 1910, Subpart S, Electrical. Baltimore, MD: Occupational Safety and Health Administration.

13 Special Installations and OSHA

Martha J. Boss and Dennis Day

CONTENTS

13.1 TUNNEL INSTALLATIONS

The provisions here apply to the installation and use of high-voltage power distribution and utilization equipment that is portable or mobile, such as substations, trailers, cars, mobile shovels, draglines, hoists, drills, dredges, compressors, pumps, conveyors, and underground excavators. Conductors in tunnels must be installed in

- Metal conduit or other metal raceway
- Metal-clad (MC) cable
- Other approved multiconductor cable

Multiconductor portable cable may supply mobile equipment. Conductors and cables must be located or guarded as protection against physical damage. An equipment grounding conductor must be run with circuit conductors inside the metal raceway or inside the multiconductor cable jacket. The equipment grounding conductor may be insulated or bare. Bare terminals of transformers, switches, motor controllers, and other equipment must be enclosed to prevent accidental contact with energized parts. Enclosures for use in tunnels must be drip-proof, weatherproof, or submersible as required by the environmental conditions.

Switch or contactor enclosures may not be used as junction boxes or raceways for conductors feeding through or tapping off to other switches, unless special designs are used to provide adequate space for this purpose. A disconnecting means that simultaneously opens all ungrounded conductors must be installed at each transformer or motor location. All non-energized metal parts of electric equipment and metal raceways and cable sheaths must be effectively grounded and bonded to all metal pipes and rails at the portal and at intervals not exceeding 305 meters (1000 feet) throughout the tunnel.

13.2 EMERGENCY POWER SYSTEMS

Emergency power systems include circuits, systems, and equipment intended to supply power for illumination and special loads in the event of failure of the normal supply. Emergency circuit wiring must be kept entirely independent of all other wiring and equipment and may not enter the same raceway, cable, box, or cabinet or other wiring except either where common circuit elements suitable for the purpose are required or for transferring power from the normal to the emergency source.

13.2.1 EMERGENCY ILLUMINATION

Emergency illumination must include all required means of egress lighting, illuminated exit signs, and all other lights necessary to provide illumination. Where emergency lighting is necessary, the system must be so arranged that the failure of any individual lighting element (e.g., a light bulb burning out) cannot leave any space in total darkness.

13.2.2 Signs

A sign must be placed at the service entrance equipment indicating the type and location of onsite emergency power sources; however, a sign is not required for individual unit equipment. If the grounded circuit conductor of the emergency power source is connected remotely, signage is required. The sign must identify all emergency and normal sources connected at that grounding location.

13.2.3 Class 1, 2, and 3 Remote-Control, Signaling, and Power-Limited Circuits

Class 1, 2, and 3 remote-control, signaling, or power-limited circuits are characterized by their usage and electrical power limitation, which differentiates them from light and power circuits. These types of circuits are classified in accordance with their respective voltage and power limitations.

A Class 1 power-limited circuit must be supplied from a source having a rated output of ≤30 volts and 1000 volt-amperes. A Class 1 remote-control circuit or a Class 1 signaling circuit must have a voltage not exceeding 600 volts; however, the power output of the source need not be limited. The power source for a Class 2 or Class 3 circuit must be listed equipment marked as a Class 2 or Class 3 power source, except that (1) thermocouples do not require listing as a Class 2 power source, and (2) a dry-cell battery is considered an inherently limited Class 2 power source, provided the voltage is 30 volts or less and the capacity is less than or equal to that available from series-connected No. 6 carbon zinc cells.

A Class 2 or Class 3 power supply unit must be durably marked where plainly visible to indicate the class of supply and the unit's electrical rating. Unless a barrier for protection against contact is used, cables and conductors of Class 2 and Class 3 circuits may not be (1) placed in any cable, cable tray, compartment, enclosure, manhole, outlet box, device box, raceway, or similar fitting; or (2) with conductors of electric light, power, Class 1, non-power-limited fire alarm, or medium-power network-powered broadband communications circuit conductors.

13.3 FIRE ALARM SYSTEMS

Fire alarm circuits must be classified either as non-power-limited or power-limited. The power sources for use with fire alarm circuits must be either power-limited or non-power-limited as follows:

- The power source of non-power-limited fire alarm (NPLFA) circuits must have an output voltage of ≤600 volts, nominal.
- The power source for a power-limited fire alarm (PLFA) circuit must be listed equipment marked as a PLFA power source.

13.3.1 Separation of Non-Power-Limited Fire Alarm Circuits

Non-power-limited fire alarm circuits and Class 1 circuits may occupy the same enclosure, cable, or raceway provided all conductors are insulated for maximum voltage of any conductor within the enclosure, cable, or raceway. Power supply and fire alarm circuit conductors are permitted in the same enclosure, cable, or raceway only if connected to the same equipment.

13.3.2 Power-Limited Circuit

Power-limited circuit cables and conductors may not be placed in any cable, cable tray, compartment, enclosure, outlet box, raceway, or similar fitting with conductors of electric light, power, Class 1, non-power-limited fire alarm, or medium-power network-powered broadband communications circuits. Power-limited fire alarm circuit conductors must be separated at least 50.8 mm (2 in.) from conductors of any electric light, power, Class 1, non-power-limited fire alarm, or medium-power network-powered broadband communications circuits unless a special and equally protective method of conductor separation is employed. Conductors of one or more Class 2 circuits are permitted within the same cable, enclosure, or raceway with conductors of power-limited fire alarm circuits provided that the insulation of Class 2 circuit conductors in the cable, enclosure, or raceway is at least that needed for the power-limited fire alarm circuits.

13.3.3 Identification

Fire alarm circuits must be identified at terminal and junction locations in a manner that must prevent unintentional interference with the signaling circuit during testing and servicing. Power-limited fire alarm circuits must be durably marked as such where plainly visible at terminations.

13.4 COMMUNICATIONS SYSTEMS

Communication systems include central-station-connected and non-central-station-connected telephone circuits, radio and television receiving and transmitting equipment (including community antenna television and radio distribution systems, telegraph, and district messenger), outside wiring for fire and burglar alarm, and similar central station systems.

13.4.1 Protective Devices

A listed primary protector must be provided (1) on each circuit run partly or entirely in aerial wire or aerial cable not confined within a block and (2) on aerial or underground circuits which due to location may be exposed to accidental contact with electric light or power conductors operating at over 300 volts to ground. In addition, where lightning exposure exists, each interbuilding circuit on premises must be protected by a listed primary protector at each end of the interbuilding circuit.

13.4.2 CONDUCTOR LOCATION

Lead-in or aerial-drop cables from a pole or other support, including the point of initial attachment to a building or structure, must be kept away from electric light, power, Class 1, or non-power-limited fire alarm circuit conductors to avoid accidental contact. A separation of at least 1.83 meters (6 feet) must be maintained between communications wires and cables on buildings and lightning conductors. Where communications wires and cables and electric light or power conductors are supported by the same pole or run parallel to each other in-span the following conditions apply:

- Where practicable, communication wires and cables on poles must be located below the electric light or power conductors.
- Communications wires and cables may not be attached to a crossarm that carries electric light or power conductors.

Indoor communications wires and cables must be separated by at least 50.8 mm (2 in.) from conductors of any electric light, power, Class 1, non-power-limited fire alarm, or medium-power network-powered broadband communications circuits. Alternatively, a special and equally protective method of conductor separation, identified for that purpose, may be employed.

13.4.3 EQUIPMENT LOCATION

Outdoor metal structures supporting antennas, as well as self-supporting antennas (e.g., vertical rods or dipole structures), must be located as far away from overhead conductors of electric light and power circuits of over 150 volts to ground as necessary to prevent the antenna or structure from falling into or making accidental contact with such circuits.

13.4.4 GROUNDING

The metal sheath of aerial cables entering buildings must be grounded or interrupted close to the entrance to the building, if potentially in contact with electric light or power conductors, by an insulating joint or equivalent device. Protective devices must be grounded in an approved manner. Masts and metal structures supporting antennas must be permanently and effectively grounded without splice or connection in the grounding conductor.

Transmitters must be enclosed in a metal frame or grill or must be separated from the operating space by a barrier, all metallic parts of which are effectively connected to ground. All external metal handles and controls accessible to the operating personnel must also be effectively grounded. Unpowered equipment and enclosures are considered to be grounded where connected to an attached coaxial cable with an effectively grounded metallic shield.

13.5   SOLAR PHOTOVOLTAIC SYSTEMS

Solar photovoltaic systems can be interactive with other electric power production sources or can stand alone with or without electrical energy storage (e.g., batteries). These systems may have alternating current (AC) or direct current (DC) output for utilization.

13.5.1   CONDUCTORS OF DIFFERENT SYSTEMS

Photovoltaic source circuits and photovoltaic output circuits may not be contained in the same raceway, cable tray, cable, outlet box, junction box, or similar fitting as feeders or branch circuits of other systems, unless the conductors of the different systems are separated by a partition or are connected together.

13.5.2   DISCONNECTING MEANS

Means must be provided to disconnect all current-carrying conductors of a photovoltaic power source from all other conductors in a building or other structure. Where a circuit grounding connection is not designed to be automatically interrupted as part of the ground-fault protection system, a switch or circuit breaker used as a disconnecting means may not have a pole in the grounded conductor.

13.6   INTEGRATED ELECTRICAL SYSTEMS

An integrated electrical system is a unitized segment of an industrial wiring system where all of the following conditions are met:

- An orderly shutdown process minimizes employee hazard and equipment damage.
- Conditions of maintenance and supervision ensure that only qualified persons service the system.
- Effective safeguards have been established and are maintained.

Note: Overcurrent devices that are critical to integrated electrical systems need not be readily accessible if located with mounting heights to ensure security from operation by unqualified persons.

13.7   SYSTEMS OVER 600 VOLTS, NOMINAL

Wiring methods for systems over 600 volts, nominal, must meet certain requirements. Aboveground conductors must be installed in one of these:

- A rigid metal conduit
- An intermediate metal conduit
- Cable trays

- A cable bus
- Other suitable raceways
- As open runs of metal-clad cable suitable for the use and purpose

In locations accessible to qualified persons only, open runs of type MV cables, bare conductors, and bare busbars are also permitted. Busbars must be either copper or aluminum. Metallic shielding components (e.g., tapes, wires, or braids for conductors) must be grounded. Conductors emerging from the ground must be enclosed in approved raceways.

13.7.1 OPEN RUNS

Open runs of non-metallic-sheathed cable, bare conductors, or busbars must be installed in locations accessible only to qualified persons. Open runs of insulated wires and cables must have a bare lead sheath or a braided outer covering supported to prevent physical damage to the braid or sheath.

13.7.2 OPEN INSTALLATIONS OF BRAID-COVERED INSULATED CONDUCTOR

The braid on open runs of braid-covered insulated conductors must be flame retardant or must have a flame-retardant applied after installation. This treated braid covering must be stripped back a safe distance at conductor terminals, according to the operating voltage.

13.7.3 INSULATION SHIELDING

Metallic and semiconductor insulation shielding components of shielded cables must be removed for a distance dependent on the circuit voltage and insulation. Stress reduction means must be provided at all terminations of factory-applied shielding. Metallic shielding components (i.e., tapes, wires, or braids, or combinations thereof) and their associated conducting and semiconducting components must be grounded.

13.7.4 MOISTURE OR MECHANICAL PROTECTION FOR METAL-SHEATHED CABLES

Where cable conductors emerge from a metal sheath and where protection against moisture or physical damage is necessary, the insulation of the conductors must be protected by a cable sheath terminating device.

13.7.5 INTERRUPTING AND ISOLATING DEVICES

Suitable barriers or enclosures must be provided to prevent contact with non-shielded cables or energized parts of oil-filled cutouts. Load interrupter switches will be used only if suitable fuses or circuits are used in conjunction with these devices to interrupt fault currents. Where these devices are used in combination, the devices must be coordinated electrically so the effects of closing, carrying, or interrupting all possible currents up to the assigned maximum short-circuit rating can be safely accomplished. Where more than one switch is installed with interconnected load terminals to provide for alternate connection to different supply conductors, each switch must be provided with a conspicuous sign reading:

WARNING—SWITCH MAY BE ENERGIZED BY BACKFEED

A means (i.e., a fuseholder and fuse designed for the purpose) must be provided to completely isolate equipment for inspection and repairs. Isolating means that are not designed to interrupt the load current of the circuit must be either interlocked with an approved circuit interrupter or provided with a sign warning against opening them under load.

13.8   PORTABLE CABLES OVER 600 VOLTS, NOMINAL

Multiconductor portable cable for use in supplying power to portable or mobile equipment at over 600 volts, nominal, must consist of No. 8 or larger conductors employing flexible stranding; however, the minimum size of the insulated ground-check conductor of Type G–GC cables must be No. 10. Cables operated at over 2000 volts must be shielded for the purpose of confining the voltage stresses to the insulation. Equipment grounding conductors must be provided. All shields must be grounded. The minimum bending radii for portable cables during installation and handling in service must be adequate to prevent damage to the cable.

Connectors used to connect lengths of cable in a run must be of a type that locks firmly together. Provisions must be made to prevent opening or closing these connectors while they are energized. Strain relief must be provided at connections and terminations. Portable cables may not be operated with splices unless the splices are of the permanent molded, vulcanized, or other approved type. Termination enclosures must be suitably marked with a high-voltage hazard warning, and terminations must be accessible only to authorized and qualified employees (OSHA, 2007).

REFERENCES

NFPA. (2005). *NFPA 70: National Electrical Code (NEC) Handbook*. Quincy, MA: National Fire Protection Association.

OSHA. (2007). Occupational Safety and Health Standards, 29 CFR, Part 1910, Subpart S, Electrical. Baltimore, MD: Occupational Safety and Health Administration.

14 Distance Requirements and OSHA

Martha J. Boss and Dennis Day

CONTENTS

Space must be provided around all electric equipment to permit ready and safe operation and maintenance. The Occupational Safety and Health Administration (OSHA) specifies both the width (30 inches or the width of the panel, whichever is greater), and the depth, which is at least 3 feet and based on the voltage of the equipment. Points of control must be located so that persons are not likely to come in contact with any live part or moving part of the equipment while turning on the lights. When normally enclosed live parts are exposed for inspection or servicing, the working space must be suitably guarded. Where electric equipment might be exposed to physical damage, enclosures or guards must be so arranged and of such strength as to prevent damage. In all cases, the working space must permit at least a 90-degree opening of equipment doors or hinged panels.

14.1 CONTACT PROTECTION

Protective shields, protective barriers, or insulating materials as necessary must be used to avoid inadvertent contact with live parts. If possible, the arc radii determined as the hazard zone around live parts should not be entered. If entry is deemed necessary, only qualified persons should enter within the arc radii boundaries. Conductive materials and equipment that are in contact with any part of a worker's body must be handled in a manner that will prevent them from contacting exposed energized conductors or circuit parts. If a worker must handle long conductive objects in areas with exposed live parts, the employer must institute work practices (use of insulation, guarding, and material handling techniques) to minimize the hazard. Suitable insulating mats or platforms must be provided so the attendant cannot readily touch live parts unless standing on the mats or platforms (1) where live parts of motors or controllers operating at over 150 volts to ground are guarded against accidental contact only by location, and (2) where adjustment or other attendance may be necessary.

If portable ladders are used where the worker or the ladder could contact exposed energized parts, they must have non-conductive siderails. Permanent ladders or stairways must be provided to give safe access to the working space around electric equipment installed on platforms, balconies, mezzanine floors, or in attics, roof rooms, or other spaces. Electrically conductive cleaning materials (steel wool, metalized cloth, silicon carbide, conductive liquid solutions) must not be used in proximity to energized parts unless procedures are followed that will prevent electrical contact. Conductive articles of jewelry and clothing (watch bands, jewelry, rings, key chains, cloth with conductive thread, or metal headgear) must not be worn if these articles might contact exposed energized parts. If equipment is exposed to physical damage from vehicular traffic, suitable guards must be provided to prevent such damage.

14.2 ELECTRICAL INSTALLATIONS

14.2.1 QUALIFIED PERSONS

Even though access is controlled by lock and key or other approved means, electrical installations in a vault, room, closet, or an area surrounded by a wall, screen, or fence should be accessible to qualified persons only.

14.2.2 UNQUALIFIED PERSONS

Electrical installations open to unqualified persons must be (1) made with metal-enclosed equipment, or (2) enclosed in a vault or an area where access is controlled by a lock. Where non-metallic or metal-enclosed equipment is accessible to unqualified personnel and the bottom of the enclosure is less than 8 feet above the floor, the door or cover must be kept locked. In addition, when the equipment is located outdoors, the nuts and bolts must be designed so easy removed by the general public is not possible.

14.3 OUTDOOR ENCLOSURES

Outdoor electric equipment must be installed in suitable enclosures and protected from accidental contact by unauthorized personnel, or by vehicular traffic, or by accidental spillage or leakage from piping systems. No architectural appurtenance or other equipment (that by its height obstructs the electrical equipment space requirements) will be located in the working space. The enclosure height must be at least 7 feet high. A wall, screen, or fence less than 7 feet in height is not considered to prevent access unless other features (three-stranded barbed wire on the top) provide a degree of isolation equivalent to a 7-foot fence. *Note:* For clearances of conductors for specific voltages and basic insulation level (BIL) ratings, see ANSI C2-1997.

14.4 INDOOR DEDICATED SPACES

The dedicated space will equal the width and depth of the equipment and must

- Extend from the floor to a height of 1.83 meters (6.0 feet) above the equipment or to the structural ceiling, whichever is lower. *Note:* A dropped, suspended, or similar ceiling that does not add strength to the building structure is not considered a structural ceiling.
- Only contain piping (sprinkler systems), ducts, or equipment foreign to the electrical installation if these accoutrements are isolated by height or physical enclosures so as to prevent damage to the electrical equipment and inadvertent contact by unauthorized persons.
- Be kept clear of foreign systems unless protection is provided to avoid damage from condensation, leaks, or breaks in such foreign systems.

Control equipment is permitted in the dedicated space if necessary to guarantee electrical safety. Sufficient space must be provided and maintained about electric equipment to permit ready and safe operation and maintenance of such equipment.

14.5 ENTRANCE AND ACCESS

Sufficient access and working space must be provided and maintained. Electrical installations having exposed live parts must be accessible to qualified persons only. Entrances to all buildings, rooms, or enclosures containing exposed live parts or exposed conductors operating at over 600 V, nominal, must be kept locked or be under the observation of a qualified person at all times. At least one entrance not less than 610 mm (24 in.) wide and 1.98 m (6.5 ft) high must be provided to give access to the working space around electric equipment. On switchboard and control panels exceeding 1.83 m (6.0 ft) in width, two entrances (one at each end) must be provided, or one if the location permits a continuous and unobstructed way of exit travel or the work space required in Table 14.1 is doubled. The entrance must be located so that the edge of the entrance nearest the switchboards and control panels is at least the minimum clear distance given in Table 14.1 away from such equipment. Where bare energized parts at any voltage or insulated energized parts above 600 V, nominal, to ground are located adjacent to such an entrance, these energized parts must be suitably guarded. Permanent ladders or stairways must be provided to give safe access to the working space around electric equipment. *Note:* These requirements do not apply to equipment on the supply side of the service point.

TABLE 14.1
Minimum Depth of Clear Working Space at Electric Equipment, 600 Volts or More

Nominal Voltage to Ground	Minimum Clear Distance for Condition					
	Condition A		Condition B		Condition C	
	Meters	Feet	Meters	Feet	Meters	Feet
601 to 2500 V	0.9	3.0	1.2	4.0	1.5	5.0
2501 to 9000 V	1.2	4.0	1.5	5.0	1.8	6.0
9001 to 25,000 V	1.5	5.0	1.8	6.0	2.8	9.0
25,000 to 75,000 V	1.8	6.0	2.5	8.0	3.0	10.0
Above 75,000 V	2.5	8.0	3.0	10.0	3.7	12.0

Source: OSHA, Occupational Safety and Health Standards, 29 CFR, Part 1910, Subpart S, Electrical, Occupational Safety and Health Administration, Baltimore, MD, 2007.

Notes: Minimum depth of clear working space in front of electric equipment with a nominal voltage to ground above 25,000 volts may be the same as that for 25,000 volts under Conditions A, B, and C for installations built before April 16, 1981. Minimum clear distances may be 0.7 m (2.5 ft) for installations built before April 16, 1981. Condition A: Exposed live parts on one side and no live or grounded parts on the other side of the working space, or exposed live parts on both sides effectively guarded by suitable wood or other insulating material. Insulated wire or insulated busbars operating at not over 300 volts are not considered live parts. Condition B: Exposed live parts on one side and grounded parts on the other side. Concrete, brick, and tile walls are considered as grounded surfaces. Condition C: Exposed live parts on both sides of the work space (not guarded as provided in Condition A) with the operator between.

TABLE 14.2

Minimum Depth of Clear Working Space at Electric Equipment, 600 Volts or Less

Nominal Voltage to Ground	Minimum Clear Distance for Condition					
	Condition A		Condition B		Condition C	
	Meters	Feet	Meters	Feet	Meters	Feet
0 to 150 V	0.9	3.0	0.9	3.0	0.9	3.0
151 to 600 V	0.9	3.0	1.0	3.5	1.2	4.0

Source: OSHA, Occupational Safety and Health Standards, 29 CFR, Part 1910, Subpart S, Electrical, Occupational Safety and Health Administration, Baltimore, MD, 2007.

Notes: Minimum clear distances may be 0.7 m (2.5 ft) for installations built before April 16, 1981. Condition A: Exposed live parts on one side and no live or grounded parts on the other side of the working space, or exposed live parts on both sides effectively guarded by suitable wood or other insulating material; insulated wire or insulated busbars operating at not over 300 volts are not considered live parts. Condition B: Exposed live parts on one side and grounded parts on the other side. Condition C: Exposed live parts on both sides of the work space (not guarded as provided in Condition A) with the operator between.

14.5.1 Switchboard and Control Panels

Switchboard and control panels exceeding 48 in. (4 ft) in width must have one entrance at each end where practical; however, the National Electrical Code (NEC 110.33(A)(1)) defines large equipment as switchboard and control panels exceeding 6 feet in width. In this case, OSHA is the more restrictive requirement.

Switchboard and control panels exceeding 1.83 m (6.0 ft) in width must have two entrances, one at each end, or one if

- Location permits a continuous and unobstructed way of exit travel.
- The work space required in Table 14.2 is doubled.

These entrances must be located so the edge of the entrance nearest the switchboards and control panels is at least the minimum clear distance given in Table 14.1 away from such equipment.

Size openings in partitions or screens must be sized and located so that people are not likely to come into accidental contact with the live parts, or bring conducting objects into contact with them.

14.5.2 Exposed Live Parts: Guards, Locks, and Signage

Where bare energized parts at any voltage or insulated energized parts above 600 V, nominal, to ground are located adjacent to an entrance, these energized parts must be suitably guarded. Entrances to all buildings/rooms or enclosures containing exposed live parts or conductors must (1) be kept locked or continually supervised by a qualified person or (2) have permanent and conspicuous warning signs that state:

DANGER—HIGH VOLTAGE—KEEP OUT

Entrances to rooms and other guarded locations containing exposed live parts must be marked with conspicuous warning signs forbidding unqualified persons to enter. When doors are open or panels are removed from compartments containing voltages over 250 volts AC or DC, DANGER labels must be attached on the equipment and plainly visible.

14.6   STORAGE AND ACCESS

Working space required around electrical equipment may not be used for storage. Thus, what may appear to be a housekeeping issue may in fact also be an electrical safety issue.

14.7   BACK OF ASSEMBLIES

Working space is not required in back of assemblies (e.g., dead-front switchboards, motor control centers) where

- No renewable or adjustable parts (e.g., fuses or switches) are on the back.
- All connections are accessible from locations other than the back.

When rear access is required to work on de-energized parts on the back of enclosed equipment, a minimum working space of 762 mm (30 in.) horizontally must be provided. Switchboards, panelboards, and distribution boards installed for the control of light and power circuits and motor control centers must be located in dedicated spaces and protected from damage.

14.8   ELECTRICAL SAFETY SPACE CONSIDERATIONS FOR ELECTRICAL EQUIPMENT RATED AT 600 VOLTS, NOMINAL, OR LESS

Working space for equipment likely to require examination, adjustment, servicing, or maintenance while energized must have

- Clearance space (around the equipment) greater than or equal to
  - 1.98 m (6.5 ft) high (measured vertically from the floor or platform)
  - 914 mm (3.0 ft) wide (measured parallel to the equipment)
- Depth of the working space in the direction of access to live parts as indicated in Table 14.2. Distances will be measured from the live parts exposed or from the enclosure front or opening if enclosed.
- Width of working space in front of the electric equipment that is the width of the equipment or 762 mm (30 in.), whichever is greater.

In existing buildings where electrical equipment is being replaced, Condition B (see Table 14.2 notes) working clearances will be permitted between dead-front switchboards, panelboards, or motor control centers located across the aisle from each other where (1) maintenance or supervision ensures that written procedures prohibit opening of equipment on both sides of the aisle at the same time, and (2) authorized, qualified persons must service the equipment. The work space must be clear and extend from the grade, floor, or platform. Where the electrical equipment exceeds 6.5 ft in height, the minimum headroom must not be less than the height of the equipment. In other words, the equipment cannot be installed on a slant platform. Both NEC and OSHA have additional requirements for larger dimension equipment.

14.8.1 Live Parts

Live parts must be guarded against accidental contact by

- Using approved cabinets or enclosures
- Locating them in a room or vault accessible only to qualified persons
- Arranging permanent, substantial partitions or screens so only qualified persons have access within reach of the live parts; any partition opening must be sized and located to prevent accidental contact
- Placing them on a suitable balcony, gallery, or platform that is elevated and located to prevent access by unqualified persons (see Table 14.3).
- Elevating them 2.44 m (8.0 ft) or more above the floor or other working surface

If normally enclosed live parts must be exposed for maintenance or inspection and this exposes these parts to passageways or general open space, the working space must be suitably guarded and ≥3 ft in front of the equipment.

TABLE 14.3
Elevation of Unguarded Live Parts above Working Space

Nominal Voltage between Phases	Elevation	
	Meters	**Feet**
601 to 7500 V	2.81	9.01
7501 V to 35,000 V	2.8	9.0
Over 35,000 V	2.8 + 9.5 mm/kV over 35 kV	9.0 + 0.37 in./kV over 35 kV

Source: OSHA, Occupational Safety and Health Standards, 29 CFR, Part 1910, Subpart S, Electrical, Occupational Safety and Health Administration, Baltimore, MD, 2007.

Notes: The minimum elevation will be 2.6 m (8.5 ft) for installations built before August 13, 2007. The minimum elevation will be 2.4 m (8.0 ft) for installations built before April 16, 1981, if the nominal voltage between phases is in the range of 601 to 6600 volts.

14.8.2   Minimum Headroom

Around service equipment, switchboards, panelboards, and motor centers, the minimum headroom of working spaces varies given the building's age:

- For installations built before August 13, 2007, the minimum headroom is 1.91 m (6.25 ft).
- For installations built on or after August 13, 2007, the minimum headroom is 1.98 m (6.5 ft), or greater than the height of any equipment that exceeds. 1.98 m (6.5 ft).

Within the height required, other equipment that is associated with the electrical installation and is located above or below the electrical equipment is not permitted to extend more than 6 inches beyond the front of the electrical equipment.

14.8.3   Work Space Width

The work space must not be less than 30 inches wide in front of the electric equipment. According to the NEC, the width of the working space in front of the equipment must be whichever is the greater of the following: (1) the width of the equipment proper, or (2) 30 inches. In all cases, the work space must permit at least a 90-degree opening of doors or hinged panels. By special permission, smaller working spaces are permitted where all uninsulated parts operate at not greater than 30 volts rms, 42 volts peak, or 60 volts DC.

14.8.4   Equipment Rated 1200 Amps and over 1.83 Meters Wide

For equipment rated 1200 amperes or more and over 1.83 m (6.0 feet) wide, containing overcurrent devices, switching devices, or control devices, one entrance not less than 610 mm (24 in.) wide and 1.98 m (6.5 ft) high at *each end* of the working space must be provided, except for the following circumstances, in which case one means of exit is permitted:

- Where the location permits a continuous and unobstructed way of exit travel
- Where the working space required in Table 14.2 is doubled

However, the entrance must be located so that the edge of the entrance nearest the equipment is the minimum clear distance given in Table 14.2 away from such equipment.

14.9   ELECTRICAL SAFETY SPACE CONSIDERATIONS FOR HIGH-VOLTAGE EQUIPMENT OVER 600 VOLTS, NOMINAL

14.9.1   Outdoor Installations with Exposed Live Parts

Outdoor electrical installations having exposed live parts must be accessible to qualified persons only. *Note:* The above requirements do not apply per OSHA to equipment on the supply side of the service point.

14.9.2   OUTDOOR INSTALLATIONS WITH ENCLOSED EQUIPMENT

For equipment accessible to unqualified employees, ventilating or similar openings in equipment must be designed such that foreign objects inserted through these openings are deflected from energized parts. Where installations are exposed to physical damage from vehicular traffic, suitable guards must be provided. Non-metallic or metal-enclosed equipment located outdoors and accessible to the general public must be designed so exposed nuts or bolts cannot be readily removed, permitting access to live parts. Where non-metallic or metal-enclosed equipment is accessible to the general public and the bottom of the enclosure is less than 2.44 m (8.0 ft) above the floor or grade level, the enclosure door or hinged cover must be kept locked. Doors and covers of enclosures used solely as pull boxes, splice boxes, or junction boxes must be locked, bolted, or screwed on. Underground box covers that weigh over 45.4 kg (100 lb) are not required to be locked, bolted, or screwed on. *Note:* The above requirements do not apply per OSHA to equipment on the supply side of the service point.

14.9.3   ENTRANCE MARKINGS SIGNAGE

Entrances to all buildings, rooms, or enclosures containing exposed live parts or conductors must be kept locked or continually supervised by a qualified person. These entrances must have permanent and conspicuous warning signs that read as follows:

DANGER—HIGH VOLTAGE—KEEP OUT

14.9.4   EQUIPMENT

14.9.4.1   Metal-Enclosed Equipment

Metal-enclosed switchgear, unit substations, transformers, pull boxes, connection boxes, and other similar associated equipment must be marked with appropriate caution signs. Vents or similar openings in metal-enclosed equipment must be designed so that foreign objects inserted through these openings are deflected from energized parts.

14.9.4.2   Mobile and Portable Equipment Systems over 600 Volts, Nominal

Mobile and portable equipment for systems over 600 V, nominal, must meet certain requirements. A metallic enclosure must be provided on the mobile machine for enclosing the terminals of the power cable. This enclosure (1) must include provisions for a solid connection to the ground wire terminals to effectively ground the machine frame, (2) must be marked with a sign warning of the presence of energized parts, and (3) must have a provision for locking so that only authorized qualified persons may open the enclosure. All energized switching and control parts must be enclosed in effectively grounded metal cabinets or enclosures. Circuit breakers and protective equipment must have the operating means projecting through the metal cabinet or enclosure so these units can be reset without locked doors being opened. The method of cable termination used must prevent any strain or pull on the cable from stressing the electrical connections.

14.9.4.3   Access to Internal Equipment Systems from 500 to 1000 Volts

Where doors are used for access to voltages from 500 to 1000 volts AC or DC, one of the following must be provided: (1) door locks, or (2) interlocks. Where doors are used for access to voltages of over 1000 volts AC or DC, provide either (1) mechanical lockouts with a disconnecting means to prevent access until voltage is removed from the cubicle, or (2) both door interlocking and mechanical door locks.

14.9.5   ACCESS TO EQUIPMENT SYSTEMS OVER 600 VOLTS, NOMINAL

If switches, cutouts, or other equipment operating at 600 V, nominal, or less are installed in a high-voltage room or enclosure where exposed live parts or exposed wiring operating at over 600 volts, nominal, are present, then the high-voltage equipment must be effectively separated from the space occupied by the low-voltage equipment by a suitable partition, fence, or screen. However, switches or other equipment operating at 600 V, nominal, or less and serving only equipment within the high-voltage vault, room, or enclosure may be installed in the high-voltage enclosure, room, or vault if accessible to qualified persons only. Conductors and equipment used on circuits exceeding 600 V, nominal, must be accessible to qualified persons only. Electrical installations located in a vault, room, or closet (or in an area surrounded by a wall, screen, or fence) with access controlled by lock and key are considered to be accessible to qualified persons only. A wall, screen, or fence must be used to enclose an outdoor electrical installation to deter access by persons who are not qualified. A fence may not be less than 2.13 m (7.0 ft) in height or a combination of 1.80 m (6.0 ft) or more of fence fabric and a 305-mm (1-ft) or more extension utilizing three or more strands of barbed wire or equivalent. *Note:* These requirements do not apply per OSHA to equipment on the supply side of the service point.

14.9.6   ACCESS TO INDOOR INSTALLATIONS BY OTHER-THAN-QUALIFIED PERSONS

For indoor installations accessible to other-than-qualified persons, the installations must be made with metal-enclosed equipment or enclosed in a vault or located in an area to which access is controlled by a lock. Metal-enclosed switchgear, unit substations, transformers, pull boxes, connection boxes, and other similar associated equipment must be marked with appropriate caution signs. Openings in ventilated dry-type transformers and similar openings in other equipment must be designed so that foreign objects inserted through these openings are deflected from energized parts. *Note:* These requirements do not apply per OSHA to equipment on the supply side of the service point.

14.9.7   WORKING SPACES AROUND LIVE PARTS
OPERATING AT OVER 600 VOLTS, NOMINAL

Working space around live parts operating at over 600 V, nominal, must meet certain requirements. Sufficient space must be provided and maintained about electric equipment to allow ready and safe operation and maintenance of such equipment.

Distances are to be measured from the live parts (if exposed) or from the enclosure front or opening (if the live parts are not exposed). *Note:* These requirements do not apply per OSHA to equipment on the supply side of the service point.

14.9.8 WORKING SPACE AND GUARDING OF EQUIPMENT OVER 600 VOLTS, NOMINAL

The minimum clear working space in the direction of access to live parts of electric equipment must not be less than specified in Table 14.1. Distances must be measured from live exposed parts or from the enclosure front or opening, if live parts are enclosed.

REFERENCES

IEEE. (1997). *IEEE C2: National Electrical Safety Code*, 7 CFR 1775.503(d)(1). New York: Institute of Electrical and Electronics Engineers.

NFPA. (2005). *NFPA 70: National Electrical Code (NEC) Handbook*. Quincy, MA: National Fire Protection Association.

OSHA. (2007). Occupational Safety and Health Standards, 29 CFR, Part 1910, Subpart S, Electrical. Baltimore, MD: Occupational Safety and Health Administration.

15 Arc Flash, Personal Protective Equipment, and Classified Locations

Gayle Nicoll, Martha J. Boss,
Reza Tazali, and Robert Zweifel

CONTENTS

An arc flash is a short circuit through the air. In an arc-flash incident, an enormous amount of concentrated radiant energy explodes outward. Arc blast or flash hazards include high temperatures (hotter than the surface temperature of the sun) over short periods of time (fractions of a second), hot gases, an intense pressure wave from the explosion (like having a hand grenade explode inches away), and shrapnel from vaporized and molten metal particles. Arc flashes are short circuits that flash from one exposed live conductor to another. Unfortunately, many times that conductor is a human being. The arc flash ionizes the air, creating electrically conductive, super-heated plasma with temperatures that can reach over 5000°F. This explosion can take less than a fraction of a second and produces a bright flash, intense heat, and a blast wave equivalent to detonating dynamite. The combination can be lethal to anyone standing in the blast area. Arcing parts, including the parts of electric equipment that in ordinary operation produce arcs, sparks, flames, or molten metal, must be enclosed or separated and isolated from all combustible material. An arc flash can occur any time electrical equipment is disconnected, inspected, or serviced. Some examples include the following:

- Accessing a damaged circuit breaker panel
- Accidentally contacting live ("hot") equipment
- Overvoltage conditions
- Insulation failure

- Corrosion of terminals
- High-amp current source close to a conducting medium
- Internal and external sparks

15.1 LOCKOUT, TAGOUT, AND ARC FLASH

De-energizing equipment does not absolve the facility from the responsibility of performing arc-flash analysis or providing the necessary personal protective equipment (PPE). Both the Occupational Health and Safety Administration (OSHA) and the National Fire Protection Association (NFPA) have basic rules that prohibit energized work. During hazardous energy control procedures, the circuit must be approached to verify that the circuit has been de-energized. Until the verification testing is completed, the circuit must be considered energized per NFPA 70E, Standard for Electrical Safety in the Workplace (NFPA, 2012); therefore, workers who approach a circuit for verification testing must wear full PPE (Tajali, 2012).

15.2 ARC-FLASH INJURIES

Arc-related injuries can range from minor to severe burns to blindness, hearing and memory loss due to the pressure wave, broken bones, or death. When a worker is exposed to an arc, the clothing worn may play a large role in the severity of the potential injury. Arc-flash accidents can result in serious injury to the worker:

- Severe burns
- Shrapnel wounds
- Vision loss
- Lung collapse
- Hearing loss
- Concussion
- Broken bones
- Death

Additionally, damage to equipment and buildings and fiduciary risk can result.

15.3 ARC FLASH AND OSHA

Arc flash is a serious and recognized hazard that falls under the jurisdiction of OSHA's 29 CFR 1910.303 and 29 CFR 1910.132–138 with regard to PPE selection. OSHA 29 CFR 1910.303 (Occupational Safety and Health Standards, Subpart S—Electrical, General) requires employers to assess the workplace prior to working on energized parts above 50 volts. If an arc-flash hazard is present, or likely to be present, then the employer must train employees and require employees to use PPE. OSHA's 29 CFR 1910.269 (Electric Power Generation, Transmission, and Distribution) requires that workers be trained in the potential hazards of electric arcs and the flames these arcs can produce by igniting other materials in the area. Once generated, arcs can be energy sources that potentiate ignition within gassy (hydrogen or methane) environments or where coal dust can serve as a combustion component.

Workers are prohibited by OSHA from wearing clothing that, in the presence of an arc, can potentially increase the extent of injury—that is, clothing that would ignite and continue to burn or clothing that would melt on the skin. Thus, workers are generally prohibited from wearing clothing materials made entirely of, or blended with, synthetic materials (e.g., acetate, nylon, polyester, rayon). OSHA mandates that all services to electrical equipment be done in a de-energized state. Working live can only be done under special circumstances.

The regulations outlined in NFPA 70E, Article 130, should be used as a tool to comply with OSHA mandates Subpart S Part 1910.333(a)(1). NFPA 70E is a voluntary standard, so many people erroneously believe compliance is voluntary; however, OSHA's electrical safety mandates, found in Subpart S Part 1910 and Subpart K Part 1926, are based on NFPA 70E. NFPA 70E is recognized as the tool that illustrates how an employer might comply with these OSHA standards. The relationship between the OSHA regulations and NFPA 70E can be described as OSHA is the "should" and NFPA 70E is the "how."

Article 110.16 of the National Electric Code (NEC) requires that all equipment worked on while energized must be identified and marked with arc-flash warning labels. Every panel, switchboard, or control panel fed by a motor control center (MCC) must be assessed for both arc-flash and shock hazards. Remember that motor loads can contribute to the available fault current. Although the MCC is the assumed final access point for motor loads, it may feed a variety of power centers. For example, an MCC may feed a 277/480-volt power panel or a 480/120–240 transformer (Littelfuse, 2007).

15.4 ARC-FLASH INVESTIGATION

In an arc-flash analysis, onsite investigation of the facility is imperative. During the analysis, electrical equipment and its control features must be catalogued. As part of the arc-flash analysis, the available fault current at each energized equipment point and approach boundaries, including the flash protection boundary, are calculated. From these calculations, the type of PPE appropriate when working within the arc-flash boundary will be determined. According to NFPA 70E-2000 Section 220.2(B)(1), the arc-flash analysis must be performed before employees are allowed to approach any exposed electrical conductor or circuit. This includes work done to service, modify, or install electrical systems. *Note:* Anyone who opens a circuit breaker box is potentially working in a live electrical work scenario. A detailed arch-flash study contains at least the following steps:

1. Data collection
2. Single-line diagram development
3. System modeling
4. Arcing short-circuit calculations
5. Time–current curve evaluation
6. Incident energy calculations
7. PPE determination

8. Arc-flash warning label location determination
9. Report and recommendations
10. Integration of data into the electrical safety program
11. Training

Although the class of equipment and system voltage are required considerations, calculations must also include the effectiveness of any energy control devices. These energy control devices are needed either to remove power that is sustaining the arc or to redirect the ground fault to earth. The IEEE 1584 (Guide for Performing Arc-Flash Hazard Calculations) method is a systematic approach that calculates the exact arc-flash energies from the electrical power system parameters. Complete data collection from the power system is required in order to generate short-circuit and coordination studies, and these can be a valuable adjunct to the arc-flash energy calculations (Tajali, 2012).

15.5 CURRENT-LIMITING DEVICES

Current-limiting protective devices can reduce incident energy by clearing the fault faster and by reducing the current seen at the arc source. These current-limiting effects are typically only applicable at high-fault currents. If the available fault current is not within the current-limiting range of the fuse, the faster clearing time may not be achieved, and a false sense of security may exist. Because the arcing fault current is often substantially less than the bolted fault current, the current-limiting effects must be evaluated carefully. Using the faster trip times associated with current-limiting devices without making adjustments for any reduced let-through current can lead to an overly optimistic assessment. An assessment based on an inaccurate evaluation of the current-limiting device ready-activation will under-predict the time that the arc can sustain. Faster trip times in part of the system design may result in lower arcing fault currents. The assumption is that the less energy released, the faster the arcing fault is cleared; however, faster trip times may also cause less power system reliability as a result of reduced selectivity and an increased chance of nuisance tripping (Walls, 2005).

15.6 PARALLEL SOURCES

Arcing faults can be fed from parallel sources. The energy at the fault location will be based on the total fault current from all of the sources. The time required to clear the fault will be based on the fault current portion that flows through each protective device. Some fault sources (e.g., utility feeds) are constant, whereas other fault sources (e.g., induction motors) are transient and only contribute current for a few cycles. The total clearing time of the overcurrent protective device depends on all of these variables (Walls, 2005).

15.7   RISK QUANTITATION AND THE 50-VOLT QUESTION

Both OSHA 29 CFR 1910.303 and NFPA 70E require an electrical shock evaluation for all energized parts operating at 50 volts or higher. OSHA and NFPA standards also require that all equipment operating at 50 volts and higher be assessed for potential shock and arc-flash hazards. All equipment operating at 50 volts and higher must be assessed for the following:

- Electrical shock, which establishes
  - Protection boundaries
  - PPE required
- Arc-flash hazards, which establishes
  - Incident energy level
  - Hazard Risk Category (HRC)
  - Flash protection boundary
  - PPE required

15.7.1   INCIDENT ENERGY

Incident energy is the amount of energy impressed on the face and body of the electrical worker. One of the units used to measure incident energy is calories per square centimeter (cal/cm^2) (Tajali, 2012). IEEE 1584 states that calculating incident energy on equipment under 240 V fed from a transformer less than 125 kVA is not necessary. This statement is based on the assumption that the fault current amperage will not be sufficiently high to sustain an arc flash. The IEEE 1584 statement, however, refers only to incident energy calculations and does not release employers from assessing all equipment down to 50 V. Substantial arc flashes can occur on equipment operating at 208 V and less if supplied by large transformers. Arcing faults that clear in one-half cycle or less, as required in current-limiting devices, may result in lower incident energy (Walls, 2005).

15.7.2   IMPEDANCE

Impedance (which has both magnitude and phase) may reduce available fault current. Due to the inverse characteristics of typical overcurrent devices, a small reduction in available fault current can sometimes result in a large increase in incident energy due to a longer trip time (Walls, 2005). Consequently, a circuit breaker may open more slowly (than that predicted by the assumed fault current) in equipment located downstream from power sources vs. equipment located closer to power sources and on the same circuit. This will significantly increase the amount of potential incident energy. Equipment downstream from a power source must be individually assessed for arc potential. For low-voltage systems, the arc fault current will be lower than the bolted fault current because of arc impedance (Tajali, 2012). *Note:* Some motors fed from the MCC may also have a disconnect switch (as opposed to an on/off switch) near the motor that must be assessed for arc-flash hazards.

15.7.3 Fault Current and Clearing

Arc-flash hazards are to be identified at each electrical equipment location where live work may occur. The available fault current and the clearing time of the circuit protection devices are both factors in determining the hazard risk. Fault calculations have historically been performed to provide a basis for conservative equipment ratings and selection. Utility contributions were often represented as an infinite source. For arc-flash hazard analysis, the available fault current must be assessed as accurately as possible. A conservatively high value may result in lower calculated incident energy than might actually be possible depending on the protective device's time–current response. The lower results would be caused by using a faster time–current response value from the protective device's time–current curve. To account for inaccuracies in the system model, tolerances should be used to estimate both minimum and maximum short-circuit currents for use in the arc-flash calculations (Walls, 2005).

The trip time is determined based on the arcing fault current that flows through the protective device when the arcing fault occurs. Due to parallel contributions and voltage transformations, the fault current through the protective device may not equal the total arcing fault current at the equipment bus. When tolerances are used, the trip time must be determined for the minimum arcing fault and the maximum arcing fault. IEEE 1584 requires calculation of trip time and incident energy for 85% and 100% of the calculated arcing fault current to account for variations in test data. The larger of the results should be used.

15.8 LIVE PARTS AND GUARDING

Article 110.16 of the National Electrical Code requires all equipment that may be worked on while energized to be identified and marked with an arc-flash warning label (Littelfuse, 2007). Per OSHA 29 CFR 1910.303(g)(2), live parts operating at 50 volts or more must be guarded against accidental contact by the following methods:

- Use approved enclosures or locate them in a room, vault, or similar enclosure that is accessible only to qualified persons.
- Arrange suitable permanent, substantial partitions or screens so that only qualified persons will have access to spaces within reach of the live parts.
- Size and locate any openings so persons are not likely to come into accidental contact with the live parts or to bring conducting objects into contact with them.
- Place live parts on a suitable balcony, gallery, or platform that is elevated to prevent access by unqualified persons, or elevate 2.44 meters (8.0 feet) or more above the floor or other working surface.
- In locations where electric equipment is likely to be exposed to physical damage, be sure enclosures or guards are arranged and of such strength to prevent damage.

If an arc is occurring, the arc can be avoided through safe radii and PPE. This requirement starts at 50 volts. Although much discussion in the industry has centered around whether an arc can be sustained, the issue really comes down to guarding.

15.9   GROUNDING TO EARTH (EARTHING)

The power that sustains an arc flash should be removed as quickly as possible. Being able to do so relies on an interplay of various control features. Ultimately, the overcurrent devices must respond correctly to discontinue the power supply. For this reason, overcurrent devices that are not functioning correctly, for one reason or another, will negate the assumptions made in arc-flash radii calculations that are reliant on their assumed disconnect times. In other words, the arc will sustain longer than anticipated.

While these overcurrent devices are in the process of disconnecting power the electric current may still be available to the frame of the equipment. For this reason, a path to ground is needed for some high-hazard equipment in order to take some of the electrical current away from the frame and into the earth. In order for this earthing path to be preferential compared to the path to a worker who is touching the frame, the earthing ground path must be continuous and of lower resistance. The resistance required is calculated based on an interplay between the earthing ground system and the circuitry overcurrent device disconnect requirements. A low-impedance path is required within the equipment grounding circuit that must be connected to a grounded conductor. This grounded conductor is associated with the power supply (service).

To correctly apply the NFPA 70E Hazard Risk Category task tables or the IEEE 1584 method, knowledge of the available short-circuit current, the opening time of the overcurrent protective device, and the available paths to ground is required. For this reason a thorough knowledge of the site's electrical system is required to prepare a proper arc-flash study. Remember that the available fault current where the arc flash occurs and the opening time of the upstream overcurrent devices determine the magnitude and time vectors of the power being provided to the arc. Impedance will affect the fault current, which will decrease farther from the power source. This lowering of the fault current may then result in a weaker current differential determination by the upstream overcurrent device, and the opening time of the upstream fuse or circuit breaker may be inappropriately delayed. Consequently, incident energy supply at the arcing site will be sustained longer. For this reason, overcurrent protective devices downstream of Hazard Risk Category 0 devices may actually be of a higher hazard risk category during an actual arcing event. Possible incident energy at all equipment supplied by overcurrent devices must, therefore, be independently analyzed for a complete arc-flash study. Impedance and clearing time must always be considered (Littelfuse, 2007).

If building grounding electrodes are not present or are not intact and the machine is not provided with a separate grounding electrode, the pathway of the ground fault current is in question. Consequently, prior to any arc-flash radii studies the status of electrical equipment (including protective devices per NEC requirements) and ground-fault pathways, if required for certain machines and equipment, must be assessed.

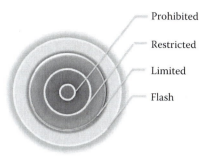

Prohibited

Restricted

Limited

Flash

FIGURE 15.1   Arc-flash boundaries.

15.10   APPROACH BOUNDARIES

Four different approach boundaries for personnel safety are defined in NFPA 70E (see Figure 15.1):

1. *Flash (protection) boundary*—This is the closest approach allowed without the use of arc protection PPE; represents the boundary between curable second-degree and incurable third-degree burns on exposed skin.
2. *Limited approach boundary*—No unqualified person may approach any exposed energized conductor. PPE is required.
3. *Restricted boundary*—Qualified persons are allowed to approach exposed, energized conductors closer than the restricted approach boundary if PPE is worn and a written work plan is approved and used.
4. *Prohibited boundary*—Crossing this boundary is the same as actually contacting the exposed energized part. In addition to the restricted boundary approach requirements, workers must perform a risk assessment before the prohibited boundary is crossed.

Equipment must be labeled with arc-flash and shock information to meet NEC requirements; this information includes the arc-flash radii determinations.

15.11   EQUIPMENT AND PERSONAL PROTECTIVE EQUIPMENT

Whereas OSHA 29 CFR 1910.132 requires PPE hazard assessments, 29 CFR 1910 Subpart S—Electrical contains the electrical safety rules. Although it is the technical basis for some of Subpart S, NFPA 70E is not incorporated by reference (see 29 CFR 1910.6). However, the 2004 edition of NFPA 70E was used by OSHA in the new Electrical Standard (2007 Final Rule). Equipment must be labeled with arc-flash and shock information to meet NEC requirements, and this information includes PPE requirements.

TABLE 15.1
Personal Protective Equipment

PPE Category	Minimum Arc Rating of PPE (cal/cm²)
0	1.2
1	4
2ª	8
3	25
4	40

Source: NFPA, *NFPA 70E: Standard for Electrical Safety in the Workplace®*, National Fire Protection Association, Quincy MA, 2012.

ª PPE Category 2 includes head and face protection.

15.11.1 NFPA 70E AND HAZARD RISK CATEGORIES

The NFPA 70E standard details Hazard Risk Categories (HRCs) called Levels; the greater the risk, the higher the HRC. The arc rating is the incident energy on a fabric that results in sufficient heat transfer through the fabric to cause the onset of a second-degree burn. Incident energy is expressed as calories per centimeter squared (cal/cm²) at a distance from the arc (the arc-radii limit). *Note:* Unprotected, second-degree burns can occur at an energy level of approximately 1.2 cal/cm², which corresponds to an HRC of 0; thus, HRC 0 does not denote zero risk! OSHA does not incorporate the NFPA 70E standard by reference in 29 CFR 1910.6; however, in making decisions as to personal protective equipment, using the NFPA 70E criteria as a reference is in keeping with OSHA training materials and interpretations. Table 15.1 shows PPE levels based on the HRCs.

15.11.2 SURFACE CHECKING

Equipment must be free of harmful physical irregularities that can be detected during tests or inspections. Surface irregularities that may be present on rubber goods due to imperfections on forms or molds or inherent difficulties in the manufacturing process and that may appear as indentations, protuberances, or imbedded foreign material are acceptable if

- The indentation or protuberance blends into a smooth slope when the material is stretched.
- Foreign material remains in place when the insulating material is folded and stretches with the surrounding insulating material.

Note: Rubber insulating equipment meeting the following national consensus standards are deemed to be in compliance. The national consensus standards include the following American Society for Testing and Materials (ASTM) specifications:

- ASTM D120-87, Specification for Rubber Insulating Gloves
- ASTM D178-93 (D178-88), Specification for Rubber Insulating Matting
- ASTM D1048-93 (D1048-88a), Specification for Rubber Insulating Blankets
- ASTM D1049-93 (D1049-88), Specification for Rubber Insulating Covers
- ASTM D1050-90, Specification for Rubber Insulating Line Hose
- ASTM D1051-87, Specification for Rubber Insulating Sleeves
- ASTM D120-87, Specification for Rubber Insulating Gloves

These standards also contain specifications for conducting the various tests required.

15.11.3 Testing

Equipment used during electrical work, including that associated with battery-powered systems and battery chargers, must be capable of withstanding the alternating current (AC) proof-test voltage or the direct current (DC) proof-test voltage. The proof test will reliably indicate that the equipment can withstand the voltage involved. The test voltage must be applied continuously for 3 minutes for equipment other than matting and must be applied continuously for 1 minute for matting.

15.11.4 Gloves

Refer to NFPA 70E for a table of glove ratings based on voltage shock hazard.

15.11.4.1 Protector Gloves

When gloves are used for HRC 0 situations under limited-use conditions and those requiring high finger dexterity, protector gloves are not required. In all other situations, protector gloves are required. In addition to the HRC, the possibility of physical damage to the gloves must be considered. Extra care is needed in visual examination of the workplace and the glove itself. Sharp objects must be avoided, as even a pinhole breach in a glove negates the effectiveness of the glove for electrical protection. Insulating gloves that have been used without protector gloves may not be used at a higher voltage until the insulating gloves have been tested.

15.11.4.2 AC Proof Test

Gloves must be capable of withstanding the AC proof-test voltage specified in ASTM standards and manufacturer's requirements after a 16-hour water soak. When the AC proof test used on gloves is 60 Hz, the proof-test current may not exceed the ASTM or manufacturer's prescribed test values at any time during the test period. When the AC proof test is at a frequency other than 60 Hz, the permissible proof-test current must be computed from the direct ratio of the frequencies. For the test, gloves (right side out) must be filled with tap water and immersed in water to a depth that is in accordance with ASTM standards and manufacturer's requirements. Water must be added to or removed from the glove as necessary so the water level is the same inside and outside the glove. After the 16-hour water soak, the 60-Hz proof-test current may exceed the values in accordance with ASTM standards and manufacturer's requirements by a factor less than 2 mA. Any exceedance above the 2-mA level indicates the glove does not pass and must be destroyed. *Note:* Equipment that has been subjected to a minimum breakdown voltage test must not be used for electrical protection.

15.11.5   Type II Insulating Equipment

Material used for Type II insulating equipment must be capable of withstanding an ozone test with no visible effects. The ozone test is used to reliably indicate that the material will resist ozone exposure in actual use. Any visible signs of ozone deterioration of the material (e.g., checking, cracking, breaks, pitting) is evidence of failure to meet the requirements for ozone-resistant material.

15.11.6   Maintenance

Electrical protective equipment must be maintained in a safe, reliable condition. The following specific requirements apply to insulating blankets, covers, line hose, gloves, and sleeves made of rubber:

- Maximum-use voltages must conform to those listed in ASTM standards and manufacturer requirements.
- Insulating equipment must be inspected for damage before each day's use and immediately following any incident that can reasonably be suspected of having caused damage. Insulating gloves must be given an air test along with the inspection.
- Insulating equipment with any of the following defects may not be used:
  - A hole, tear, puncture, or cut
  - An embedded foreign object
  - Any texture changes, such as swelling, softening, hardening, or becoming sticky or inelastic
  - Any other defect that damages the insulating properties
- Insulating equipment previously used for ozone cutting or ozone checking must not be used. The ozone can produce interlacing cracks that cause rubber failure when mechanical stress occurs.
- Insulating equipment found to have other defects that might affect the insulating properties of the equipment must be removed from service and returned for testing.
- Insulating equipment must be cleaned as needed to remove foreign substances.
- Insulating equipment must be stored in such a location and in such a manner as to protect the equipment from light, temperature extremes, excessive humidity, ozone, and other injurious substances and conditions.
- Electrical protective equipment must be subjected to periodic electrical tests. Test voltages and the maximum intervals between tests must be in accordance with ASTM standards and manufacturer's requirements. The test method will reliably indicate whether the insulating equipment can withstand the voltages involved. Standard electrical test methods considered as meeting this requirement are given in the following national consensus standards:
  - ASTM D120-87, Specification for Rubber Insulating Gloves
  - ASTM D1048-93, Specification for Rubber Insulating Blankets
  - ASTM D1049-93, Specification for Rubber Insulating Covers
  - ASTM D1050-90, Specification for Rubber Insulating Line Hose

- ASTM D1051-87, Specification for Rubber Insulating Sleeves
- ASTM F478-92, Specification for In-Service Care of Insulating Line Hose and Covers
- ASTM F479-93, Specification for In-Service Care of Insulating Blankets
- ASTM F496-93b, Specification for In-Service Care of Insulating Gloves and Sleeves

Insulating equipment failing to pass inspections or electrical tests may not be used by employees, except as follows:

- Rubber insulating line hose may be used in shorter lengths with the defective portion cut off.
- Rubber insulating blankets may be repaired using a compatible patch that results in physical and electrical properties that are equal to those of the blanket.
- Rubber insulating blankets may be salvaged by severing the defective area from the undamaged portion of the blanket. The resulting undamaged area may not be smaller than 560 mm by 560 mm (22 in. by 22 in.) for Class 1, 2, 3, and 4 blankets.
- Rubber insulating gloves and sleeves with minor physical defects (small cuts, tears, or punctures) may be repaired by the application of a compatible patch. In addition, rubber insulating gloves and sleeves with minor surface blemishes may be repaired with a compatible liquid compound. The patched area must have electrical and physical properties equal to those of the surrounding material. Repairs to gloves are permitted only in the area between the wrist and the reinforced edge of the opening.
- Repaired insulating equipment must be retested before the repaired equipment may be used by employees. The employer will certify that equipment has been tested. The certification will identify the equipment that passed the test and the date the equipment was tested. *Note:* Marking equipment and entering the results of the tests and dates of testing into logs are two acceptable means of meeting this requirement.

15.12   NFPA 70E AND FLAME-RESISTANT CLOTHING

Flame-resistant (FR) clothing made from 100% cotton or wool may be acceptable if the weight of the clothing is appropriate for the flame and electric arc conditions. As heat levels increase, these materials will not melt but can ignite and continue to burn. The amount of heat required to ignite these materials is dependent upon the weight, texture, weave, and color of the material. This type of clothing does not, however, meet the requirements of OSHA 29 CFR 1910.269, as the clothing can ignite (and continue to burn) under the electric arc and flame exposure conditions found in the workplace.

The FR clothing industry has developed a heat energy rating system for FR fabrics. The heat energy is measured in cal/cm² and the FR criterion is defined as clothing that protects the worker at the various heat energy levels. Guidance for calculating the heat energy of an arc is contained in Section 130.7(C) and Appendix D of NFPA 70E (NFPA, 2012). FR clothing must be cared for as instructed by the manufacturer. Clothing that is damaged often requires special repair techniques. For example, using common nylon thread may reduce the value of the clothing's FR protection, so thread as specified by the clothing manufacturer must be used.

15.13 CHEMICAL EXPOSURE

Chemical exposures to electrolytes used in batteries require the use of barrier PPE to protect workers. This PPE includes splash shields, goggles, and/or safety glasses with side shields, as well as chemically resistant aprons, gloves, and boots. Dielectric fluids may contain polychlorinated biphenyls (PCBs). In the absence of testing, PCB content should be assumed and barrier PPE provided to protect workers.

15.14 LABELING

When the incident energy, flash boundary, and required PPE have been determined, labels can be generated and displayed on each piece of equipment. In addition, the method of calculating the data to support the labeling information must be documented. NFPA 70E 130.5(C) lists new requirements for labeling (NFPA, 2012). Electrical equipment (switchboards, panelboards, control panels, meter socket enclosures, MCCs) is to be marked with a label that includes the (1) available incident energy and corresponding working distance (i.e., arc-flash radii), and (2) minimum required level and rating of clothing/PPE given the highest HRC for the equipment (see Figure 15.2).

FIGURE 15.2 Sample arc-flash hazard warning label.

OSHA addresses labeling through training requirements. 29 CFR 1910.303(e) requires employers to mark electrical equipment with descriptive markings, including the equipment's voltage, current, wattage, or other ratings. In 29 CFR 1910.335(b), OSHA requires employers to use alerting techniques (safety signs and tags, barricades, attendants) to warn employees of hazards that could cause injury due to electric shock, burns, or failure of electric equipment parts. Further, 29 CFR 1910.335(b)(1) requires the use of safety signs, safety symbols, or accident prevention tags as required by 29 CFR 1910.145 to warn employees about electrical hazards (e.g., electric arc-flash hazards).

15.15   TRAINING

In 29 CFR 1910.332, employers are required to identify all actual and potential electrical hazards and to train and qualify their employees in safe work practices and standard operating procedures to reduce hazards and increase worker safety. Workers should be trained on arc-flash safety at least yearly, and this training must be documented.

Arc-flash assessments should be integrated into the plant's ongoing preventive maintenance and documentation programs. Equipment that is properly assessed and labeled may be moved at some point in the future and require a reassessment. Plant single-line drawings must be updated when systems are changed or if the electric utility changes the service coming into the building.

15.16   HAZARDOUS (CLASSIFIED) LOCATIONS AND OSHA

There are requirements for electric equipment and wiring used in locations that are classified depending on the properties of the flammable vapors, liquids, or gases or combustible dusts or fibers that may be present therein and the likelihood that a flammable or combustible concentration or quantity is present. Each room, section, or area must be considered individually in determining its classification. All areas designated as hazardous (classified) locations under the Class and Zone system and areas designated under the Class and Division system established after August 13, 2007, must be properly documented. This documentation must be available to those authorized to design, install, inspect, maintain, or operate electric equipment at the location. Factors in determining the classification and boundaries for each location include the following:

- Quantity of flammable material that might escape in case of accident
- Adequacy of ventilating equipment
- Total area involved
- Record of the industry with respect to explosions or fires

15.17   CLASS I LOCATIONS

Class I locations have flammable gas or vapor airborne concentrations that produce explosive or ignitable conditions.

15.17.1 CLASS I, DIVISION 1

Class I, Division 1, locations have concentrations of flammable gases or vapors that are present under normal operating conditions; exist frequently because of repairs, maintenance operations, or leakage; or are present due to faulty operation of equipment or processes that may also cause simultaneous failure of electric equipment. This classification usually includes locations where volatiles are transferred from one container to another or are contained within (1) interiors of spray booths and/or in the vicinity of spraying and painting operations; (2) pen tanks or vats; (3) drying rooms or compartments for evaporation; (4) fat and oil extraction equipment; (5) cleaning and dyeing plants; (6) gas generator rooms and manufacturing plants; (7) inadequately ventilated pump rooms; (8) interiors of refrigerators and freezers; or (9) other areas where ignitable concentrations occur during normal operations.

15.17.2 CLASS I, DIVISION 2

Class I, Division 2, locations have ignitable concentrations of liquids, gases, or vapors that are

- Handled, processed, or used
- Normally confined within closed containers
- Present within closed systems that will only be breached by accidental rupture or breakdown, or due to abnormal operation
- Normally prevented by positive mechanical ventilation and will only become hazardous if this ventilating equipment fails
- Adjacent to Class I, Division 1, locations where transfer can occur due to lack of (1) adequate positive-pressure ventilation from a source of clean air, or (2) effective safeguards against ventilation failure

The following are not normally considered hazardous locations:

- Piping without valves, checks, meters, and similar devices
- Storage locations containing only sealed containers

In non-hazardous locations, electrical conduits and enclosures separated from process fluids by a single seal or barrier are classified as Division 2.

15.17.3 CLASS I, ZONE 0

Class I, Zone 0, locations have ignitable concentrations of flammable gases or vapors that are present continuously or for long periods of time.

15.17.4 CLASS I, ZONE 1

Class I, Zone 1, locations have ignitable concentrations of flammable gases or vapors that

- Are present under normal operating conditions
- Occur frequently due to repair, maintenance operations, or leakage
- Occur where faulty operation of equipment or processes could cause the release of ignitable concentrations of flammable gases or vapors and simultaneous failure of electric equipment which then becomes an ignition source
- Are adjacent to a Class I, Zone 0, location where transfer can occur due to (1) inadequate positive-pressure ventilation from a source of clean air, or (2) ineffective safeguards against ventilation failure.

15.17.5   CLASS I, ZONE 2

Class I, Zone 2, locations have the following conditions:

- Ignitable concentrations of flammable gases or vapors are not likely to occur in normal operation, or if they do they will exist for only a short period.
- Volatile flammable liquids, flammable gases, or flammable vapors are handled, processed, or used; however, the liquids, gases, or vapors are normally confined within closed containers or closed systems that will only be breached by accidental rupture or breakdown or as the result of abnormal operation.
- Ignitable concentrations of flammable gases or vapors normally are prevented by positive mechanical ventilation and will only become hazardous if this ventilating equipment fails.
- The location is adjacent to a Class I, Zone 1, location where transfer can occur due to (1) inadequate positive-pressure ventilation from a source of clean air, or (2) ineffective safeguards against ventilation failure.

15.17.6   CLASS II LOCATIONS

Class II locations have combustible dust.

15.17.6.1   Class II, Division 1

Class II, Division 1, locations have combustible dust that is (1) normally present or airborne in quantities that can produce explosive or ignitable mixtures; (2) present when mechanical failure occurs or when abnormal equipment operation causes explosive or ignitable mixtures to occur and may also cause simultaneous failure of electric equipment or operation of protection devices that potentiate ignition; or (3) electrically conductive.

Combustible dusts that are electrically non-conductive include dusts produced in the handling and processing of grain and grain products, pulverized sugar and cocoa, dried egg and milk powders, pulverized spices, starch and pastes, potato and wood flour, oil meal from beans and seed, dried hay, and other organic materials that may produce combustible dusts when processed or handled. Dusts containing magnesium or aluminum are particularly hazardous, and the use of extreme caution is necessary to avoid ignition and explosion.

15.17.6.2 Class II, Division 2

Class II, Division 2, locations have combustible dust that is not normally airborne in quantities that can produce explosive or ignitable mixtures or that accumulate sufficiently to interfere with the normal operation of electric equipment or other apparatus. Such dust, however, may become airborne as a result of infrequent malfunctioning of equipment and accumulate near electric equipment in sufficient quantities to interfere with the safe dissipation of heat from electric equipment and may also be ignitable due to abnormal operation or failure of electric equipment. This classification includes locations where dangerous concentrations of suspended dust, while not likely, could accumulate about or in electric equipment.

15.17.7 CLASS III LOCATIONS

Class III locations are hazardous due to easily ignitable fibers or flyings that are not likely to become airborne in quantities sufficient to produce ignitable mixtures.

15.17.7.1 Class III, Division 1

Class III, Division 1, locations have easily ignitable fibers or materials producing combustible flyings that are handled, manufactured, or used. Such locations usually include some parts of rayon, cotton, and other textile mills; combustible fiber manufacturing and processing plants; cotton gins and cotton-seed mills; flax-processing plants; clothing manufacturing plants; woodworking plants and establishments; and industries involving similar hazardous processes or conditions. Easily ignitable fibers and flyings include rayon, cotton (including cotton linters and cotton waste), sisal or henequen, istle, jute, hemp, tow, cocoa fiber, oakum, baled waste kapok, Spanish moss, excelsior, and other similar materials.

15.17.7.2 Class III, Division 2

Class III, Division 2, locations have easily ignitable fibers that are stored or handled. These locations do not include areas where such fibers are present due to manufacturing.

15.18 ELECTRICAL INSTALLATIONS

Equipment, wiring methods, and installations of equipment in hazardous (classified) locations must be intrinsically safe. Equipment and associated wiring approved as intrinsically safe are permitted in any hazardous (classified) location for which the equipment and associated wiring have been approved. Electrical installations must be approved for the hazardous (classified) location. Equipment will be approved for the class of location and the ignitable or combustible properties of the specific gas, vapor, dust, or fiber that will be present. Electrical installations must also be safe for the hazardous (classified) location and of a type and design that the employer demonstrates will provide protection from the hazards arising from the combustibility and flammability of the vapors, liquids, gases, dusts, or fibers involved. *Note:* NFPA 70, the National Electrical Code, lists or defines hazardous gases, vapors, and dusts by groups, which are characterized by their ignitable or combustible properties.

15.19 MARKING

Equipment must be marked to show the class, group, and operating temperature or temperature range, based on operation in a 40°C ambient temperature environment. The temperature marking may not exceed the ignition temperature of the specific gas or vapor to be encountered; however, the following provisions modify this marking requirement for specific equipment:

- Equipment of the non-heat-producing type (e.g., junction boxes, conduit, fittings) and equipment of the heat-producing type having a maximum temperature less than or equal to 100°C (212°F) need not have a marked operating temperature or temperature range.
- Fixed lighting fixtures marked for use in Class I, Division 2, or Class II, Division 2, locations only need not be marked to indicate the group.
- Fixed general-purpose equipment in Class I locations, other than lighting fixtures, that is acceptable for use in Class I, Division 2, locations need not be marked with the class, group, division, or operating temperature.
- Fixed dust-tight equipment, other than lighting fixtures, that is acceptable for use in Class II, Division 2, and Class III locations need not be marked with the class, group, division, or operating temperature.
- Electric equipment that is suitable for ambient temperatures exceeding 40°C (104°F) must be marked with both the maximum ambient temperature and the operating temperature or temperature range at that ambient temperature.

15.20 CONDUITS

All conduits must be threaded and wrench-tight. Where making a threaded joint tight is impracticable, a bonding jumper must be utilized.

15.21 EQUIPMENT IN DIVISION 2 LOCATIONS

Equipment that has been approved for a Division 1 location may be installed in a Division 2 location of the same class and group. General-purpose equipment or equipment in general-purpose enclosures may be installed in Division 2 locations if the employer can demonstrate that the equipment does not constitute a source of ignition under normal operating conditions.

15.22 PROTECTION TECHNIQUES FOR HAZARDOUS LOCATIONS

The following are acceptable protection techniques for electric and electronic equipment in hazardous (classified) locations where approved:

- *Explosion-proof apparatus*—Equipment in Class I, Division 1 and 2, locations
- *Dust ignition-proof*—Equipment in Class II, Division 1 and 2, locations

- *Dust-tight*—Equipment in the Class II, Division 2, and Class III locations
- *Purged and pressurized*—Equipment in any hazardous (classified) location
- *Nonincendive circuit*—Equipment in Class I, Division 2; Class II, Division 2; or Class III, Division 1 or 2, locations
- *Nonincendive equipment*—Equipment in Class I, Division 2; Class II, Division 2; or Class III, Division 1 or 2, locations
- *Nonincendive component*—Equipment in Class I, Division 2; Class II, Division 2; or Class III, Division 1 or 2, locations
- *Oil immersion*—Current-interrupting contacts in Class I, Division 2, locations
- *Hermetically sealed*—Equipment in Class I, Division 2; Class II, Division 2; and Class III, Division 1 or 2, locations

15.23   CLASS I, ZONE 0, 1, AND 2 LOCATIONS

Employers may use the zone classification system as an alternative to the division classification system for electric and electronic equipment and wiring for all voltage in Class I, Zone 0, Zone 1, and Zone 2 hazardous (classified) locations where fire or explosion hazards may exist due to flammable gases, vapors, or liquids.

15.24   LOCATION AND GENERAL REQUIREMENTS

Locations are classified depending on the properties of the flammable vapors, liquids, or gases that may be present and the likelihood that a flammable or combustible concentration or quantity is present. Where pyrophoric materials are the only materials used or handled, these locations need not be classified. Each room, section, or area must be considered individually in determining its classification. All threaded conduit must be threaded with an NPT (National Pipe Thread Taper) standard conduit cutting die that provides 3/4-in. taper per foot. The conduit must be made wrench-tight to prevent sparking when fault current flows through the conduit system and to ensure the explosion proof or flameproof integrity of the conduit system where applicable.

Equipment that is to be provided with threaded entries for field wiring connections must be installed as follows:

- For equipment with threaded entries for NPT threaded conduit or fittings, listed conduit, conduit fittings, or cable fittings will be used.
- For equipment with metric threaded entries, such entries will be identified as being metric, or listed adaptors to permit connection to conduit of NPT threaded fittings will be provided with the equipment. Adapters will be used for connection to conduit or NPT threaded fittings.

15.25   PROTECTION TECHNIQUES AND ZONE CLASSIFICATION

One or more of the following protection techniques must be used for electric and electronic equipment in hazardous (classified) locations classified under the zone classification system and as approved:

- Flameproof ("d")—Equipment in Class I, Zone 1, locations
- Purged and pressurized—Equipment in Class I, Zone 1 or Zone 2, locations
- Intrinsic safety—Equipment in Class I, Zone 0 or Zone 1, locations
- Type of protection ("n")—Equipment in Class I, Zone 2, locations; type of protection is further subdivided into nA, nC, and nR.
- Oil immersion ("o")—Equipment in Class I, Zone 1, locations
- Increased safety ("e")—Equipment in Class I, Zone 1, locations
- Encapsulation ("m")—Equipment in Class I, Zone 1, locations
- Powder filling ("q")—Equipment in Class I, Zone 1, locations

15.26   SPECIAL PRECAUTIONS

Equipment construction and installation must ensure safe performance under conditions of proper use and maintenance. Classification of areas and selection of equipment and wiring methods should be under the supervision of a qualified registered professional engineer. In instances of areas within the same facility classified separately,

- Class I, Zone 2, locations may abut but not overlap Class I, Division 2, locations.
- Class I, Zone 0 or Zone 1, locations may not abut Class I, Division 1 or Division 2, locations.

A Class I, Division 1 or Division 2, location may be reclassified as a Class I, Zone 0, Zone 1, or Zone 2, location only if all of the space that is classified because of a single flammable gas or vapor source is reclassified.

Low ambient conditions require special consideration. Electric equipment may not be suitable for use at temperatures lower than 20°C (4°F) unless approved for use at such lower temperatures. However, at low ambient temperatures, flammable concentrations of vapors may not exist in a location classified Class I, Zone 0, 1, or 2, at normal ambient temperature.

15.27   LISTING AND MARKING

Equipment that is listed for a Zone 0 location may be installed in a Zone 1 or Zone 2 location of the same gas or vapor. Equipment that is listed for a Zone 1 location may be installed in a Zone 2 location of the same gas or vapor. Equipment approved for Class I, Division 1 or Division 2, must, in addition to other required marking, be marked with the following:

- Class I, Zone 1, or Class I, Zone 2 (as applicable)
- Applicable gas classification groups
- Temperature classification

Equipment meeting one or more of the protection techniques must be marked with the following in the order shown:

1. Class, except for intrinsically safe apparatus
2. Zone, except for intrinsically safe apparatus
3. "AEx"
4. Protection techniques
5. Applicable gas classification groups
6. Temperature classification, except for intrinsically safe apparatus

An example of such a required marking is "Class I, Zone 0, AEx ia IIC T6" (OSHA, 2007).

REFERENCES

ASTM. (1998). ASTM D1049: Standard Specification for Rubber Insulating Covers. West Conshohocken, PA: American Society for Testing and Materials.

ASTM. (2011). ASTM D1050: Standard Specification for Rubber Insulating Line Hose. West Conshohocken, PA: American Society for Testing and Materials.

IEEE. (2002). IEEE 1584: Guide for Performing Arc-Flash Hazard Calculations. New York: Institute of Electrical and Electronics Engineers.

Littelfuse. (2007). *Misconceptions About Arc-Flash Hazard Assessments*. Chicago, IL: Littelfuse, Inc.

Morley, L.A. (1990). *Mine Power Systems*, Information Circular 9258, NTIS No. PB 91-241729. Pittsburgh, PA: U.S. Department of the Interior, Bureau of Mines.

NFPA. (2005). *NFPA 70: National Electrical Code (NEC) Handbook*. Quincy, MA: National Fire Protection Association.

NFPA. (2012). *NFPA 70E: Standard for Electrical Safety in the Workplace®*. Quincy MA: National Fire Protection Association.

OSHA. (2007). Occupational Safety and Health Standards, 29 CFR, Part 1910, Subpart S, Electrical. Baltimore, MD: Occupational Safety and Health Administration.

Tajali, R. (2012). *Arc-Flash Analysis: IEEE Method versus the NFPA 70E Tables*. Palatine, IL: Schneider Electric.

Walls, G. (2005). *Understanding Arc-Flash Requirements*, Revision 3. Virginia Beach, VA: Professional Power Systems.

16 Electrical Testing

Martha J. Boss and Gayle Nicoll

CONTENTS

Inspection and testing technologies are those tests that give results that can be used for acceptance criteria but are not normally used for trending. Most tests in this category can be classified as a go/no-go test (i.e., the equipment either passes or fails the test). In both instances, the sustainability of the equipment can be measured and appropriate actions taken. Those actions collectively define the stewardship of the equipment and the facility where the equipment is used.

16.1 INSULATION POWER FACTOR TESTING

Power factor, sometimes referred to as *dissipation factor*, is the measure of the power loss through the insulation system to ground. This measurement is a dimensionless ratio that is expressed as a percent of the resistive current flowing through insulation relative to the total current flowing. To measure this value, a known voltage is applied to the insulation, and the resulting current and current/voltage phase relationship is measured. Figure 16.1 shows the phase relationships of the resulting currents. This test is not destructive, will not deteriorate or damage insulation, and is recommended for inclusion in any commissioning program. Usually, I_R is very small compared to I_T because most insulation is capacitive in nature. As a comparison, consider the similarities between a capacitor and a piece of electrical insulation. A capacitor is two current-carrying plates separated by a dielectric material. An electrical coil,

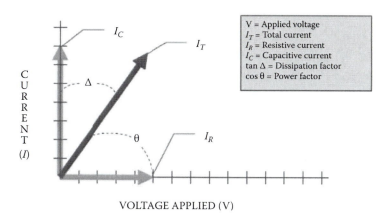

FIGURE 16.1   Power factor phase relationships.

such as that found in a transformer or motor, is a current-carrying conductor with an insulation material protecting the conductor from shorting to ground. The conductor of the coil and ground is similar to the two conducting plates in the capacitor. The insulation of the coil is like the dielectric material of the capacitor. The dielectric material prevents the charge on each plate from "bleeding through" until the voltage level of the two plates exceeds the voltage capacity of the dielectric. The coil insulation prevents the current from flowing to ground, until the voltage level exceeds the voltage capacity of the insulation.

The resistive current (I_R) is in phase with the applied voltage (V), and I_C is leading the voltage by a phase angle of 90 degrees. The total current is the resultant combination of both I_R and I_C. The tangent of the angle between the applied voltage and resultant current is the *dissipation factor*, and the cosine of the angle between the resultant current and the capacitive current is called the *power factor*.

As the impedance of the insulation changes due to physical damage, insulation shorts, moisture, contamination, or aging, the ratio between I_C and I_R will become less. The resulting phase angle between the applied voltage and resultant current then becomes less, and the power factor will rise. Consequently, the power factor test is used for making routine comparisons of the condition of an insulation system and for acceptance testing to verify that the equipment has been manufactured and installed properly. The test is not destructive, and regular follow-up maintenance testing will not deteriorate or damage the insulation. The power factor can be measured in three ways:

- *Grounded specimen test (GST)*—A voltage is applied to the circuit under test and all leakage current flowing through the insulation to ground is detected. Grounding leads are provided for the return path. This test provides a good overall test of the circuit.
- *Ungrounded specimen test (UST)*—A voltage is applied to the circuit under test and a direct measurement on only a portion of the circuit can be accomplished. Any currents flowing to ground are not measured. This test is good for isolating a reading for troubleshooting.

TABLE 16.1

Typical Power Factors

Equipment Type	Percent (%) Power Factor at 20°C
Oil-filled transformer:	
New, high-voltage (115 kV and up)	0.25–1.0
15 years old, high-voltage	0.75–1.5
Medium-voltage distribution	1.5–5.0
Dry type transformer (>600 V)	1.0–7.0
Oil circuit breakers (5 kV and up)	0.5–2.0
Air circuit breakers (5 kV and up)	0.5–2.0
Oil-paper cables:	
Solid (up to 27 kV), new condition	0.5–1.5
High-voltage oil-filled or pressurized (up to 69 kV)	0.2–0.5
Rotating machines stator windings (2–13.8 kV)	2.0–8.0
Capacitors (resistor out of circuit)	0.2–0.5
Bushings:	
Solid or dry	3.0–10.0
Compound filled (up to 15 kV)	5.0–10.0
Compound filled (15–46 kV)	2.0–5.0
Oil-filled (below 110 kV)	1.5–4.0
Oil-filled (110 kV and up)	0.3–3.0

Source: Adapted from NASA, *Reliability Centered Building and Equipment Acceptance Guide*, National Aeronautics and Space Administration, Washington, DC, 2004.

- *Grounded specimen test with guard*—A voltage is applied to the circuit under test and all leakage current flowing through the insulation to ground is detected; however, an additional lead is attached to the circuit and any leakage current up to that part of the circuit will be bypassed by the measuring circuit. This test is good for troubleshooting.

Table 16.1 contains typical insulation power factor values of electrical equipment. All power factor test results must be corrected to 20°C for comparison purposes. Test set manufacturers' instruction manuals should be consulted for the appropriate correction tables. Power factor readings should not be taken under the following conditions:

- *High humidity*—Humid conditions over 75% can result in excessive surface leakage currents on exposed surfaces.
- *Freezing ambient temperatures*—Power factor tests are very sensitive to moisture; however, frozen water becomes non-conducting and defects or degradation can remain hidden. This temperature limitation does not apply to insulation oils.
- *Dirty or contaminated surfaces*—Poor surface conditions result in excessive leakage currents that will be added to the losses and may give a false impression of the test.

16.2   INSULATION RESISTANCE TESTING

An insulation resistance test is a non-destructive direct current (DC) test used to determine insulation resistance to ground. A DC voltage is applied to the equipment under test, resulting in a small current flow. The test set then calculates the resistance. Insulation resistance is generally accepted as a reliable indication of the presence of contamination or degradation; however, test results vary greatly due to environmental conditions, specifically temperature. All readings must be corrected to 20°C for comparisons to be accurate. Under ideal conditions, modern insulation systems can be expected to have life cycles in excess of 100,000 operating hours. The total current through and along the insulation is made up of three components:

- The capacitance charging current actually "charges" the insulation up to the applied voltage, just like a capacitor, thus the term *capacitance charging*. It starts high and very rapidly falls off to zero after the insulation has been fully charged.
- The dielectric absorption current results from absorption within imperfections (or differences) in the insulation that cause voltage polarization. It starts high and decreases rapidly with time.
- The leakage current is the most important component of the insulation resistance test when the condition of the insulation under test is being evaluated. The path of this current can be either through the insulation itself to ground or over leakage surfaces.

Whereas the other two currents are essentially temporary charging currents, the leakage current actually represents a current loss. Theoretically, the leakage current should be constant with time for any single voltage value. This is an indication that the insulation under test can withstand the voltage being applied. A steady increase in the leakage current with time at the same applied voltage indicates an abnormality in the insulation, and the test should be stopped and the equipment examined. Actual test results are dependent on the length of the conductor being tested, the temperature of the insulating material, and the humidity of the surrounding environment at the time of the test. Insulation resistance tests are performed to establish a trending pattern and a deviation from the baseline information obtained during maintenance testing.

16.2.1   INSULATION RESISTANCE CORRECTION FACTORS

Taking the insulation resistance test a step further, the resistance readings can be recorded at 15-, 30-, 45-, and 60-second intervals for up to 10 minutes. These readings can be plotted on log–log graph paper. The resulting curves should have a smooth rise over the test time. For condition monitoring purposes, the best application for the insulation resistance test is the polarization index (PI) or the dielectric absorption ratio (DAR). The PI is the ratio of the 10- minute resistance to the 1-minute resistance. The DAR is the 30-second reading divided by the 1-minute reading. The resulting values are dimensionless and do not require any temperature correction. They are purely numeric and offset the fact that previous test information might not be available.

16.2.2 Leakage Current

Leakage current increases faster in insulation with the presence of moisture or contamination. The PI and DAR values will be lower for insulation that is in poor condition. In a condition monitoring program using insulation resistance, the PI is the value that is trended.

16.3 HIGH-POTENTIAL TESTS

High-potential (Hi-Pot) testing is a high-voltage DC standard acceptance and investigative test that (1) shows excessive leakage current in equipment, (2) is used to verify that insulation systems in new equipment can withstand designed voltage levels, (3) is used for new and repaired electrical transmission and distribution equipment and new or rewound motors, and (4) is a potentially destructive test that is, therefore, not used for periodic testing. In repaired equipment, if the leakage current continues to increase at a constant test voltage, this repair is not to the proper standard and will probably fail prematurely. For new equipment, if the equipment will not withstand the appropriate test voltage, the insulation system or construction method is inadequate for long-term service reliability. DC test voltages are applied to discover gross problems (e g., incorrectly installed accessories, mechanical damage). Manufacturers of the equipment, cables, and accessories to be tested must be consulted before applying the test voltage.

16.4 AIRBORNE ULTRASONIC TESTS

A relatively inexpensive device, the ultrasonic noise detector can be used to locate liquid and gas (pressure and vacuum) leaks. When a fluid or gas moves from a high-pressure region to a low-pressure region, the fluid or gas produces ultrasonic noise due to turbulent flow. The detector translates the ultrasonic noise in the frequency range of 20 to 100 kHz to the audible range, allowing an inspector to identify the source of the leak. Electrical arcing and discharges create sounds in the ultrasonic frequency ranges, sometimes long before a catastrophic failure occurs. Corona discharge is normally associated with high-voltage distribution systems and is produced as a result of a poor connection or insulation problem. Corona discharges, loose switch connections, and internal arcing in dead-front electrical connections are conditions that can all be discovered using ultrasonic test devices. For electrical systems, ultrasonic tests are often used in conjunction with infrared testing, as corona occurs in the ultraviolet region of the spectrum. Ultrasonic test kits generally consist of a receiver "gun" with variable frequency selection, a set of ear-isolation headphones, various contact probes, and various tone generators.

16.4.1 Airborne Ultrasonic Performance Survey

Electrical equipment should be surveyed for indications of arcing or electrical discharge, including corona. Piping systems should be surveyed for indications of leakage. For switchyards, the operator should use a parabolic concentrator, as the minimum distance to any live circuit will be at least 13 feet. A fixed frequency should be used to listen for a crackling sound. When inspecting electrical panels, the operator should use a rubber concentrator ("bootie") placed over the receiver to narrow the inspection area and help block out surrounding noises.

Airborne ultrasonic testing can be subjective and dependent on perceived differences in noises. To maximize the usefulness of this technology, care should be taken when setting test equipment controls for frequency ranges, sensitivity, and scale. Additionally, the operator should be cognizant of the fact that electrical loading and the presence of moisture (high humidity) may affect the ultrasonic signal. Any defects or exceptions noted by the use of airborne ultrasonics should be corrected and a re-survey of the repaired areas required to ensure that proper corrective action has been taken.

16.4.2 PULSE ECHO ULTRASONIC PERFORMANCE SURVEY

Material thickness measurements should be taken on a representative sample of all material where a thickness is specified.

16.5 PARTIAL DISCHARGE ANALYSIS

Partial discharge (PD) analysis is an online technology designed to monitor the condition of insulation in machines and cables above 4000 VAC. A partial discharge is an incomplete, or partial, electrical discharge that occurs between insulation and either other insulation or a conductor. These discharges create a high-frequency signal that partial discharge monitoring systems are designed to detect. PD analysis typically is performed for very large power generation equipment and large drive units (e.g., those associated with wind tunnel operations).

16.6 INFRARED THERMOGRAPHY

Infrared thermography (IRT) is the application of infrared detection instruments to identify temperature differences (thermogram) in equipment. The test instruments used are non-contact, line-of-sight, thermal measurement and imaging systems. As a non-contact technique, IRT is useful in identifying hot and cold spots in energized electrical equipment where standoff temperature measurement is necessary. IRT inspections are identified as either qualitative or quantitative. The quantitative inspection is interested in accurate measurement of the temperature of the item of interest. The qualitative inspection identifies relative differences, hot and cold spots, and deviations from normal or expected temperature ranges. Qualitative inspections are significantly less time consuming than quantitative because the thermographer is not concerned with highly accurate temperature measurement. What the thermographer does identify is highly accurate temperature differences between like components.

Infrared thermography can be used to identify installation defects, latent manufacturing defects, and safety hazards in electrical systems (e.g., transformers, motor control centers, switchgear, switchyards, power lines) (see Figure 16.2). For meaningful infrared data to be recorded the loading of the circuit should be at least 50%. In mechanical systems, IRT can identify blocked flow conditions in transformer cooling radiators and in pipes.

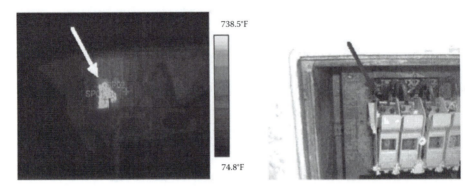

FIGURE 16.2 Infrared image of a failed electrical connection.

Table 16.2 shows typical temperature differences relative to given baseline (ΔT) criteria used in both industry and the military for in-service electrical equipment. For new equipment, any temperature rise over the reference temperature should be investigated and repaired. The indicated values are for equipment at 50% loading or greater. As the loading becomes less, the ΔT values become less. The infrared imager should be a focal plane array camera that is (NASA, 2004)

- Self-contained with a minimum of 2 hours of battery capacity
- Capable of using a temperature range of −20 to 300°C
- Sensitive to 0.2°C over all temperature ranges
- Accurate to within ±2%
- Capable of storing up to 12 images for later use
- Linked to a video recorder interface

TABLE 16.2
Typical Temperature Differentials Relative to Given Baseline Criteria

| Criticality | Temperature above Reference | | Condition |
	MilSpec	Industry	
Nominal	10–25°C	0–10°C	Nominal possibility of permanent damage; repair next maintenance period.
Intermediate	25–40°C	10–20°C	Possibility of permanent damage; repair soon.
Serious	40–70°C	20–40°C	Probability of permanent damage to item and surrounding area; repair immediately.
Critical	Over 70°C	Over 40°C	Failure imminent.

Source: Adapted from NASA, *Reliability Centered Building and Equipment Acceptance Guide,* National Aeronautics and Space Administration, Washington, DC, 2004.

16.6.1 SURVEY

For the survey, use predefined relative difference limits, hot and cold spots, and deviations from normal or expected temperature ranges consistent with manufacturer's design data. Perform a thermographic survey to detect uneven heating indicative of loose or dirty terminal connections or other internal electrical connections to the individual components. This survey may also indicate internal corrosion or other flaws. Localized heating may be indicative of flaws in windings or insufficient ventilation of the surrounding area. The survey should qualitatively verify good terminal connections, internal connections to the individual components, and relative equivalent temperatures of the adjacent components during operation. Acceptable deviations in temperature will vary with the size and type of the electrical panel; however, these limits should not exceed ±5°C. The temperature range should be ≤11.3°C ambient for all of the terminal connections or internal connections to the individual components.

16.7 OHM/RESISTANCE TESTING

Ohm/resistance testing determines whether grounding systems are working properly. Grounding that goes to earth is described as earthing in this text. The earth is actually a poor conductor compared to metals used as conductors (copper, aluminum). The waters of earth influence this conductivity. Waters with high mineral content are good conductors, and pure water is a poor conductor. Soil layering and type influence the efficacy of various grounding systems due to differences in soil resistivity.

Testing to determine whether a path to an earthing ground is present requires tracing a charge from application point to the supposed earthing ground. The testing should take into account that ambient temperatures and water loading of soil porosities will influence the test results. One of the complexities involved in testing for ground continuity is the issue of whether building frames and/or plumbing features can be used as a path to ground. Depending on the age of the building, grounding pathways may effectively be transferred via the building frame.

16.7.1 BUILDING FRAMES

Discontinuities in the metal frame structures may cause paths to ground to also be discontinuous and to fail. The result can be that the ground fault being transferred exposes workers to the risk of electric shock where the path to ground becomes discontinuous. For this reason, questions arose during promulgation of 29 CFR 1910 Subpart S—Electrical as to whether the building frame should be an allowed pathway to an earthing ground. During ohm/resistance testing, various component building sections may have various ages, so understanding the history of the Occupational Safety and Health Administration (OSHA) rulemaking and code requirements (National Electrical Code [NEC] and local) is important. In finalizing the current Subpart S—Electrical, OSHA found that requiring changes to installations that had been permitted by the NEC for many years before 1978 was impracticable and did not believe that the substantial cost to employers of changing grounding connections was worth the reduction in risk associated with moving from the use of the structural metal frame of a building to a separate equipment grounding conductor.

Consequently, the final rule in 29 CFR 1910.304(g)(8)(iii) continues to allow use of the grounded structural metal frame of a building as the equipent grounding conductor for equipment secured to, and in metallic contact with, the metal frame in installations made before April 16, 1981. However, the final rule requires such grounds to be replaced any time work is performed on the branch circuit. OSHA's rationale is that, because the circuit is de-energized, no increased risk during the installation of a new equipment grounding conductor would occur. Additionally, OSHA believed that the costs of installing an acceptable equipment grounding conductor in such cases would be minimized. Local building codes and the current National Electrical Code may or may not agree with OSHA in this regard and may take precedence. Consequently, research as to code requirements is important when determining the scope of an ohm/resistance study.

16.7.2   Water Pipes

Various commenters were noted in the Preamble to 29 CFR 1910 Subpart S—Electrical as opposing the use of water pipes. Their rationale included the following points:

- Section 250.52 of the NEC states that an interior metal water pipe more than 1.52 meters (5 feet) from the point of entrance of the water pipe into the building is no longer allowed to serve as part of the grounding electrode system.
- Using an isolated equipment grounding conductor (e.g., cold-water pipe) may increase the risk of reactance along the equipment grounding conductor when an alternating current (AC) fault occurs.
- Using a water pipe to ground equipment violates 2002 NEC Section 300.3(B), which requires all circuit conductors to be grouped together so magnetic fields are offset and reluctance is minimized.
- Plastic pipe makes water pipes an unreliable ground and using water pipes to ground electric equipment can pose hazards to employees working on the piping system.
- Water pipes cannot be counted upon to serve the same function as an equipment grounding conductor, which is to prevent electrocution due to malfunctioning equipment on the branch circuit by allowing large amounts of current to flow and trip the overcurrent device.
- Use of water pipes as equipment grounding conductors is actually more likely to cause an electrocution in the event that a plumber, pipefitter, or similar professional working on the water piping system would break a pipe connection involved in a fault, thereby exposing themselves to the full lethal circuit voltage and providing a path for current to flow. Plumbers, pipefitters, or similar professionals are not required to de-energize and lock out electrical circuits in order to work on plumbing systems.

- Using non-metallic pipe in all or part of a plumbing system would cause metallic parts of equipment or sections of the water piping to become energized if a tool or equipment were to malfunction and expose anyone (plumber, pipefitter, general plant employee) to an electrocution hazard from simple contact with the piping system. Employees working on water pipes used in this manner can be exposed to hazardous differences in electrical potential across an open pipe.

In the Preamble, OSHA agrees with these comments and states that water pipes used as equipment grounding conductors are not reliable and do not have low impedance. However, OSHA has allowed grounded cold-water pipes to be used for grounding branch circuit extensions since 1972. Because very few reported accidents had come to the attention of OSHA, the agency did not believe that the risk to employees and/or the substantial cost to employers of rerunning these branch circuit extensions was worth the reduction in risk. Their rationale can be summarized as follows:

- To redo a branch circuit extension, an employee would need to deenergize the existing circuit and run new conductors back to a point where an acceptable connection to the ground is available. (Section 250.130(C) of the 2002 NEC lists acceptable grounding points.)
- The risk of inadvertently contacting an energized part during the recircuiting process is likely to be at least as high as the risk of electric shock caused by using the water pipe as an equipment grounding conductor.
- Knowledge as to which branch circuit receptacles are grounded to a water pipe may be lacking; thus, employees may be introduced to hazards in the process of tracing the existing wiring installation.

Consequently, the final rule (1) allows using a grounded cold-water pipe as the equipment grounding conductor on branch circuit extensions in existing installations and (2) requires equipment grounding connections to be replaced any time work is performed on the branch circuit. In such cases, the circuit would have to be deenergized anyway, and no increased risk during the installation of a new equipment grounding conductor was anticipated by OSHA. The prior standard had permitted using a grounded cold-water pipe as an equipment ground only for extensions of branch circuits that did not have an equipment grounding conductor (OSHA, 2007).

16.7.3 EFFECTIVE

The final rule in 29 CFR 1910.304(g)(5) requires a equipment grounding conductor to be permanent, continuous, and effective. The 2002 edition of NEC defines "effectively grounded" in Article 100 as (1) intentionally connected to earth through a ground connection or connections of sufficiently low impedance and (2) having sufficient current-carrying capacity to prevent the buildup of voltages that may result in undue hazards to connected equipment or to persons. This same definition appears in Part I of the 2000 edition of National Fire Protection Association (NFPA) 70E. To maintain consistency with the NEC and NFPA 70E, OSHA adopted the NEC

definition of "effectively grounded" in 29 CFR 1910.399 and has applied that definition in the final rule to all voltages. The term "effectively grounded" (or the equivalent) is used in final 29 CFR 1910.304(b)(2)(ii), (g)(5), (g)(8)(ii), and (g)(8)(iii); in
29 CFR 1910.305(c)(5); and in 29 CFR 1910.308(a)(6)(ii), (a)(7)(viii), (e)(4)(ii), and
(e)(4)(iii).

16.7.4 CONSEQUENCE

When determining whether an ohm/resistance path is continuous, attachments to
building frames and water pipes must be considered as possible continuous conductors to earthing grounds. The emphasis should be on the term "possible," as path to
ground must be proved. All potential metal structures must be of sufficient size and
width/cross-section that their resistance is low.

16.7.5 ELECTRODES

Electrodes are chosen during the design phase. The intent is to provide an electrode
to earth contact resistance that is consistent with design requirements that take into
consideration overall circuitry, earthing grounds, motor usage, and circuitry (service/
supply) grounds. Variables in electrode conformation and placement include depth,
diameter, and numbers (per a given area). However, over time electrodes can deteriorate or be damaged, so testing must include an evaluation of electrode integrity given
original and current design/facility parameters. If the electrodes are damaged or of
insufficient size the resistance will be overly large, resulting in a grounding system
that is ineffective.

16.7.6 TESTING

The order of testing varies; however, for this discussion, earthing grounds are tested
first. The area within the earth that receives the fault current is effectively large,
so resistance within the earth is assumed to be low. Various rods are driven into
the earth and voltage is applied. An ammeter is used to measure current flow and
potential difference (voltage) between the rods. From these values, the resistance is
calculated at various distances from a given point. The number of measuring points
and relationship between measurements is dependent on knowledge of the overall
grounding system (intended) conformation. In general, the distance between measuring stakes inserted in the soil should be three times the stake depth (length).
Some measuring devices automatically calculate ground resistance. Buried conductive metal structures can create anomalies and may require additional measurements
be taken using various stake distancing conformations. Types of testing for buildings
may include the following:

- *Fall of potential* for testing the entire ground system and portions of this
  system
- *Selective* for testing individual resistances of the ground system and portions of this system

- *Stakeless* for measuring separate and individual grounds (earthing electrode and their immediate conductor attachments, including water pipes and structural frames)

The choice of testing intervals and the methodology used will vary given the chosen equipment, grounding system design parameters, and conditions at the time of testing.

16.7.7 RESISTANCE GOALS

Effective grounding relies on correct ground resistance. Various standards and codes provide various thresholds of choice, ranging from less than 5 ohms to less than 25 ohms impedance to ground. Factors such as impedance (capacitive and inductive) may also require consideration. The information provided by such tests should show that a continuous and effective path to earthing ground is present (OSHA, 2007).

16.8 ARC-FLASH STUDY SPECIFICATIONS

Arc-flash studies can take varying forms depending on the electrical system. The complexity and extent of coordinated power system protection depend on the type of buildings or facilities required, on the load demand of facilities, and on the quantity and types of facilities to be constructed. The coordination required to ensure power system protection depends on the complexity and extent of the power system. The protective devices must protect electrical power system conductors or equipment against sustained overloads, inrush conditions, electrical faults, or other abnormal power system or equipment operating conditions, in accordance with UFC 3-520-01,[*] IEEE Std. 242-2001,[†] and IEEE Std. 141-1993.[‡] The requirements for fuses, circuit breakers, protective relays, or other protective devices associated with the arc-flash testing must be evaluated. The decision logic is that if these protective devices do not work properly the arc will sustain past anticipated time limits and the arc-flash radii may increase. Consequently, an evaluation of all protective devices is an essential component of an arc-flash study. The coordinated power system protection is assumed to be based on life safety, economics, simplicity, and the electrical power availability. The functional use of the facilities or utilities requires the following:

- Maximum power service with a minimum of power interruptions should be provided.
- The protective device operating speed must be such that it minimizes damage to electrical equipment, prevents injury to personnel, and lessens nuisance tripping.
- DC power sources must have protective devices with proper closing and tripping functions.
- Overcurrent protective devices and frame grounding (to earth) must be coordinated.

---

[*] UFC 3-520-01, Interior Electrical Systems, with Change 1; replaced by UFC 3-520-01 with Change 2.
[†] IEEE 242-2001 (errata, 2003), Recommended Practice for Protection and Coordination of Industrial and Commercial Power Systems (IEEE Buff Book).
[‡] IEEE 141-1993, IEEE Recommended Practice for Electric Power Distribution for Industrial Plants.

Short-circuit studies, load-flow studies, motor-starting analysis, and coordination studies may all be required for a complete arc-flash hazard analysis in accordance with the InterNational Electrical Testing Association (NETA) Acceptance Testing Specifications (ATS).

Work must be performed is in conformance with the following standards: IEEE C2,[*] IEEE 1584,[†] NEMA Z535.4,[‡] NETA ATS,[§] NFPA 70,[¶] and NFPA 70E.[**] Documentation must include verifying system coordination and recommended ratings and settings of protective devices per IEEE C2, IEEE 1584, NEMA Z535.4, NETA ATS, NFPA 70, and NFPA 70E.

16.8.1 PROPOSED TEST PLAN

A proposed test plan should be developed to include the following:

- Complete field test procedure
- Tests to be performed
- Test equipment required
- Tolerance limits
- Complete testing and verification of the ground-fault protection equipment

All power system analysis software data files necessary to restore and edit the model should be preserved throughout the study and in some cases will be provided to the client thereafter.

16.8.2 FIELD TESTING

During testing, use safety devices (e.g., arc-flash personal protective equipment, electrically insulating rubber gloves, protective barriers, danger signs) to protect and warn personnel in the test vicinity. Replace any devices or equipment damaged due to improper test procedures or handling. Field tests may include adjustments for each component.

16.8.2.1 Molded-Case Circuit Breakers

Visually inspect circuit breakers. Verify current ratings and adjustable settings incorporated in accordance with the coordination study.

---

[*] IEEE C2-2012 (errata, 2012; INT 1-4 2012; INT 5-6 2013), National Electrical Safety Code.
[†] IEEE 1584-2002 (Am 1 2004; Int 1-3 2008), Guide for Performing Arc-Flash Hazard Calculations.
[‡] NEMA Z535.4-2011, American National Standard for Product Safety Signs and Labels.
[§] NETA ATS (2013), Standard for Acceptance Testing Specifications for Electrical Power Equipment and Systems.
[¶] NFPA 70-2014 (AMD 1 2013; errata, 2013; AMD 2 2013), National Electrical Code.
[**] NFPA 70E-2012 (errata, 2012), Standard for Electrical Safety in the Workplace.

16.8.2.2   Power Circuit Breakers

Visually inspect the circuit breaker and implement settings in accordance with the coordination study. Conduct current injection tests. These tests are typical for NETA acceptance testing. Test with approved settings installed in accordance with NETA ATS and the test plan. Document the results in the field test reports and determine the following:

- Long-time pick-up value and delay by primary current injection
- Short-time pick-up value and delay by primary current injection
- Instantaneous pick-up value by primary current injection
- Ground-fault pick-up value and delay by primary current injection

Test trip unit functions by secondary current injection or in accordance with the manufacturer's requirements.

16.8.2.3   Protective Relays

Visually inspect protective relays. Implement relay settings in accordance with the coordination study. Conduct current injection tests. Test with approved settings installed in accordance with NETA ATS and the test plan. Document the results in the field test reports and determine the following:

- Long-time pick-up value and delay by secondary current injection
- Short-time pick-up value and delay by secondary current injection
- Instantaneous pick-up value by secondary current injection
- Ground-fault pick-up value and delay by secondary current injection

Data should include calibration and testing procedures, as well as frequency of calibration, inspection, adjustment, cleaning, and lubrication.

16.8.3   Performance Test Reports

A written description of procedures (e.g., diagrams, instructions, and precautions required to properly install, adjust, calibrate, and test the devices and equipment) must be provided. Submit performance test reports showing adjustments made and field tests performed to prove compliance with performance criteria. Indicate in each test report the final position of controls.

16.8.3.1   Equipment Data

Data should include the following:

- Manufacturer's time–current characteristic curves for individual protective devices
- Recommended settings of adjustable protective devices
- Recommended ratings of non-adjustable protective devices

Review protective device submittals of equipment to be provided and indicate any options or modifications required to achieve the requirements of this section. Submit Certificates of Conformance for Devices and Equipment, documenting that all devices or equipment met the requirements.

16.8.3.2   Project/Site Conditions

Site conditions must be detailed in the report. Unusual site conditions that can contribute to variances include the following:

- Service conditions for altitudes above 1005 m (3300 ft) for most apparatus
- Ambient temperature ranges of –30 to –40°C
- Frequency of 50 Hz vs. the generally predicted 60 Hz
- Fungus within electrical devices in tropical areas
- Seismic conditions (seismic requirements should be shown on the drawings with details conforming to UFC 3-310-04[*])
- Hazardous (classified) locations
- Humidity and ventilation issues

The short-circuit values for the connection points on the electrical system must be determined through either site knowledge or testing. Existing protective device settings that will affect arc radii must be evaluated, including device time–current curves. For certain tasks (e.g., working in manholes, elevated work areas, or underneath equipment), trained personnel may not be able to distance themselves from the arc flash within 2 seconds. Discuss those situations. Analyze other unusual situations and the rationale for their subsequent hazard determinations.

16.8.4   Coordination Study

Demonstrate through the study that the maximum possible degree of selectivity has been obtained between devices. Discuss the consistency of protection for equipment and conductors given the probability of damage from overloads and fault conditions. Include a description of the coordination of the protective devices, such as (1) which devices may operate in the event of a fault at each bus, (2) the logic used to arrive at device ratings and settings, (3) situations where system coordination is not achievable due to device limitations (an analysis of any device curves that overlap), (4) coordination between upstream and downstream devices, and (5) relay settings.

Provide recommendations to improve or enhance system reliability by reducing the incident energy level, and detail where such changes would involve additions or modifications. Provide composite coordination plots on log–log graph paper, providing separate plots for phase and ground faults. Provide applicable cable and transformer damage curves on phase–fault plots; limit the number of protective device curves on any plot to five for clarity.

---

[*] UFC 3-310-04 (2012), Seismic Design for Buildings.

16.8.5   Fault Current and Arc-Flash Analysis

Fault current and arc-flash analysis includes the following:

- Main electric supply substations
- The source bus and system buses where fault availability is 10,000 amperes (symmetrical) for 600-volt-level distribution buses
- The source bus and extended through the secondary side of transformers for medium-voltage distribution feeders
- The source bus and extended through outgoing breakers and outgoing medium-voltage feeders, down to the individual protective devices for medium-voltage radial taps
- Outgoing medium-voltage feeders through the secondary side of transformers
- Nearest upstream device in the existing source system and the device's extension through the downstream devices at the load end

16.8.5.1   Determination of Facts

Available fault capacity at the power source or a fault capacity on which to base the analysis must be defined and documented in the report. Field inspections to determine and document the time–current characteristics, features, and nameplate data for each existing protective device may be required if onsite proven documentation of same is not available. The power generation utility may be able to provide an estimate of fault current availability at the site, and the fault current availability may then be one basis for fault current studies.

16.8.5.2   Single-Line Diagram

A single-line diagram to show the electrical system buses, devices, transformation points, and all sources of fault current (including generator and motor contributions) is required. A fault-impedance diagram or a computer analysis diagram may also be needed. Be sure that each bus, device, or transformation point has a unique identifier. If a fault impedance diagram is provided, show

- Impedance data
- Location of switches, breakers, and circuit-interrupting devices
- Available fault data and protective device interrupting rating

16.8.5.3   Fault Current Analysis

Perform the fault current analysis in accordance with methods described in IEEE 242 and IEEE 399,[*] including (1) single-line drawings based on existing hardware, and (2) specialized computer-aided engineering software designed for fault current analysis with single-ended substation source operation, double-ended substation source operation, and generator source operation. Utilize actual hardware data in fault calculations. Be sure that bus characteristics and transformer impedance are correctly characterized. Perform the analysis and generate a separate study report for each operational scenario.

---

[*] IEEE 399-1997, IEEE Recommended Practice for Industrial and Commercial Power Systems Analysis.

16.8.5.4 Fault Current Availability

At each voltage transformation point and at each power distribution bus, analyze as appropriate:

- Balanced three-phase fault
- Bolted line-to-line fault
- Line-to-ground fault current values

Show the maximum and minimum values of fault available at each location in tabular form on the single-line diagram or in the textual report.

16.8.6 ARC-FLASH WARNING LABELS

During the study, as information is finalized, labels should be provided. These labels are to be weatherproof and detailed as required by NFPA 70 and NFPA 70E. Equipment to be labeled includes (but is not limited to) medium-voltage switches, transformers, switchgear main breakers, switchgear bus-tie breakers, switchgear feeder breakers, switchgear cable compartments, switchboards, panelboards, motor control centers, enclosed breakers, safety switches, automatic transfer switches, motor starters, control panels, and other equipment likely to require examination, adjustment, servicing, or maintenance while energized.

16.8.6.1 Label Format

Label format should be of the NFPA 70E detailed format type. This format includes different colors and formatting per NEMA Z535.4 for different hazard levels and includes the following components:

- Flash hazard boundary
- Incident energy
- Hazard category and personal protective equipment (PPE)
- Shock voltage
- Minimum insulated glove rating
- Limited approach boundary distance
- Restricted approach boundary distance
- Prohibited approach boundary distance

16.8.6.2 Label Content

The arc-flash label content must be based on the operational scenario, fault location, and fault type (arcing or bolted) that results in the highest incident energy.

16.8.7 REPORT

The final report should be organized with separate sections with all applicable content for all operating scenarios. Utilize single-line diagrams whenever practicable. Include tabulated arc-flash data for all equipment requiring an arc-flash warning label. In the textual narrative describe the analyses performed and the basis and methods used.

16.8.7.1 Study Options

Indicate which study options have been chosen and the tabular input, calculations, and software utilized to generate the short-circuit analysis and arc-flash hazard analysis results. Options may include the following:

- Standard used for arc-flash calculations, with IEEE 1584 being the preferred standard
- 240-volt exceptions (report Category 0 if transformer is less than 125 kVA)
- Maximum arcing time, with 2 seconds being preferred (fully dependent on task performed)

In the context of the textual reports, discuss all evaluations and the preferential decision logic used to include

- Motor fault contributions, with 5 cycles preferred (50-hp or greater motors are to be evaluated)
- Level miscoordination checked, with 5 levels being preferred
- Miscoordination ratio, with 80% being preferred (be sure that the cleared fault threshold matches the miscoordination ratio)
- Flash boundary calculation adjustments above 1 kV, trip time ≤0.1 second, with 1.5 cal/m² being preferred

Properly categorize all equipment types in the arc-flash evaluation. Discuss whether switchgear and panelboards have the proper gap distance. Include the short-circuit study utilized (comprehensive being preferred). Discuss the fault types analyzed; three-phase, single line to ground, line to line, and line to line to ground are all preferred.

16.8.7.2 Technical Data and Analysis

The technical data and analysis should include documented utility company data (e.g., system voltages, fault megavolt-ampere [MVA], system X/R ratio, time–current characteristic curves, current transformer ratios, relay device numbers and settings, existing power system data). Information from the local utility company should include an area load flow and fault study. The analysis should also include the desired method of coordinated power system protection, as well as descriptive and technical data for existing devices and new protective devices being proposed, including manufacturers' published data, nameplate data, time–current curves, and definition of the fixed or adjustable features of the existing or new protective devices. Other data that should be included are as follows:

- Time–current characteristic curves and protective device ratings and settings
- Time–current characteristics curves for each bus in the system as required to ensure coordinated power system protection between protective devices or equipment
- Recommended ratings and settings of all protective devices
- All calculations performed for the analyses, including the computer analysis programs utilized, the name of the software package, the developer, and the version number

16.8.7.3 Node Analysis

Include all information input to define nodes (e.g., cable data, conduit type, circuit length, transformer impedance, bus impedance, generator impedances). Provide and analyze the following:

- Fault short-circuit levels and reactance-to-resistance (X/R) ratios
- Ground-fault short-circuit levels and X/R ratios
- Load flow levels
- Arc-flash energy (for both bolted and arcing short-circuit levels)
- Motor starting study results

16.8.7.4 Protective Devices

Provide protective device setting sheets as separate pages suitable for use by installing technicians, separate from other report analysis and data. Include recommended changes to existing protective device settings and settings for all new protective devices. Provide all information to field install the settings, including settings or features not used or turned off.

16.8.7.5 Equipment Evaluation Report

The equipment evaluation report (EER) shows the available interrupt current (AIC)/short-circuit current rating (SCCR) for all equipment evaluated and the required rating for the application where the equipment is installed. The EER must identify underrated and marginally rated equipment. Underrated equipment is defined as equipment with actual AIC/SCCR values that do not meet the required AIC/SCCR for the application/installation. Marginally rated equipment is defined as equipment within 90 to 100% of the required AIC/SCCR for the application or installation (NASA, 2011).

16.9 ELECTROMAGNETIC FIELD TESTING

Electromagnetic (EM) radiation is a term given to a wide range of waves that have two components: an electric field and a magnetic field. These fields propagate at 90° to each other in the far field, an area that is a certain distance away from the origin of the fields. A familiar form of EM radiation is visible light. The entire EM spectrum ranges from low-energy radiowaves to high-energy gamma rays.

16.9.1 Fields

Electricity and magnetism are inherently linked. Electricity is the motion of electrons. Whenever an electron moves, creating electricity, a magnetic field is also produced. The opposite effect is also true, as electric fields can be generated by magnets.

Whenever electricity is flowing through a wire, the electricity generates electromagnetic fields (EMFs) perpendicular to the flow of the electricity. The electric field is gyrating up and down, which can be shown to be similar to a sinusoidal wave. At the same time, the magnetic field is similarly going side to side, which can also be shown as a sinusoidal wave. When an electrical current flows through a wire, an EMF emanates parallel out from the wire; therefore, whenever electricity-carrying wires are active an EMF is present. The stronger the electrical current, the stronger the magnetic field. Similarly, the larger and more powerful the magnet, the stronger the electric field. The larger the amount of current applied, the larger the magnitude of the EMF generated. The EMF strength is also proportional to proximity, as the closer a person is to the source of the EMF, the more EMF exposure the person will receive.

A collection of EM radiation waves creates an electromagnetic field. EMF studies usually focus on lower energy EM waves. Electromagnetic fields with frequencies below visible light are categorized based on their frequency. The frequencies are used for different applications and electronic items (e.g., radiowaves, television broadcasting, microwaves, cell phones, tablets). When people refer to EMF studies, the lower energy EM waves are usually the primary focus of these studies.

16.9.2 DETECTION

Electromagnetic field measurements are made with special instruments designed to detect various types and frequencies of the EMF. EMFs span a wide range of frequencies and energies, from extremely low frequency (ELF) to low frequency (LF) to radiofrequency (RF). EMF surveys generally focus on the ELF, very low frequency (VLF), and LF sections. In order to detect this wide range, several different instruments are used. Most EMF instruments are capable of measuring the flux density or field strength of magnetic and the field strength of electric fields in three dimensions.

16.9.3 EMF SURVEYS

In an EMF study, sources of EMFs are identified and a series of detectors are used to measure the field strength systematically in real time. The measurements are made using industry guidelines, and they incorporate a fixed grid pattern to characterize the field strength over the entire area. From this information, a map of the affected area, identification of various relevant EMFs, and recommended shielding or other mitigation can be determined. EMF studies are conducted for many reasons and may address sources of EMFs that are hidden (e.g., an electrical cable passing behind and through a wall of the building) and EMFs from different electrical or magnetic sources that can interact. The different EMF sources can be additive and result in an elevated field strength within various areas.

Electromagnetic fields can affect equipment performance. EMFs from a single source (i.e., electrical conductors or wires) may adversely affect the performance of other equipment. EMFs can affect people with medical conditions; for example, an elevated EMF can interact with cardiac pacemakers, causing the pacemakers to malfunction in the elevated EMF area. Also, EMFs may affect health in ways as yet unknown or proven. External EMF sources can affect a facility. Power transmission lines, power substations, and transformers can all generate EMFs. If a facility is close to a source of high electricity, EMF risks may be present. The combination of EMF strength, duration of exposure, and distance from the source must all be considered when identifying risks. The duration of exposure is quantified as the duty factor.

16.9.4 SURVEY ASSUMPTIONS

All survey assumptions must be documented, as various site conditions will affect measurements. For a simple survey with extraneous EMFs, the following assumptions are typical:

- Magnetic fields are generated by circuitry induction.
- As the electric field varies (increases or decreases), the strength of the magnetic field can be assumed to vary.
- The stronger the electric field, the stronger the magnetic field. Because shielding can affect electrical field measurement, the primary measurements may be for the magnetic field ranges rather than for electrical field determinants.

16.9.5 EMF STANDARDS

Several organizations have developed voluntary guidelines for EMF exposure. The International Commission on Non-Ionizing Radiation Protection (ICNIRP) addresses exposures at 60 Hz. The magnetic field limits are set at 4200 milligauss (mG) for occupational exposure and 833 mG for the general public. The ICNIRP has made a series of recommendations for limiting EMF exposure to human beings based on the epidemiological data available from verifiable research studies (ICNIRP, 1997). Based on ICNIRP's work, the European Union has adopted these same standards for EMF exposure of personnel. Although the guidelines are voluntary, the levels are designed to prevent undue health risks associated with EMF exposure.

The American Conference of Governmental Industrial Hygienists (ACGIH) provides that occupational exposures should not exceed 10,000 mG. The ACGIH additionally recommends that workers with pacemakers should not be exposed to 1000 mG. The ACGIH 10,000-mG guideline level and the ICNIRP level of 4200 mG are intended to prevent effects (e.g., induced currents in body cells, unwanted nerve stimulation).

The International Committee on Electromagnetic Safety (ICES) is sponsored by the Institute of Electrical and Electronics Engineers Standards Association (IEEE-SA). ICES and its predecessor organizations have been developing American national standards for the safe use of electromagnetic energy near workers or other people; these standards define safety levels and address the assessment of exposure to EMF. The first in a series of standards for safety levels with respect to human exposure to electromagnetic energy at radiofrequencies was published in 1966, followed by a full revision published in 2005 and amended in 2010 (see Table 16.3).

The differentiation between *occupational exposure* and *general public exposure* is important to note. Occupational exposure limits have been set for people who work in professions specifically dealing with electricity and/or magnetism. General public exposure limits have been set for people who may come into contact with areas having higher than normal EMF levels. When developing the standards, the ICNIRP assumed that occupational workers would be trained in how to mitigate hazards and in safe operational handling practices. In contrast, general public exposure limits are lower or more stringent than occupational worker limits because the general public

TABLE 16.3

Summary of ICNIRP Electromagnetic Field Exposure Limits

Frequency	Electric Field Strength (V/m)	Magnetic Field (µT)
Occupational: 0.025–0.82 kHz	$500/f$	$25/f$
Occupational: 60 Hz	8333	416
Public: 0.025–0.82 kHz	$250/f$	$5/f$
Public: 60 Hz	416	83
Public: 50 Hz	500	100

- Will generally not know if or when exposure to EMFs is occurring
- Will not know what safety precautions to take
- Is not trained on how to mitigate EMF exposure
- Consists of a variety of diverse populations with varying susceptibilities to EMF exposure

In addition, the ICNIRP (1997) states that people with pacemakers or other electronic implants should have even lower EMF exposures.

16.9.6 EXAMPLE CALCULATIONS

Calculations are based on information obtained from the device manufacturer (i.e., transmitted power per port, number of antennae, and expected antenna gain). The equivalent isotropically radiated power (EIRP) can then be calculated using the Federal Communications Commission (FCC) equations. The aggregate EIRP is calculated using the following equation:

$$\text{EIRP (milliwatts, mW)} = PT + GA + 10 \times \log(AN)$$

where
PT = Transmit power per port based on design and manufacturer's information.
GA = Antenna gain.
AN = Number of antennas.
EIRP in microwatts (µW) = EIRP in milliwatts (mW) × 1000.

Power density (P_{df}) is calculated as a function of distance from the EMF:

$$P_{df} = D_f \times \frac{\text{EIRP}}{4\pi r^2}$$

where
D_f = Duty factor, obtained from research evaluations of the type activity the end user is engaged in.
r = Straight-line distance to the emitting device in centimeters.

The time-weighted average can be calculated using Equation 2 in FCC Technical Bulletin 65 (FCC, 1997):

$$\sum S_{exp}t_{exp} = S_{limit}t_{avg}$$

where

S_{exp} = Power density level of exposure (W/cm²).

t_{exp} = Allowable time of exposure for S_{exp}.

S_{limit} = Appropriate power density maximum permissible exposure (MPE) limit (W/cm²).

t_{avg} = Appropriate MPE averaging time.

16.9.7 INSTRUMENTATION

Many types of instrumentation are available for measurement of EMF. The examples contained below are representative of isotropic EMF and magnetic field measuring systems. *Note:* The frequency ranges described herein have been chosen to be representative of those systems being tested.

16.9.7.1 Broadband Radiofrequency Isotropic Field Probes

Broadband radiofrequency (RF) isotropic field probes are used for exposure measurements in the vicinity of industrial RF sources. Radiofrequency radiation is electromagnetic radiation in the frequency range of 3 kHz to 300 MHz. An example probe assembly consists of a spherical casing containing a sensor on one end and instrumentation electronic housings on the other end. The sensor and instrumentation electronics (housing) operate together and are calibrated as a unit. The measurement capability and frequency response depend on which electronic housing and sensor are selected. Field strength is determined in each of three axes (x, y, z). For each axis, the probe measures the radiofrequency signal level and generates a linearized reading. After vector addition, the resultant readout is typically displayed in amperes per meter (A/m) for magnetic fields and volts per meter (V/m) for electric fields.

16.9.7.2 Extremely Low Frequency Survey Meter for EMFs

The extremely low frequency (ELF) power frequency field strength measurement system is used to evaluate EMFs associated with 50- or 60-Hz sources, including (1) electric power transmission and distribution, and (2) electrically operated equipment and appliances. Both electric and magnetic fields are sensed and measured. Manual and automatic (via microprocessor programming) range changing can be used to best measure the desired frequency range. Automatic zeroing of the instrument is another important feature.

Readouts for electric fields are in volts per meter. Electric fields are detected by a displacement current sensor. The displacement current sensor has two thinly separated conductive disks, which are connected together electrically. When it is immersed in an electric field, the charge is redistributed among the two parallel disks so the electric field between the two disks remains at zero. This redistribution of charge is measured as a displacement current that indicates external electric field strength.

Readouts for magnetic fields are in gauss (G) or tesla (T). A magnetic field is determined by measuring the voltage developed across the terminals of an induction coil, with a sensitivity of typically 0.1 mG to 20 G. Surrounding the circular displacement current sensing disks is a coil consisting of several hundred turns of fine-gauge wire. When placed in an alternating magnetic field, a current is induced in the coil that is proportional to the applied magnetic field strength. The magnetic field strength is determined by measuring the voltage developed across the terminals of the coil. The outputs of both field transducers are measured with a true root-mean-square (RMS) detector. True RMS detection provides accurate evaluation of fields having waveform variations.

16.9.7.3   ELF Magnetic Field Meter

The ELF magnetic field meter measures the flux density of magnetic fields in the lower frequency ranges (i.e., 5 Hz to 2 kHz). The three-axis flux density detector is responsive to either sinusoidal or complex magnetic field waveforms and computes the RMS value of magnetic flux density. The sensor consists of three multi-turn loops connected to the instrumentation readout package via a 1.2-meter-long cable. The microprocessor in the meter continually computes the magnitude of the field (flux density value) independent of the probe orientation. The field sensor is electrically shielded. Both ambient magnetic fields and high-level fields found near high-current-carrying conductors can be measured.

16.9.8   RADIOFREQUENCY AND MICROWAVE SURVEYS

Two basic types of survey instruments are available for RF and microwave surveys: broadband and narrowband. Broadband equipment is the most common, where all sources of RF or microwave energy are detected together (over a certain frequency range) and the total field strength is detected. Narrowband instruments have the ability to separate emissions from each other and can display the total energy while also identifying the frequency or frequencies of the emissions that form the total. When the frequency of the emitter is known, such as for industrial generators, then the lower cost and easier to use broadband equipment is normally sufficient. In multiple emitter environments, such as rooftops in most cities, narrowband equipment is more useful as this equipment can display the field strength of each emitter to directly compare against federal regulations.

The Selective Radiation Meter (SRM-3006; Narda Safety Test Solutions) is a narrowband measuring system for safety analysis and environmental measurements and high-frequency electromagnetic fields in the frequency range from 9 kHz to 6 GHz. Because signals with frequencies of this magnitude are very difficult to sample digitally, the SRM-3006 uses a combination of analog and digital signal processing. The SRM is used for measuring absolute and limit values of electromagnetic fields including high-frequency electromagnetic fields and EMFs due to radio (AM, FM, or digital) and television (analog or digital) broadcasts. Digital video broadcasting–terrestrial (DVB-T) is the digital video broadcasting European-based consortium standard for the broadcast transmission of digital terrestrial television.

TETRA BOS is a digital radio network being established by the German Federal Agency for Digital Radio and Security Authorities and Organisations (BDBOS) to provide a nationwide digital voice and data communications network for all security authorities and organizations within the Federal Republic of Germany.

The SRM is also used to measure mobile telecommunications, radar, and wireless communications electromagnetic fields. The Global System for Mobile Communications (GSM) is a European Telecommunications Standards Institute (ETSI) standard to describe protocols for second-generation digital cellular networks used by mobile phones. The Universal Mobile Telecommunications System (UMTS) is a third-generation mobile cellular network system based on the GSM standard. UMTS specifies a complete network system. The Worldwide Interoperability for Microwave Access (WiMAX) wireless communications standard is based on providing up to 1 gigabits for fixed stations. Wireless local area networks (WLANs) link two or more devices using a wireless distribution method and provide a connection via an access point (AP) to the wider Internet. Most modern WLANs are based on IEEE 802.11 standards.

In unknown field environments (e.g., those around so-called shared sites) where several providers of mobile telephone services share a common antenna site, the SRM-3006 displays total field level and contributions made by the individual services, either as absolute values or as a percentage of the permitted limit level. Each service can be resolved down to its individual channels, and the contribution made by each channel to the overall field emission can be measured using the SRM-3006. In the same way, the value can be integrated over the frequency band of the service and the total value displayed, again as an absolute value or in terms of the relevant limit value.

The SRM-3006 includes all of the typical functions of a spectrum analyzer and as such can be universally applied. High measurement speed at small resolution bandwidths (RBWs) are possible. The complete SRM-3006 measuring system is comprised of the basic SRM-3006 unit and the three-axis antennas. Single-axis antennas, covering various applications and frequency ranges, can also be used. All antennas can be mounted directly on the basic unit or connected to the basic unit using a special radiofrequency cable.

REFERENCES

FCC. (1997). Evaluating compliance with FCC guidelines for human exposure to radiofrequency electromagnetic fields. *OET Bulletin 65*, Edition 97-01, August.

ICNIRP. (1997). *ICNIRP Guidelines: Guidelines on Limits of Exposure to Broad-Band Incoherent Optical Radiation (0.38 to 3 μM)*. Munich, Germany: International Commission on Non-Ionizing Radiation Protection.

NASA. (2004). *Reliability Centered Building and Equipment Acceptance Guide*. Washington, DC: National Aeronautics and Space Administration.

NASA. (2011). *NASA/KSC Guide Specifications: Coordinated Power System Protection and Arc-Flash Analysis*, Section 26 05 63.00 98. Washington, DC: National Aeronautical and Space Administration.

NFPA. (2005). *NFPA 70: National Electrical Code (NEC) Handbook*. Quincy, MA: National Fire Protection Association.

NFPA. (2012). *NFPA 70E: Standard for Electrical Safety in the Workplace®*. Quincy MA: National Fire Protection Association.

OSHA. (2007). Occupational Safety and Health Standards, 29 CFR, Part 1910, Subpart S, Electrical. Baltimore, MD: Occupational Safety and Health Administration.

17 Electromagnetic Shielded Equipment, Instrumentation, and Facilities, Part One

Randy Boss and Martha J. Boss

CONTENTS

Electromagnetic fields (EMFs) are invisible lines of force that surround all electrical devices. Electric fields are produced by voltage and increase in strength as the voltage increases. The electric field strength is measured in units of volts per meter (V/m). Magnetic fields result from the flow of current through wires or electrical devices and increase in strength as the current increases. Magnetic fields are measured in units of gauss (G) or tesla (T). Most electrical equipment has to be turned on (i.e., current must be flowing) for a magnetic field to be produced. Electric fields, on the other hand, are present even when the equipment is switched off, as long as the equipment remains connected to the source of electric power. Electric fields are shielded or weakened by materials that conduct electricity (including trees, buildings, and human skin). Magnetic fields, on the other hand, pass through most materials and are, therefore, more difficult to shield. Both electric and magnetic fields decrease as the distance from the source increases.

Equipment can run on either alternating current (AC) or direct current (DC). Equipment using the building's electrical supply from the utility company is provided with AC electric current that reverses direction in the electrical wiring—or alternates—60 times per second, which is equivalent to 60 hertz (Hz). If the equipment uses batteries, then electric current flows in one direction only, from the batteries to the equipment, and is termed direct current. This produces a "static" or stationary magnetic field, also known as a *direct current field*. Some battery-operated equipment can produce time-varying magnetic fields as part of their normal operation. AC electric power produces magnetic fields that can generate weak electric currents in humans. These are called *induced currents*. Much of the research on how EMFs may affect human health has focused on AC-induced currents. This chapter presents the engineering requirements when EM shielding is required. These requirements are minimum specification standards that would require modifications to be site specific.

17.1 FACILITY EM SHIELDING

Facilities must meet or exceed specified minimum attenuation decibel (dB) levels. The EM shielding system may include the following:

- Welded steel or bolted EM shield
- EM shielded doors for access into the facility
- Electrical and electronic penetrations of the shield
- EM filter/surge arrester assemblies, including their EM enclosures
- EM shielded pull boxes and junction boxes
- EM shielded conduit runs
- Special protective measures for mission-essential equipment outside the EM shield
- Structural penetrations.
- Mechanical and utility penetrations (e.g., air ducts, gas, water)
- Instrumentation and control
- Equipment door/access panels

Testing should include a shielded enclosure leak detection system (SELDS), IEEE 299,[*] and other methods of shielded enclosure testing.

17.2 FILTER AND ELECTRICAL WORK CERTIFICATION AND SAMPLING

Filter and electrical work must comply with NFPA 70,[†] UL 486A-486B,[‡] and UL 1283.[§] The label and listing of the Underwriters Laboratories or other nationally recognized testing laboratory will be acceptable evidence that the material or equipment conforms to the applicable standards of that agency. In lieu of the label or listing, a certificate may be furnished from an acceptable testing organization adequately equipped and competent to perform such services. The certificate will state that the items have been tested and conform to the specified standard. Field samples should include samples of shielding sheet installation, shielding fastening, doors, 30- and 100-ampere power filters, communication filters, waveguides, and penetration mechanisms.

17.3 WELDING

Welding of EM shielding material and sheet steel should occur at an ambient temperature of 10°C (50°F) minimum to 32°C (90°F). Shielding should not be installed until the building has been weather enclosed. Sheet steel welding is not to be performed in direct sunlight.

---

[*] IEEE 299-2006, Standard Method for Measuring the Effectiveness of Electromagnetic Shielding Enclosures.
[†] NFPA 70-2014 (AMD 1 2013; errata, 2013; AMD 2 2013), National Electrical Code.
[‡] UL 486A-486B (2013), Wire Connectors.
[§] UL 1283 (2005), Electromagnetic Interference Filters.

17.4   MAINTENANCE SUPPLIES AND PROCEDURES

Maintenance supplies must be sufficient for a 3-year period or 50,000 open/close cycles, whichever is greater, for each EM shielded door. The maintenance instructions required to maintain the door through the cycle count must be prominently displayed near the door. In addition, the following must be provided:

- Extra EM power filters and extra communications filters of each type installed
- A set of finger stock and EM gaskets (if used) for each hinged EM shielded door provided
- A set of manufacturer recommended spare parts for EM shielded doors of each style installed
- A set of tools required to maintain the doors
- A set of special tools, calibration devices, and instruments required for operation, calibration, and maintenance of the equipment
- Readily available environmentally safe lubricants, cleaning solvents, or coatings in sufficient quantities to last for 6 months

17.5   OPERATING AND MAINTENANCE MANUALS

All manufacturers' written instructions for operation and maintenance of the EM shielding system must be provided in electronic open format. The manual must provide the following:

- A complete set of assembly drawings, including penetration locations and installation details
- The EM shielding construction specification
- Shield penetration schedule
- Power/signal filter schedule
- Test plan
- The prepared preventive maintenance instructions for periodic inspection, testing and servicing, lubrication, alignment, calibration, and adjustment events normally encountered

Complex preventive maintenance events will be extracted from or will refer to detailed vendor or manufacturer data. This information will be derived from an evaluation of engineering data considering local environmental conditions, manufacturer recommendations, estimated operating life for the specific application and use of the equipment, and types of job skills available at the operating site.

Spare parts data must be approved and verified by the shielding specialist prior to submission. The data should include a complete list of recommended parts and supplies with current unit prices and source of supply. Hardness critical items (HCIs) requiring periodic inspection to maintain EM shield integrity should be listed. Hardness critical items are those components or construction features that singularly and collectively provide specific levels of high electromagnetic pulse (HEMP) protection (e.g., EM shields, surge arresters, EM shielded doors, shield welding, electrical filters, honeycomb waveguides, waveguides below cutoff). Each major item of equipment must have the manufacturer's name, address, type or style, model or serial number, and catalog number on a plate secured to the item of equipment. Equipment and materials of the EM shielding must be designed and built to facilitate testing and maintenance.

17.6 EM SHIELDING EFFECTIVENESS

The EM shielded enclosure complete with all filters, doors, and/or waveguides must have EM shielding effectiveness attenuation. Minimum magnetic field attenuation is 20 dB at 14 kHz, increasing linearly to 50 dB or 100 dB at 200 kHz or 1 MHz, respectively. Minimum electric field and plane wave attenuation is 50 dB or 100 dB from 14 kHz to 1 GHz or 10 GHz.

17.7 EM SHIELDING ENCLOSURE REQUIREMENTS (WELDED CONSTRUCTION)

17.7.1 WELDED SHIELDING ENCLOSURE

A complete metal enclosure includes floor, walls, ceiling, doors, penetrations, welds, and the embedded structural members forming a continuous EM shielded enclosure. Shielding sheets and closures must be 3.416 mm thick (10-gauge) hot-rolled steel conforming to ASTM A568/A568M.[*] Steel plates, channels, or angles a minimum of 6 mm (1/4 in.) thick must be used to reinforce shield sheets for attachments of ducts, waveguides, conduit, pipes, and other penetrating items. Furring channels used to attach shielding sheets to walls or floors must be the minimum gauge of the shielding steel. The shielding sheet steel gauge may be thicker to reduce labor and welding effort only if structurally tolerable with the existing design. Steel must be free of oil, dents, rust, and defects.

17.7.2 METAL MEMBERS

Structural steel shapes, plates, and miscellaneous metal must conform to ASTM A36/A36M.[†]

17.7.3 STEEL AND WELDING MATERIAL

Welding materials must comply with the applicable requirements of AWS D1.1/D1.1M[‡] and AWS D9.1M/D9.1.[§] Steel and welding material must conform to AISC 325 (AISC, 2011). Welding electrodes must conform to AWS D1.1/D1.1M for metal inert gas (MIG) welding method. Weld filler metal must conform to American Water Society (AWS) A5.18/A5.18M.[¶]

---

[*] ASTM A568/A568M-13, Standard Specifications for Steel, Sheet, Carbon, Structural, and High-Strength, Low-Alloy, Hot-Rolled and Cold-Rolled, General Requirements for.

[†] ASTM A36/A36M-12, Standard Specification for Carbon Structural Steel.

[‡] AWS D1.1/D1.1M:2010 (errata, 2011), Structural Welding Code—Steel.

[§] AWS D9.1M/D9.1:2012, Sheet Metal Welding Code.

[¶] AWS A5.18/A5.18M:2005, Specification for Carbon Steel Electrodes and Rods for Gas Shielded Arc Welding.

17.7.4 FASTENERS

Self-tapping screws must not be used for shielding attachment. Power-actuated drive pins must be zinc-coated steel, Type I, pin size No. 4 to secure steel sheets to concrete surfaces and to light-gauge furring channels. The drive pins must conform to ASTM A227/A227M[*] Class 1 for materials and ASTM B633[†] for plating.

17.7.5 MISCELLANEOUS MATERIALS AND PARTS

Miscellaneous materials and parts may include bolts and anchors, supports, and braces, as well as lugs, rebars, and brackets to assemble work. Holes for bolts and screws must be drilled or punched with precise matching of holes. Thickness of metal and details of assembly and supports must provide ample strength and stiffness. The materials must be galvanically similar.

17.7.6 PENETRATIONS

Penetrations of the shield (including bolts or fasteners) must be sealed with puddle welds or full circumferential EM welds. Structural penetrations (e.g., beams, columns, other metallic structural elements) must be provided with continuously welded or brazed seams and joints between the penetrating element and the shield. Non-metallic structural elements must not be allowed to penetrate the electromagnetic barrier.

17.7.7 PENETRATION PLATES (WELDED CONSTRUCTION)

The penetration plate must be the central location for treatment of penetrations. The panel must be constructed of 6-mm (1/4-in.) thick ASTM A36/A36M steel plate welded to the shield. Waveguide, conduit, and piping penetrations must be circumferentially welded at the point of penetration to the inner surface of the penetration plate. Penetration plates must extend at least 150 mm (6 in.) beyond all other penetrations.

17.7.8 FLOOR FINISH

Floor EM shielding must be covered by a reinforced cast-in-place concrete slab that is 100 mm (4 in.) thick.

17.8 EM SHIELDING ENCLOSURE REQUIREMENTS (BOLTED CONSTRUCTION)

17.8.1 PANEL CONSTRUCTION

Flat steel sheets must be laminated to each side of a 19 mm (3/4 in.) thick structural core of either plywood or hardboard. Panels must have a flame spread rating of <25 when tested according to ASTM E84.[‡] Flat steel must conform to ASTM A653/

---

[*] ASTM A227/A227M-06 (R2011), Standard Specification for Steel Wire, Cold-Drawn for Mechanical Springs.

[†] ASTM B633-13, Standard Specification for Electrodeposited Coatings of Zinc on Iron and Steel.

[‡] ASTM E84-13, Standard Test Method for Surface Burning Characteristics of Building Materials.

A653M* with G-60 coating, minimum 0.5512 mm (26 gauge) thick and zinc-coated phosphatized. Plywood must conform to APA L870† for exterior, sound-grade hardwood, Type I. Hardboard must conform to AHA A135.4,‡ Class 4, SIS, for standard type hardboard. Adhesive for laminating steel sheets to structural core must be a waterproof type that maintains a permanent bond for the lifetime of the enclosure.

17.8.2 FRAMING

Panels must be joined and supported by specially designed framing members that clamp the edges of the panels and provide continuous, uniform, and constant pressure for contact to connect the shielding elements of the panels. The walls must be self-supporting from floor to ceiling with no bracing. Deflection of walls under a static load of 335 N (75 lb) applied normally to the wall surface at any point along the framing members must be ≤1/250 of the span between the supports. Ceilings must be

- Self-supporting from wall to wall
- Supported by adjustable, non-conducting, isolated hangers from the structural ceiling above
- Designed to have a deflection under total weight, including ceiling finish, of ≤1/270 of the span

A one-piece factory prewelded corner section or trihedral corner framed with a brass machine cast corner cap assembly must be provided at all walls and floor or ceiling corner intersections. All modular enclosures must be designed for ease of erection, disassembly, and reassembly.

17.8.3 CHANNEL

The framing–joining system members will consist of 3-mm (1/8-in.) thick zinc-plated steel channels having a minimum 16-mm (5/8-in.) overlap along each side of the contacting surface. Screw fasteners must be spaced at 75- or 100-mm (3- or 4-in.) intervals. Screw fasteners must be either zinc- or cadmium-plated steel, minimum size 6 mm (1/4 in.), with a pan or flat Phillips head. Fasteners must be heat treated and hardened with a minimum tensile strength of 931 MPa (135,000 psi).

17.8.4 SOUND TRANSMISSION CLASS

Enclosure panels will have a sound transmission class (STC) of 30 dB minimum when tested according to ASTM E90.§

---

* ASTM A653/A653M-13, Standard Specification for Steel Sheet, Zinc-Coated (Galvanized) or Zinc-Iron Alloy-Coated (Galvannealed) by the Hot-Dip Process.
† APA L870-2010, Voluntary Product Standard, PS 1-09, Structural Plywood.
‡ AHA A135.4, Basic Hardboard; 24 CFR 3280.304(b)(1).
§ ASTM E90-09, Standard Test Method for Laboratory Measurement of Airborne Sound Transmission Loss of Building Partitions and Elements.

17.8.5 PENETRATION PLATES (BOLTED CONSTRUCTION)

Plates must be a minimum of 3-mm (1/8-in.) thick ASTM A36/A36M steel plate, sized 450 by 450 mm (18 by 18 in.) and must have a 6-mm (1/4-in.) thick extruded brass frame for mounting to the shielded enclosure wall panel. Penetration plates must extend at least 150 mm (6 in.) beyond all other penetrations.

17.9 EM SHIELDED DOORS

Material in shielded doors and frames must be steel conforming to ASTM A36/A36M or ASTM A568/A568M and must be stretcher-leveled and installed free of mill scale. Metal must be thicker where indicated or required for its use and purpose. Metal thresholds of the type for proper shielding at the floor must be provided. Fire-rated shielded doors and assemblies must meet NFPA 80[*] and NFPA 80A[†] requirements and must bear the identifying label of a nationally recognized testing agency qualified to perform certification programs. The EM shielded doors must be provided by a single supplier who has been regularly engaged in the manufacture of these items for at least the previous 5 years. The assemblies must be supplied complete with a rigid structural frame, hinges, latches, and parts necessary for operation. The products supplied must duplicate assemblies that have been in satisfactory use for at least 2 years. The door frame must be steel suitable for welding or bolting to the surrounding structure and shield. The EM filters, EM waveguide penetrations for door systems, and miscellaneous material must be provided for a complete system. The enclosure door must be non-sagging and non-warping. The EM shielded door must provide a shielding effectiveness 10 dB or 20 dB greater than the minimum EM shielding effectiveness requirements. The door must have a clear opening that is 915 mm (36 in.) wide and 2135 mm (84 in.) high. The door and frame assembly must have a sound rating of STC 30 minimum. Testing must be performed in accordance with ASTM E90.

17.9.1 DOOR LATCH

The door latch must be lever controlled with roller cam action requiring ≤67 N (15 lb) of operating force on the lever handle for both opening and closing. The door must be equipped with a latching mechanism having a minimum of three latching points that provides proper compressive force for the EM seal. The mechanism must be operable from both sides of the door and must have permanently lubricated ball or thrust bearings as required at points of pivot and rotation.

17.9.2 HINGES

Doors must be equipped with a minimum of three well-balanced adjustable ball-bearing or adjustable radial-thrust bearing hinges suitable for equal weight distribution of the shielded doors. Hinges must allow adjustment in two directions. The force necessary to move the doors must be ≤22 N (5 lb).

---

[*] NFPA 80-2013, Standard for Fire Doors and Other Opening Protectives.
[†] NFPA 80A-2012, Recommended Practice for Protection of Buildings from Exterior Fire Exposures.

17.9.3 THRESHOLD PROTECTORS

Threshold protectors must be furnished for each EM shielded door. Protectors will consist of portable ramps that protect the threshold when equipment carts or other wheeled vehicles are used to move heavy items across the threshold. The ramps may be asymmetrical to account for different floor elevations on each side; however, the slope of the ramp must be ≤4:1 on either side. Ramps must be designed to support (1) 227 kg (500 lb) vertical force applied to an area 75 by 13 mm (3 by 1/2 in.) for a personnel door, and (2) 907 kg (2000 lb) vertical force applied to an area 75 by 13 mm (3 by 1/2 in.) for an equipment double-leaf door. *Note:* The force is applied (during testing) to the contact area between the threshold and the door. Mounting brackets, convenient to the entry, must be provided to store the ramp.

17.9.4 FREQUENCY OF OPERATION

With proper maintenance, door assemblies will function properly through 100,000 cycles and 15-year service life minimum, without the shielding effectiveness decreasing below the overall shield required attenuation.

17.9.5 ELECTRIC INTERLOCKING DEVICES

Electric interlocking devices must be provided for vestibules equipped with shielded doors at each end. Electric interlocking devices must be provided so that shielded doors at the ends of the vestibule cannot be opened at the same time during normal operation. A manual override must be provided to allow emergency egress, and an audible alarm must be provided to indicate that doors at each end of the vestibule are open. The low-piezoelectric-type alarm, in a tamperproof enclosure, must continue to sound while both doors are open. The sound intensity must be 45 dBA minimum at 3.05 m (10 ft). Lights must be provided on the side of each door outside the vestibule to indicate that the other door is open. Interlock systems may be integrated into a cypher lock system. The interlock system must be powered by an uninterruptible power source and must be failsafe in an unlocked condition in the event of a power failure.

17.9.6 ELECTRIC CONNECTIVITY

Electric connectivity for sensors, alarms, and electric interlocking devices must be installed in accordance with the door manufacturer's instructions. Detail drawings must show the complete EM shielded enclosure:

- Installation details and sequence
- Methods necessary to ensure shield integrity under all columns and other structural members
- Doors and filters
- Location, number, and method of penetrating the shielding material
- Fabrication details for penetrations of the shielding material and the complete EM shielded enclosure

17.9.7　THRESHOLD ALARM

A press-at-any-point ribbon switch must be applied to the threshold. The switch must enunciate an alarm whenever pressure is applied to the threshold of the EM shielded door.

17.9.8　HOLD OPEN AND STOP DEVICE

Each EM shielded door leaf must be provided with a hold open and stop device permanently attached to the door leaf. Shielded doors must have a fastener plate welded onto the door. The device must not interfere with the finger stock. Drilling or tapping of the shielded door is not allowed.

17.9.9　EMERGENCY EXIT HARDWARE

Emergency exit EM shielded doors must be equipped with single-motion egress hardware. The force required to latch and unlatch emergency exit hardware on EM shielded doors must meet life safety code NFPA 101[*] requirements. Field alterations or modifications to panic hardware are not allowed.

17.9.10　FINISH

Electromagnetic shielded doors must be factory prime painted with zinc chromate primer. Doors may be factory finish painted or galvanized. Touch up any damaged finish.

17.9.11　DOOR COUNTER

A door operation counter must be provided on the enclosure interior.

17.9.12　FIRE AND SOUND RATINGS

Fire ratings and STC sound ratings must be as required by the door finish schedule on the drawings or in the specifications.

17.10　EM SHIELDED, LATCHING TYPE DOORS

Doors must be steel laminated type. Steel doors must be a minimum of 3.416 mm (10-gauge) thick steel sheet electrically and mechanically joined by welded steel frames overlapping joints with continuous EM welds. Laminated types must have substantially the same construction as enclosure panels; however, laminated types must have steel faces electrically and mechanically joined by channels or overlapping seams that are continuously seam welded or soldered along all joined surfaces. The closure seal must have an extruded brass channel containing a recess into which

---

[*] NFPA 101-2012 (Amendment 1, 2012), Life Safety Code.

two sets of beryllium copper condition heat-treatable (HT) strips are placed in accordance with ASTM B194* or stainless steel 430 (magnetic type) series contact fingers and a closed cell foam rubber air seal are fitted. This seal must be easily removed and/or replaced without the use of special tools or application of solders. The door must mate to the frame to allow the insertion of a brass knife edge between the two rows of the radiofrequency finger stock, to obtain optimum conductivity and electromagnetic shielding. High-temperature silver solder must be used to attach the brass knife edge components to the door panels and the frame. The fingers that form a contact between the door and its frame must be protected from damage due to physical contact and must be concealed within the door and frame assembly.

17.11   EM SHIELDED, PNEUMATIC SEALING DOORS

Pneumatic sealing mechanisms must achieve EM shielding by using pressure to force the door panel against the frame surface. Contact areas of door and frame must be a peripheral strip ≥75 mm (3 in.) wide completely around the door with a tinned or highly conductive noncorrosive surface. After the door is in a closed position, the pneumatic sealing mechanism will exert pressure in ≤10 seconds. The sealing mechanism release must be actuated in ≤5 seconds. Manual override operation must be less than or equal to a maximum of 155 N (35 lb). When the door is sealed, the attenuation around the edges must meet the EM shielding effectiveness requirements of this specification. Swinging doors must have a threshold of zinc-plated steel that is ≥9.5 mm (3/8 in.) thick. The door must be provided with a pneumatic system that maintains a nominal sealing pressure of 240 kPa (35 psi).

17.11.1   DOOR AND ENCLOSURE

Doors must be designed for long life and reliability without the use of EM gaskets, EM finger stock, or other sealing devices other than the direct metal-to-metal contact specified. The EM sealing device must be failsafe upon loss of air pressure and readily allow manual opening of the door. For either normal or failsafe operation, the maximum time to reach the open position must be no more than 7 seconds. The enclosure design must include a provision for removing the door for routine maintenance without disturbing the door's alignment and EM sealing properties. A label must be attached to pneumatic doors warning against painting of the mating surfaces.

17.11.2   CONTROL PANEL

The inside and outside of the shielded enclosure must contain a control panel including the necessary opening and closing pneumatic valves. The outside control panel must also have a pressure regulator and filter. The door air supply must be capable of quick opening from inside the enclosure to allow escape when opening pneumatic valves fail or malfunction.

---

* ASTM B194-08, Standard Specification for Copper-Beryllium Alloy Plate, Sheet, Strip, and Rolled Bar.

17.11.3   Air System for Pneumatic Sealing

A complete air system including compressor, filter alarm, tank, lines, air filter, dryer, air control valves, and controls must be provided. Air tank capacity must be sized so that the air volume and pressure are sufficient to operate the door through ten complete cycles after the loss of normal power.

17.12   EM SHIELDED, MAGNETIC SEALED DOOR TYPE

An EM seal must be formed by a solid metal-to-metal contact around the periphery of the door frame. The materials at the contact area must be compatible and corrosion resistant. The contact force for the door EM seal must be provided by electromagnets. When the electromagnet is energized, the door leaf must be pulled in to ensure a solid and continuous contact with the door frame. When the electromagnet is de-energized, the door leaf must be free to swing. The EM shielded doors may use electromagnets or a combination of permanent magnets and electromagnets.

17.13   EM SHIELDED, SLIDING TYPE DOOR

A sliding shielded door must be of the size and operating direction indicated. Clear openings indicated on the drawings must not require dismantling of any part of the door. The door must be manually operable from either side, inside or outside, with a maximum pull (force) of 155 N (35 lb) to set the shielded door in motion. Shielded door face panels and frames must be constructed of reinforced steel suitable for achieving the specified attenuation. Frames must be constructed of steel shapes welded together to form a true rectangular opening. In the sealed position, the shielded doors will provide the minimum shielding effectiveness specified without any derating. The doors must be designed for long life and reliability and must not use EM gaskets, EM finger stocks, or other sealing devices other than the specified direct metal-to-metal contact. A label must be attached to sliding doors warning against painting of the mating surfaces.

17.14   POWER OPERATORS

Power operators must be pneumatic electric type conforming to NFPA 80 and the requirements specified. Readily adjustable limit switches must be provided to automatically stop the door in the door's full open or closed position. All operating devices must be suitable for the hazardous class, division, and group defined in NFPA 70.

17.14.1   Pneumatic Operators

Pneumatic operators must be of a heavy-duty industrial type designed to operate the door at ≥0.2 m/s (2/3 fps) or more than 0.3 m/s (1 fps) with air pressure defined per kPa (psi). A pressure regulator must be provided if the operator is not compatible with available air pressure. Dryer, filter, and filter alarm must be provided. Pneumatic piping must be provided up to the connection with building compressed air but ≤6 m (20 ft) from door jambs. Operators must have provisions for immediate emergency manual operation of the door in case of failure. The operator must open, close, start, and stop the door smoothly. Control must be electrical, conforming to NEMA ICS 2[*] and NEMA ICS 6.[†] Enclosures must be Type 12 (industrial use) or Type 7 or 9 in hazardous locations. Electric controls must be pneumatic with pushbutton wall switches, ceiling pull switches, and rollover floor treadle.

17.14.2 Electric Operators

Electric operators must be heavy-duty industrial type designed to operate the door at ≥0.2 m/s (2/3 fps) or >0.3 m/s (1 fps). Electric controls must be pneumatic with pushbutton wall switches, ceiling pull switches, and rollover floor treadle. Electric power operators must be complete with an electric motor, brackets, controls, limit switches, magnetic reversing starter, and other accessories. The operator must be designed so that the motor may be removed without disturbing the limit switch timing and without affecting the emergency operator. The power operator must be provided with a slipping clutch coupling to prevent stalling of the motor. Operators must have provisions for immediate emergency manual operation of the door in case of electrical failure. Where controls differ from motor voltage, a control voltage transformer must be provided inside as part of the starter. Control voltage must be 120 volts or less.

17.14.2.1 Motors

Drive motors must conform to NEMA MG 1,[‡] must be high-starting torque reversible type, and must be of sufficient output to move the door in either direction from any position at the required speed without exceeding the rated capacity. Motors must be suitable for operation (1) at 120, 208, 277, or 480 volts with 60 Hz; (2) at 220, 240, or 380 volts with 50 Hz, single three phase; and (3) with across-the-line starting. Motors must be designed to operate at full capacity over a supply variation of ±10% of the motor voltage rating.

17.14.2.2 Controls

Each door motor must have an enclosed reversing across-the-line type magnetic starter with thermal overload protection, limit switches, and remote control switches. The control equipment must conform to NEMA ICS 2; enclosures must conform to NEMA ICS 6 and must be Type 12 (industrial use), Type 7 or 9 in hazardous locations, or as otherwise indicated. Each wall control station must be of the three-button type, with the controls marked and color coded:

- Open—white
- Close—green
- Stop—red

---

[*] NEMA ICS 2-2000 (R2005; errata, 2008), Standard for Controllers, Contactors, and Overload Relays Rated 600 V.
[†] NEMA ICS 6-1993 (R2011), Enclosures.
[‡] NEMA MG 1-2011 (errata, 2012), Motors and Generators; 10 CFR 431.

When the door is in motion and the stop control is pressed, the door must stop instantly and remain in the stop position. From the stop position, the door must be operable in either direction by the open or close controls. Controls must be of the full-guarded type to prevent accidental operation.

17.14.3 LEADING EDGE SAFETY SHUTDOWN

Leading edges of the door with operators must have a safety shutdown switch strip the entire length of the leading edge. The safety strip must be press-at-any-point ribbon switches. Activation of the strip must shut down the operator and release the door, with reset required to continue door operation.

17.15 EM SHIELDED DOOR FACTORY TEST

Test data must be provided on at least one shielded door of each type to verify that the EM shielded doors of the design supplied have been factory tested for compliance with this specification. Test doors must not be reused (on an actual installation).

17.15.1 SWINGING DOOR STATIC LOAD TEST

The door must be mounted and latched to the door's frame, then set down in a horizontal position to open downward with only the frame rigidly and continuously supported from the bottom. A load of 195 kg/psm (40 lb/psf) must be applied uniformly over the entire surface of the door for at least 10 minutes. The door is not acceptable if this load causes breakage, failure, or permanent deformation that causes the clearance between the door leaf and stops to vary more than 1.6 mm (1/16 in.) from the original dimension.

17.15.2 SWINGING DOOR SAG TEST

The door and its frame must be installed normally and opened 90°. Two 45-kg (100-lb) weights, one on each side of the door, must be suspended from the door within 130 mm (5 in.) of the outer edge for at least 10 minutes. The door is not acceptable if this test causes breakage, failure, or permanent deformation that causes the clearance between the door leaf and door frame to vary more than 1.6 mm (1/16 in.) from the original dimension.

17.15.3 DOOR CLOSURE TEST

Each door design must be operated 100,000 complete open/close cycles. The door is not acceptable if the closure test causes any breakage, failure, or permanent deformation that causes the clearance between the door and door frame to vary more than 1.6 mm (1/16 in.) from the original dimension.

17.15.4 HANDLE-PULL TEST

The door must be mounted and latched to its frame. The handle must have a force of 1100 N (250 lb) applied outward (normal to the surface of the door) at a point within 50 mm (2 in.) of the end of the handle. The door is not acceptable if this test causes any breakage, failure, or permanent deformation exceeding 3 mm (1/8 in.).

17.15.5 DOOR ELECTROMAGNETIC SHIELDING TEST

The EM shielded door must be factory tested in accordance with the requirements of this specification both before and after the mechanical tests described above.

17.16 ELECTROMAGNETIC FILTERS

A filter must be provided for each power, data, and signal line penetrating the enclosure. These lines include power lines, lines to dummy loads, alarm circuits, lighting circuits, and signal lines (e.g., telephone lines, antenna lines, HVAC control fire alarm). Filters and electrical surge arresters (ESAs) must be enclosed in metallic cases that protect the filter elements from moisture and mechanical damage. All external bonding or grounding surfaces must be free from insulating protective finishes. All exposed metallic surfaces must be suitably protected against corrosion by plating, lead-alloy coating, or other means. The finish (1) must provide good electrical contact when used on a terminal or as a conductor, (2) must have uniform texture and appearance, (3) must be adherent, and (4) must be free from blisters, pinholes, and other defects. The filter and ESAs must also meet the requirements of UL 1283. Insertion loss in the stop band between the load side of the filter and the power supply side must be equal to or greater than the EM shielding attenuation specified. The filter used for 400 Hz must be provided with a power factor correcting coil to limit the reactive current to 10% maximum of the full-load current rating. Each filter unit must be capable of being mounted individually and will include one filter for each phase conductor of the power line and the neutral conductor. The signal filters should include one filter for each conductor.

17.16.1 ENCLOSURE

Filter units must be mounted in an EM-modified NEMA Type 1 enclosure in accordance with NEMA ICS 6 and meet the requirements of UL 1283. Enclosures must be made of corrosion-resistant steel of 1.9837 mm (14 gauge) minimum thickness with welded seams and galvanized bulkhead cover plates. The enclosure non-conductive surfaces must be finished with a corrosion-inhibiting primer and two coats of baked or finish enamel. The input compartment should house the individual line filters and the input terminals of the filters and mounting for the surge arrester. Live parts must be spaced in accordance with NFPA 70. Filter leads must be copper. Filter enclosures must be shielding effectiveness tested in accordance with IEEE 299 and Table 17.1.

TABLE 17.1

Filter Rated Current and Short-Circuit Full Load

Filter Rated Current, RMS Amperes	Short-Circuit Full Load Amperes Symmetrical
0–100	10,000
101–400	14,000

Source: NASA, *Electromagnetic (EM) Shielding*, Section 13 27 54.00 10, National Aeronautics and Space Administration, Washington, DC, 2007.

Test leads and coaxial connectors through the enclosure must be provided for high-altitude electromagnetic pulse (HEMP) testing. The imbedded configuration must be used for filter enclosures as required by MIL-STD-188-125-1.[*]

17.16.2 FILTER UNIT MOUNTING

Each filter unit must be mounted individually in an enclosure containing one filter for each penetrating conductor. One end of the individual filter case must be attached to the radiofrequency (RF) barrier plate between the two compartments to provide a RF tight seal between the RF barrier plate and the filter case. The terminals of the filters should project through openings in the RF barrier plate into the inner terminal compartment. The case of each filter must be attached to both the enclosure and to the barrier plate to prevent undue stress being applied to the RF seal between the filter case and the RF barrier plate. Individual filters must be removable from the enclosure. Like filters must be interchangeable.

17.16.3 CONDUIT CONNECTIONS TO ENCLOSURES

The load terminal and input compartments must have no knockouts, and each compartment must have weldable threaded conduit hubs. The hubs must be circumferentially EM welded in place and must be sized and located as required for the conduits indicated.

17.16.4 ACCESS OPENINGS AND COVER PLATES

Enclosures must have separate clear front access cover plates on terminal and power input compartments. Access cover plates must be hinged with EM gaskets and have a maximum of 75-mm (3-in.) bolt spacing. The design should include thick cover plates and folded enclosure edges to prevent enclosure deformation, bolt spacers to

---

[*] MIL-STD-188-125-1 (1998; Notice 1, 2005), Department of Defense Interface Standard: High-Altitude Electromagnetic Pulse (HEMP) Protection for Ground-Based C4I Facilities Performing Critical, Time-Urgent Missions (Part I—Fixed Facilities).

prevent uneven gasket compression, and gasket mounting to facilitate replacement. All gasket contact areas must be tin-plated using the electrodeposited Type I method in accordance with ASTM B545.[*] Nuts and bolts must be permanently fastened to the enclosure by welding or captive attachments.

17.16.5 OPERATING TEMPERATURE

The filter and ESA assembly must be rated for continuous operation, with filters at rated voltage and full-load currents, in ambient temperatures from –55 to +65°C (–67 to +150°F) (measured outside the EM filter enclosure). Filter components must be suitable for continuous full-load operation at a temperature from –55 to +85°C (–67 to +155°F).

17.16.6 SHORT-CIRCUIT WITHSTAND

Filters must be labeled and built to have standard short-circuit withstand ratings in accordance with UL 1283. The minimum ratings must be as shown in Table 17.1.

17.16.7 FILTER CONNECTIONS

Individual filters must have prewired standoffs and solderless lugs. The lugs must be of the hexagonal head bolt or screw type and must conform to UL 486A-486B. Live parts must be spaced in accordance with NFPA 70.

17.17 INTERNAL ENCAPSULATED FILTERS (FILTER UNITS)

17.17.1 FILTER CONSTRUCTION

Individual filters must be of a heavy-duty type sealed in a steel case. After the filter is filled with an impregnating or encapsulating compound, the seams of the filter must be welded. When a solid potting compound is used to fill the filter, the filters may be mechanically secured and sealed with solder. Either hermetically sealed impregnated capacitors must be used or the complete filter assembly must be vacuum impregnated. Individual filter cases must be fabricated of ≥2 mm (14-gauge) thick steel and finished with a corrosion-resistant plating or one coat of a corrosion-resistant primer and two coats of finish enamel. The filter must be filled with an impregnating or potting compound that is chemically inactive with respect to the filter unit and case. The compound, either in the state of original application or as a result of having aged, must have no adverse effect on the performance of the filter. The same material must be used for impregnating as must be used for filling. Filter terminals must be copper that can withstand the pull requirements specified and measured in accordance with Section 17.16 (Electromagnetic Filters).

---

[*] ASTM B545-13, Standard Specification for Electrodeposited Coatings of Tin.

17.17.2 Ratings

Filters must be provided in the current, voltage, and frequency ratings indicated on the drawings. Filter current must be measured. Filter voltage must be (1) 120, 208, 277, or 480 volts with 60 Hz; and (2) 230, 250, or 400 volts with 50 Hz. The passband frequencies must be suitable for use with 50- or 60-Hz and 400-Hz power source and signal-line filters.

17.17.3 Voltage Drop

Voltage drop through the filter at operation frequency must be ≤2% of the rated line voltage when the filter is fully loaded with a resistive load (unity power factor). Voltage drop measurements must be in accordance with Section 17.20.1 (Voltage Drop Measurements).

17.17.4 Input Elements

Filters must be provided with inductive inputs. If inductive input is used, an ESA is required to protect the filter. The inductor must ensure firing potential for the preceding ESA and limit the current through the filter capacitor. The input inductor must be designed to withstand at least a 10,000-volt transient.

17.17.5 Drainage of Stored Charge

Filters must be provided with bleeder resistors to drain the stored charge from the capacitors when power is shut off. Drainage of the stored charge must be in accordance with NFPA 70.

17.17.6 Insertion Loss

Insertion loss must meet or exceed the levels complying with EM shielding effectiveness attenuation requirements when measured in accordance with MIL-STD-220.* Insertion loss measurements must be performed in accordance with MIL-STD-220 and Section 17.16 (Electromagnetic Filters).

17.17.7 Operating Temperature Range

Individual filters mounted in the filter enclosure operating at full load amperage and rated voltage must be ≤+85°C (185°F) based on an ambient temperature of 65°C (150°F) outside the filter enclosure. Continuous operation from −55 to +85°C (−67 to +185°F) must be demonstrated according to Section 17.20.3 (Filter Life at High Ambient Temperature). Filters must also withstand temperature cycling as specified in Section 17.16 (Electromagnetic Filters). The filter will remain at rated voltage and full-load current until temperature equilibrium is reached or 24 hours, whichever is greater.

---

* MIL-STD-220C (2009), Test Method Standard: Method of Insertion Loss Measurement.

17.17.8 CURRENT OVERLOAD CAPABILITY

Filters must be capable of operating at 140% of rated current for 15 minutes, 200% of rated current for 1 minute, and 500% of rated current for 1 second when tested in accordance with Section 17.20.5 (Overload Test).

17.17.9 REACTIVE SHUNT CURRENT

The reactive shunt current drawn by the filter operating at rated voltage must be ≤30% of the rated full-load current when measured in accordance with Section 17.20.6 (Reactive Shunt Current Measurements).

17.17.10 DIELECTRIC WITHSTAND VOLTAGE

Filters must be provided that conform to the minimum values of dielectric withstanding voltage. Filter dielectric withstand voltage test must be in accordance with Section 17.20.7 (Dielectric Withstand Voltage Test). HEMP filters must be capable of operating continuously at full-rated voltage and of withstanding an overvoltage test of 2.8 times the rated voltage for 1 minute. In addition, each filter must be capable of withstanding a 20-kV or 4-kA peak transient pulse of approximately 20-ns pulse width at full operating voltage, without damage.

17.17.11 INSULATION RESISTANCE

The insulation resistance between each filter terminal and ground must be >1 megohm when tested in accordance with Section 17.20.8 (Insulation Resistance Test).

17.17.12 PARALLEL FILTERS (CURRENT SHARING)

Where two or more individual filters are electrically tied in parallel to form a larger filter, the filters will equally share the current. Equally sharing is defined to be within 5% of the average current. The tests must be in accordance with Section 17.16 (Electromagnetic Filters).

17.17.13 HARMONIC DISTORTION

Harmonics generated by the insertion of a filter will not increase line voltage distortion more than 2.5% when measured with a unity power factor in accordance with Section 17.16 (Electromagnetic Filters).

17.18 MARKING OF FILTER UNITS

Each filter case must be marked with HCI tags and with the rated current, rated voltage, manufacturer's name, type of impregnating or potting compound, operating frequency, and model number. In addition, individual filter cases, the filter enclosures, and supply and load panelboards of filtered circuits must be marked by the manufacturer with the following:

> WARNING: Before working on filters,
> terminals must be temporarily grounded.

to ensure discharge of capacitors. Nameplates and warning labels must be securely attached.

17.19   MINIMUM LIFE

Filter assemblies must be designed for a minimum service life of 15 years. Filter schedules should include voltage, amperage, enclosure type (low, high, bandpass), location, cut-off frequency, bandpass frequencies, and electrical surge arresters (ESAs). Data and/or calculations for the design of EM doors including scheduling of EM penetrations must be considered in minimum service life assessments.

17.20   POWER AND SIGNAL LINE FACTORY TESTING

Factory test report data must be provided for each filter configuration, voltage, and amperage that will show the ability of filters to meet the specified requirements. Filter test reports must be based on prior tests of the same filter assembly design and components. Test data must include the following:

- Voltage drop measurements
- Insertion loss measurements
- Filter life test
- Thermal shock test
- Overload test
- Reactive shunt current measurements
- Dielectric withstand voltage
- Insulation resistance test
- Current sharing
- Harmonic distortion
- Terminals pull test

17.20.1   VOLTAGE DROP MEASUREMENTS

The voltage drop measurements on both AC and DC filters must be performed with the components mounted in the filter/ESA assembly enclosure or mounted on a metal plate by the same holding method that must be used for mounting in the enclosure. For AC-rated filters, measurements must be made by using expanded scale-type meters. For DC-rated filters, measurements must be made by using a DC meter when the filter is carrying rated current and rated voltage.

17.20.2 Insertion Loss Measurements

Insertion loss measurements for power filters must have the following modifications. The filters must be installed in the filter/ESA assembly enclosure. The load current power supply will operate at the rated voltage of the filters and must be capable of providing any current from no-load through rated full-load current. The radiofrequency signal generator must be a swept continuous-wave (cw) source. The buffer networks must be modified to permit valid measurements over the entire frequency band on which insertion loss requirements are specified (14 kHz to 1 Ghz). The receiver or network analyzer must be capable of operating over the entire frequency band on which insertion loss requirements are specified (14 kHz to 1 Ghz). Sensitivity must be adequate to provide a measurement dynamic range at least 10 dB greater than the insertion loss requirement. The load impedance must be resistive and must be capable of dissipating the rated full-load filter current. Insertion loss measurements must be made at 20, 50, and 100% of the filter full-load operating current. Insertion loss measurements for communication/signal line filters must be performed the same as for power filters except that the insertion loss measurements are required at load impedance equal to the image impedance of the filter. No-load insertion loss measurements must be performed over the frequencies defined in the EM shielding effectiveness attenuation requirements for both power and communication filters. Testing must be load to source for TEMPEST. Testing must be source to load for HEMP.

17.20.3 Filter Life at High Ambient Temperatures

This test is conducted to determine the effects on electrical and mechanical characteristics of a filter resulting from exposure of the filter to a high ambient temperature for a specified length of time while the filter is performing its operational function. Surge current, total resistance, dielectric strength, insulation resistance, and capacitance are measurement types that would show the deleterious effects due to exposure to elevated ambient temperatures. A suitable test chamber must be used that will maintain the temperature at the required test temperature and tolerance.

Temperature measurements must be made within a specified number of unobstructed millimeters (inches) from any one filter or group of like filters under test. This test must be made in still air. Specimens must be mounted by their normal mounting means. When groups of filters are to be tested simultaneously, the mounting distance between filters must be as specified for the individual groups; otherwise, the mounting distance must be sufficient to minimize the chances of the temperature on one filter affecting the temperature of another. Filters fabricated of different materials are not to be tested simultaneously. The test temperature must be at least 85°C (185°F) and can be as high as 87°C (189°F). The length of the test must be for 5000 hours. Specified measurements must be made prior to, during, and after exposure.

17.20.4 Thermal Shock Test

This test is conducted to determine the filter's resistance to heat exposures at extremes of high and low temperatures and to the shock of alternate exposures to these extremes. Suitable temperature-controlled systems must be used to meet the

---

TABLE 17.2

Filter Weight and Minimum Exposure Time

Filter Weight	Minimum Exposure Time
1 oz. and below	15 minutes
Above 1 oz. to 4.8 oz.	30 minutes
Above 4.8 oz. to 3 lb	1 hour
Above 3 lb to 30 lb	2 hours
Above 30 lb to 300 lb	4 hours
Above 300 lb	8 hours

Source: NASA, *Electromagnetic (EM) Shielding*, Section 13 27 54.00 10, National Aeronautics and Space Administration, Washington, DC, 2007.

---

temperature requirements and test conditions. Environmental chambers must be used to meet test requirements and to reach specified temperature conditions. Filters must be placed so that no obstruction occurs to the flow of air across and around the filter. The filter must be subjected to the specified test conditions. The first five cycles must be run continuously. After five cycles, the test may be interrupted after the completion of any full cycle and the filter allowed to return to room ambient temperature before testing is resumed.

One cycle consists of steps 1 through 4 of the applicable test condition for dual environmental test chambers (one low-temperature and one high-temperature test chamber) and steps 1 and 3 for single-compartment test chambers where both high and low temperatures are achieved without moving the filter. The test conditions are as follows:

1. $-55°C$ $0°$ and $-3°$ (which is -55 to $-58°C$)
2. $25°C$ $+10°$ and $-5°$ (which is 20 to $35°C$)
3. $85°C$ $+3°$ and $-0°$ (which is 85 to $88°C$)
4. $25°C$ $+10°$ and $-5°$ (which is 20 to $35°C$)

The effective total transfer time from the specified low temperature to the specified high temperature must be ≤ 5 minutes. The exposure time in air at the extreme temperatures is a function of the weight of the filter. The minimum exposure time per the weight of the filter must be as shown in Table 17.2.

17.20.5 OVERLOAD TEST

Filters must be mounted in the filter/ESA assembly enclosure or mounted on a metal plate by the same holding method used for mounting in the enclosure. A specified current must then be applied for a specified period of time. After the filter has returned to room temperature, the insulation resistance and voltage drop must be measured. The insulation resistance must be measured using the method in Section 17.16 (Electromagnetic Filters). AC voltage drop measurements must be made by using expanded scale-type meters that will enable voltage differences of <1 volt to be read. DC voltage drop measurements must be made by using a DC reading meter when the filter is carrying rated current and rated voltage. The insulation resistance and the voltage drop must be measured after each separate overload test. Filters must also be visually examined for evidence of physical damage after each test.

17.20.6 REACTIVE SHUNT CURRENT MEASUREMENTS

The reactive shunt current measurements must be performed with the filters mounted in the filter/ESA assembly enclosure or mounted on a metal plate by the same holding method that must be used for mounting in the enclosure. The filter must be terminated in the inner compartment in an open circuit. Rated AC voltage must be applied between the filter outer compartment terminal and the enclosure or metal plate. The AC current into the outer compartment terminal must be monitored. The measured current is equal to the filter reactive shunt current.

17.20.7 DIELECTRIC WITHSTAND VOLTAGE TEST

The dielectric withstand voltage test (high-potential, overpotential, voltage breakdown, or dielectric strength test) consists of the application of a voltage higher than the rated voltage for a specific time between mutually insulated portions of a filter or between insulated portions and ground. Repeated application of the test voltage on the same filter is not recommended, as even overpotential less than the breakdown voltage may injure the insulation. When subsequent application of the test potential is specified in the test routine, succeeding tests must be made at reduced potential. When an alternating current potential is used, the test voltage must be 60 Hz and approximate a true sine wave in form. All AC potentials must be expressed as root-mean-square values. The kVA rating and impedance of the source must permit operation at all testing loads without serious distortion of the waveform and without serious change in voltage for any setting. When a direct current potential is used, the ripple content must be ≤5% rms of the test potential. A voltmeter must be used to measure the applied voltage to an accuracy of 5%. When a transformer is used as a high-voltage source of AC, a voltmeter must be connected across the primary side or across a tertiary winding provided that the actual voltage across the filter is within the allowable tolerance under any normal load condition. Unless otherwise specified, the test voltage must be DC and must be as shown in Table 17.3.

The duration of the DC test voltages must be 5 seconds minimum and 1 minute maximum after the filter has reached thermal stability at maximum operating temperature produced by passage of rated current. The test voltage must be applied between the case (ground) and all live (not grounded) terminals of the same circuit connected together. The test voltage must be raised from zero to the specified value as uniformly as possible, at a rate of approximately 500 volts (rms or DC) per second. Upon completion of the test, the test voltage must be gradually

TABLE 17.3
Source and Voltage

Source	Voltage
DC rated only	2.5 times rated voltage
For filters with AC and DC ratings	2.5 times rated DC voltage
AC rated only	4.2 times rated rms voltage

Source: NASA, *Electromagnetic (EM) Shielding*, Section 13 27 54.00 10, National Aeronautics and Space Administration, Washington, DC, 2007.

reduced to avoid voltage surges. The changing current must be 50 mA maximum. During the dielectric withstanding voltage test, the fault indicator must be monitored for evidence of disruptive discharge and leakage current. The sensitivity of the breakdown test equipment must be sufficient to indicate breakdown when at least 0.5 mA of leakage current flows through the filter under test. The test must be performed with the components mounted in the filter/ESA assembly enclosure. Filters for AC circuits must be tested with an AC source, while filters for DC circuits must be tested with a DC source. After the test, the filter must be examined and measurements must be performed to include insulation resistance measurements to determine the effect of the dielectric withstanding voltage test on specific operating characteristics.

17.20.8 INSULATION RESISTANCE TEST

This is a test to measure the resistance offered by the insulating members of a filter to an impressed direct voltage tending to produce a leakage current through or on the surface of these filters. Insulation resistance measurements must be made on an apparatus suitable for the characteristics of the filter to be measured (such as a megohm bridge, megohm-meter, insulation resistance test set, or other suitable apparatus). The test must be performed with the components mounted in the filter/ESA assembly enclosure or mounted on a metal plate by the same holding method that must be used for mounting in the enclosure. The bleeder resistor must be disconnected. The direct potential applied to the specimen must be the largest test condition voltage (100, 500, or 1000 volts + 10%) that does not exceed the rated peak AC voltage or the rated DC voltage. A separate DC power supply may be used to charge the filters to the test voltage. The measurement error at the insulation resistance value required must be ≤10%. Proper guarding techniques must be used to prevent erroneous readings due to leakage along undesired paths. Insulation resistance measurements must be made between the mutually insulated points or between insulated points and ground. The insulation resistance value must be read with a megohmmeter and recorded after the reading has stabilized. When more than one measurement is specified, subsequent measurements of insulation resistance must be made using the same polarity as the initial measurements.

17.20.9   CURRENT SHARING

Testing must be performed with the filters mounted in the filter/ESA assembly enclosure or mounted on a metal plate by the same holding method that must be used for mounting in the enclosure. The filter inner compartment terminals must be loaded with a resistor equal in value to the rated operating voltage divided by the sum of the current ratings of the devices in parallel. The resistor must be capable of dissipating the total current. Rated operating voltage must be applied at the filter outer compartment terminals. The current into each filter outer compartment terminal must be monitored.

17.20.10   HARMONIC DISTORTION TEST

Harmonic distortion measurements must be made using a spectrum analyzer having a dynamic range of 70 dB and a frequency range from 10 kHz to 1.7 GHz. Total harmonic distortion must be measured at the input and output terminals of the filter when operating at 25, 50, and 100% of rated full-load current.

17.20.11   TERMINALS PULL TEST

This test determines whether the design of the filter terminals can withstand the mechanical stresses during installation or disassembly of equipment. Testing must be performed with the components mounted in the filter/ESA assembly enclosure or mounted on a plate by the same holding method that must be used for mounting in the enclosure. The force applied to the terminal must be 89 N (20 lb). The point of application of the force and the force applied must be in the direction of the axes of the terminations. The force must be applied gradually to the terminal and then maintained for a period of 5 to 10 seconds. The terminals must be checked before and after the pull test for poor workmanship, faulty designs, inadequate methods of attaching of the terminals to the body of the part, broken seals, cracking of the materials surrounding the terminals, and changes in electrical characteristics (e.g., shorted or interrupted circuits). Measurements are to be made before and after the test.

REFERENCES

AISC. (2011). *Steel Construction Manual*. Chicago, IL: American Institute of Steel Construction.

NASA. (2007). *Electromagnetic (EM) Shielding*, Section 13 27 54.00 10. Washington, DC: National Aeronautical and Space Administration.

NFPA. (2005). *NFPA 70: National Electrical Code (NEC) Handbook*. Quincy, MA: National Fire Protection Association.

NFPA. (2012). *NFPA 70E: Standard for Electrical Safety in the Workplace®*. Quincy MA: National Fire Protection Association.

OSHA. (2007). Occupational Safety and Health Standards, 29 CFR, Part 1910, Subpart S, Electrical. Baltimore, MD: Occupational Safety and Health Administration.

18 Electromagnetic Shielded Equipment, Instrumentation, and Facilities, Part Two

Randy Boss and Martha J. Boss

CONTENTS

18.1   ELECTRICAL SURGE ARRESTERS

18.1.1   POWER AND SIGNAL LINE ESA

Electrical surge arresters (ESAs) must be metal oxide varistors (MOVs) or spark gaps. When a spark gap is specified, the ESA must be enclosed within a metal case. Discharges must be contained within the case; no external corona or arcing is permitted. ESAs must be factory installed with minimum lead lengths within the outer compartment. For all power filter/ESA assemblies, the ESAs must be installed a minimum of 75 mm (3 in.) apart, with terminals at least 75 mm (3 in.) from a grounded surface. For telephone filter/ESA assemblies, the ESAs must have a minimum clearance spacing of 25 mm (1 in.), and terminals must be at least 75 mm (3 in.) from a grounded surface. Each phase, neutral, and telephone circuit conductor must be connected through an ESA to the ground bus.

The ESA must be installed in the power input compartment of the filter in a separate electromagnetic (EM) shielded enclosure. ESA units within the filter/ESA assembly must be individually replaceable, and like ESAs must be interchangeable. ESA terminals must be able to withstand the 89-N (20-lb) pull test. Live parts must be spaced in accordance with NFPA 70. ESA leads must be copper. Individual ESAs must be marked with hardness critical item (HCI) tags and must be marked with the manufacturer's name or trademark and part number. The ESA must meet the requirements of IEEE C62.11,[*] IEEE C62.41.1,[†] IEEE C62.41.2,[‡] and UL 1449.[§]

18.1.2 Wiring

The ESAs must be located so that leads of minimum length connect the ESA ground terminal to the enclosure. The total lead length connecting the ESA to the filter and the ESA ground terminal to the enclosure must be <300 mm (12 in.). Powerline ESA wiring must be No. 4 AWG minimum. Communication/signal line ESA wiring must be of the same or heavier gauge) than the communication/signal line conductor.

18.1.3 Voltage Characteristics

Measurements of voltage (MOV) at 1 mA DC current and spark-gap DC breakdown voltage must be made. Testing must be performed with the ESAs mounted in the filter/ESA assembly enclosure or mounted on a metal plate by the same holding method that must be used for mounting in the enclosure. A variable DC power supply must be connected between the ESA terminal and the enclosure (or plate). The applied DC voltage must be increased at a rate not to exceed 10% of the rated firing voltage per second. The (MOV) voltage at 1 mA DC current is the power supply output voltage when the output current is 1 mA. The spark-gap DC breakdown voltage is the applied voltage just prior to breakdown (indicated by a rapid decrease in the voltage across the device).

The power supply must be de-energized immediately after the value has been recorded. MOV direct current breakdown voltage at 1 mA DC current must be at least 340, 500, or 1000 volts and <425 or 1500 volts. MOV testing must be in accordance with IEEE C62.33.[¶] Spark-gap DC breakdown (sparkover) voltage must be at least 500 or 1000 volts and <1500 or 3000 volts. Spark-gap impulse sparkover voltage of the ESA must be <4000 volts. This voltage must be on surges of either polarity having a rate of rise of 1000 V/ns. Testing of the ESA impulse sparkover voltage must be performed with the spark gaps mounted in the filter/ESA assembly enclosure or mounted on a metal plate by the same holding method that must be used for mounting in the enclosure. The pulse generator must be connected between the spark-gap terminal and the enclosure (or plate) with a minimum inductance connection. The pulse generator must be capable of providing a ramp voltage of 1 kV/ns to a peak voltage that is at least twice the open circuit impulse sparkover voltage.

---

[*] IEEE C62.11-2012, Standard for Metal-Oxide Surge Arresters for Alternating Current Power Circuits (>1 kV).

[†] IEEE C62.41.1-2002 (R2008), Guide on the Surges Environment in Low-Voltage (1000 V and Less) AC Power Circuits.

[‡] IEEE C62.41.2-2002, Recommended Practice on Characterization of Surges in Low-Voltage (1000 V and Less) AC Power Circuits.

[§] UL 1449 (2006), Surge Protective Devices.

[¶] IEEE C62.33-1982 (R2000), Standard for Test Specifications for Varistor Surge-Protective Devices.

Voltage across the spark gap must be monitored on an oscilloscope or transient digitizing recorder capable of at least 1-ns resolution. The peak transient voltage during the pulse is the impulse sparkover voltage. Response time must be <4 ns. Clamping voltage of the ESA must be <900 volts at a current pulse of 10 kA. ESA clamping voltage measurements must be performed with the ESAs mounted in the filter/ESA assembly enclosure or mounted on a metal plate by the same holding method that must be used for mounting in the enclosure. The pulse generator must be connected between the ESA terminal and the enclosure (or plate) with a minimum inductance connection. The pulse generator must be capable of providing a 10-kA current pulse on an 8- by 20-μs waveshape into the ESA. Current through the ESA and voltage across the ESA must be monitored on oscilloscopes or transient digitizing recorders. The asymptotic voltage during the 10-kA portion of the pulse is the clamping voltage.

18.1.4   ESA Extinguishing Characteristics

The ESA must extinguish and be self-restoring to the normal non-conductive state within one-half cycle at the operating frequency. The ESA extinguishing test must be performed with the ESA mounted in the filter/ESA assembly enclosure or mounted on a metal plate by the same holding method that must be used for mounting in the enclosure. The extinguishing test uses an AC power source connected between the ESA terminal and ground that must be at the rated voltage and frequency capable of providing at least 25 amperes into a short-circuit load. A pulse generator capable of providing a short pulse that will fire the ESA must also be connected across the ESA. Voltage across the ESA must be monitored on an oscilloscope or transient digitizing recorder. A series of ten pulses must be injected. Performance of the ESA is satisfactory if the arc extinguishes (indicated by reoccurrence of the sinusoidal waveform) within 8.5 ms after the start of each pulse.

18.1.5   ESA Extreme Duty Discharge Current

The ESA must be rated to survive the extreme duty discharge current of a single 8- × 20-μs pulse with a 10 to 90% rise time of 8 μs and fall time to a value of 36.8% of peak in 20 μs. The following conditions apply:

- The ESA for high-voltage power lines (above 600 volts) must have an extreme duty discharge capability of ≥70 kA.
- The ESA for low-voltage power lines (below 600 volts) to building interiors, area lighting, and external HVAC equipment must have an extreme duty discharge capability of ≥50 kA.
- The ESA for control circuits (e.g., interior alarms, indicator lights, door access controllers, HVAC controls, telephones) must have an extreme duty discharge capability of ≥10 kA.

The ESA extreme duty discharge test must be performed with the ESA mounted in the filter/ESA assembly enclosure or mounted on a metal plate by the same holding method that must be used for mounting in the enclosure.

A pulse generator must be connected between the ESA terminal and the enclosure (or plate) with a minimum inductance connection. The pulse generator must be capable of supplying an 8- $\times$ 20-μs waveshape and only a single pulse is required. Current through the ESA and voltage across the ESA must be monitored on oscilloscopes or transient digitizing recorders. The ESA must be visually monitored during the test and after the pulse should be inspected for charring, cracks, or other signs of degradation or damage. Tests must be on a prototype only. The DC breakdown voltage test must be repeated.

18.1.6   MINIMUM OPERATING LIFE

The ESA operating life tests must be performed with the ESA mounted in the filter/ESA assembly enclosure or mounted on a metal plate by the same holding method that must be used for mounting in the enclosure. A pulse generator must be connected between the ESA terminal and the enclosure (or plate) with a minimum inductance connection. The pulse generator must be capable of supplying repetitive 4-kA current pulses, with a 50 ns $\times$ 500 ns waveshape, to the ESA. A series of ten pulses is required. Current through the ESA and voltage across the ESA must be monitored on oscilloscopes or transient digitizing recorders. The ESA must be visually monitored during the series of pulses for indications of external breakdown. The ESA must be able to conduct 2000 pulses at a peak current of 4 kA and a 50 ns $\times$ 500 ns waveform. Inspection for charring, cracks, or signs of degradation should be done after the test. The DC breakdown voltage test must be repeated.

18.1.7   OPERATING TEMPERATURE

The ESA must be rated for continuous operation in ambient temperatures from –25 to +125°C (–12 to +255°F).

18.2   ESA TESTING

The ESA factory test data must be submitted that will show the ability to meet the requirements herein, based on prior tests of the same ESA assembly components and design. Testing must be performed with the ESA mounted in the filter/ESA assembly enclosure or mounted on a metal plate by the same holding method that must be used for mounting in the enclosure. The pulse generator must be connected between the ESA terminal and the enclosure (or plate) with a minimum inductance connection. Current through the ESA and voltage across the ESA must be monitored on oscilloscopes or transient digitizing recorders. Test data should include the following:

- Breakdown voltage
- Impulse sparkover voltage
- Clamping voltage
- Extinguishing
- Extreme duty discharge
- Surge life

18.3 WAVEGUIDE ASSEMBLIES

Waveguide-below-cutoff (WBC) protection must be provided for all piping, ventilation, and fiberoptic cable penetrations and microwave communications barrier penetrations of the high electromagnetic pulse (HEMP) electromagnetic barrier. These WBC penetrations must be protected with cutoff frequencies and attenuation no less than the EM shielding effectiveness values listed herein. The cutoff frequencies must be ≥1.5 times the highest frequency of the shielding effectiveness. For 1 GHz, the maximum rectangular linear diagonal dimension must be 100 mm (4 in.) and the maximum circular diameter must be 100 mm (4 in.). The length-to-cell cross-section dimension ratio of the waveguide must be a minimum of 5:1 to attain 100 dB or 3:1 to attain 50 dB. Penetration locations must be arranged to facilitate installation and testing by minimizing the number of locations. Waveguides of each assembly type must be factory tested in accordance with IEEE 299.[*]

18.3.1 WAVEGUIDE-TYPE AIR VENTS

Each ventilation WBC array must be a honeycomb-type air vent with a core fabricated of corrosion-resistant steel. Waveguide construction must include heavy frames to dissipate the heat of welding to the shield. A welded WBC array must be constructed from sheet metal or square tubes. Array cells must be formed by welding the sheets at intersections or welding adjacent tubes along the entire length of the WBC section. The maximum cell size must be 100 mm (4 in.) on a diagonal. The length of the WBC section must be at least five times the diagonal dimension of the cells. Air vents must be a permanent part of the shielded enclosure and will have a shielding effectiveness equal to that of the total enclosure. Static pressure drop through the vents must be <3.4 gpscm (0.01 in. of water) at an air velocity of 305 m/s (1000 fpm). Waveguides for air vents (honeycomb) will have access doors in duct work for maintenance. The frame of the honeycomb panel must be welded or bolted into the penetration plate with continuous circumferential EM welds or with bolts 75 mm (3 in.) on-center. Welds for the fabrication and installation of honeycomb waveguide panels are primary shield welds and must be inspected on a regular schedule. Acceptance testing of all honeycomb panels must be included with the final acceptance test. Conductors (e.g., wires and louver operating rods) must not pass through the waveguide openings.

18.3.2 PIPING PENETRATIONS

All piping penetrations of the HEMP barrier, including utility piping, fire mains, vent pipes, and generator and boiler exhausts, must be made with piping WBC sections. The WBC material must be steel with a composition suitable for welding to the HEMP shield. The minimum wall thickness must be 3.2 mm (0.125 in.). The maximum inside diameter must be 100 mm (4 in.), or a metallic honeycomb insert with a maximum cell dimension of 100 mm (4 in.) must be installed. The WBC section must have an unbroken length of at least five diameters to form a minimum cutoff

---

[*] IEEE 299-2006, Standard Method for Measuring the Effectiveness of Electromagnetic Shielding Enclosures.

frequency of 1.5 times the highest frequency of the shield effectiveness. The piping WBC section must be circumferentially welded or brazed to the HEMP shield, pipe sleeve, or a penetration plate as shown on the drawings. Generator and boiler exhausts must be constructed and configured as a WBC or WBC array. The circumferential penetration welds are primary shield welds and must be inspected and tested on a regular basis.

18.3.3 WAVEGUIDE PENETRATIONS

Waveguide penetrations for dielectric fibers or hoses must be implemented in the same manner as piping penetrations. Dielectric hoses or pipes must be converted to metal waveguide piping before penetrating the shield. Conductors (e.g., wires and fiber cable strength members) must not pass through the waveguide opening.

18.4 GROUNDING STUD

The enclosure must have a 13-mm (1/2-in.) diameter stud circumferentially welded to each side of the shielding penetration plate.

18.5 PENETRATION PLATES

Penetration plates must be a minimum of 6 mm (1/4 in.) thick. The penetration plate must overlap the shield penetration cutout dimension by a minimum of 150 mm (6 in.) on each side. The penetration plate must be welded or bolted to the HEMP shield with continuous circumferential EM welds or with bolts 75 mm (3 in.) on-center.

18.6 GALVANIZING

Galvanizing, when practical and not otherwise indicated, must be hot-dipped processed after fabrication. Galvanizing must be in accordance with ASTM A123/A123M* or ASTM A653/A653M,† as applicable. Exposed fastenings must be galvanically compatible material. Electrolytic couples and dissimilar metals that tend to seize or gall must be avoided.

18.7 EM SHIELDED CABINETS AND PULL BOXES

Cabinets and pull boxes must be modified NEMA 1 in accordance with NEMA ICS 6‡ and must (1) be made of corrosion-resistant steel that is ≥2 mm thick (14-gauge) with welded seams and galvanized bulkhead cover plates, and (2) have access cover plates hinged with EM gaskets and 75-mm (3-in.) maximum bolt

---

* ASTM A123/A123M-13, Standard Specification for Zinc (Hot-Dip Galvanized) Coatings on Iron and Steel Products.
† ASTM A653/A653M-13, Standard Specification for Steel Sheet, Zinc-Coated (Galvanized) or Zinc-Iron Alloy-Coated (Galvannealed) by the Hot-Dip Process.
‡ NEMA ICS 6-1993 (R2011), Enclosures.

spacing. Design must include thick cover plates, folded enclosure edges, and bolt spacers to prevent uneven gasket compression and enclosure deformation. The gasket must be easy to replace. Gasket contact areas must be tin-plated using the electrodeposited Type I method in accordance with ASTM B545.* Conduit hub must be circumferentially EM welded to the enclosure. The cabinets must be finished with a corrosion-inhibiting primer and two coats of baked or finish enamel. Cabinets must be provided with mounting brackets for wall mounting or legs for floor mounting. Cabinets and boxes of each type must be factory tested in accordance with IEEE 299.

18.8 QUALITATIVE MONITORING SYSTEM

A built-in shield monitoring system for shielded enclosure leak detection system (SELDS) testing must be provided. The system must consist of either multiple injection points or a surface loop system. Driving conductors must be brought to a single lockable EM shielded connection box, located outside the shield in a controlled space.

18.9 INSTALLATION

During installations, cleaners, solvents, coatings, finishes, physical barriers, and door threshold protectors must be provided as required to protect the shielding system from corrosion, damage, and degradation. Inspection must be completed before a finish or concrete topping coat is installed. Handle shielding steel without causing damage. Penetrations of the shield, including fasteners and mounting bolts, may void the shielding capability. Clean and buff contacting surfaces to ensure firm contact with shielding steel.

18.9.1 CONTROL OF WARPING

Keep warping of steel shielding plates during installation and welding within 1 mm in 1 m (1/8 in. in 10 ft). Use embeds, drive pins, and/or anchor bolts or ties to hold plates in place during welding. Other techniques (e.g., skip welding) may also be used to reduce warpage. Fasteners, drive pins, and other shield penetrations must be sealed with full-penetration circumferential EM welds.

18.9.2 PLACEMENT OF FLOOR SHIELD

Placement of the floor shield should not begin until at least 14 days after the floor slab has been poured. The placement of the floor shield should utilize the shingle overlap method. Individual floor sheet must be attached on the top and one side only with air-pressure drive tools to the floor. Floor shielding sheets must be overlapped 50 mm (2 in.) at joints, bent and laid flat on the concrete floor without voids or gaps, and sealed with continuous EM welds at all seams and joints. The floor shield installation should start at the center of the space.

---

* ASTM B545-13, Standard Specification for Electrodeposited Coatings of Tin.

18.9.3 Placement of Overslab

Placement of the overslab occurs over any portion of the floor shield. Both visual and SELDS testing of the shielding within the area to be covered must first be successfully completed, any defects repaired and retested, and full test results analyzed prior to placement of the overslab. A vapor barrier must be placed over the floor shield.

18.9.4 Welding

The shielding work must be provided in accordance with the performance criteria discussed herein. Shielding steel structurally welded to the steel frame must be welded in accordance with AWS D1.1/D1.1M[*] and AWS D1.3/D1.3M.[†] EM shielding seams must be sealed EM-tight by the metal inert gas (MIG) method, using electrodes structurally and electrically compatible with the adjacent steel sheets. Sheet steel must be welded to support steel by plug or tack welding at 300 mm (12 in.) on-center, and then sheet seams must be continuously EM welded to seal the enclosure. Slag inclusions, gas pockets, voids, or incomplete fusion is not allowed anywhere along welded seams.

Weld failures must be corrected by grinding out such welds and replacing them with new welds. A qualified welder must perform the welding, both structural and EM sealing. Weldments critical to shielding effectiveness are shown on the drawings and must be performed in the manner shown on the drawings. Where both structural integrity and shielding quality are required for a given weldment, both criteria must be met simultaneously. Brazing will conform to the documents discussed above, where practical, and will also conform to requirement of the AWS *Brazing Handbook* (AWS, 2007). Structural, mechanical, or electrical systems penetrations must be sealed by providing a continuous solid perimeter weld, or braze to the shield as specified. All shield joints and seams must have a minimum 50-mm (2-in.) overlap and must be sealed with a continuous solid weld. After testing, the contracting officer must inspect and approve the installation prior to covering by other trades.

18.9.5 Wall Shielding Attachment

Continuous 1.613-mm thick (16-gauge) furring channels spaced ≤600 mm (≤24 in.) on-center must be secured to steel wall studs by using self-tapping sheet metal screws. The steel sheets must be tack welded to the furring strips every 400 mm (16 in.) on-center horizontally and 600 mm (24 in.) on-center vertically. A continuous full-penetration EM weld must be made to join the sheets and form the shield. Welds must not form dimples or depressions causing fish mouths at the edge of the sheet.

18.9.6 Formed Closures

Formed closures must be installed where indicated and/or necessary to completely close all joints, openings, enclosures of pipe chases, and structural penetrations, columns, and beams.

---

[*] AWS D1.1/D1.1M:2010 (errata, 2011), Structural Welding Code—Steel.
[†] AWS D1.3/D1.3M:2008 (errata, 2008), Structural Welding Code—Sheet Steel.

18.9.7 Sequence of Installation

Erection of the steel must be sequenced to prevent steel sheet warpage. Install shielding components that have passed initial testing (Part 1, as described in Section 18.13.3.1) before construction of any features that would limit access for repairs to the shield.

18.9.8 Door Assemblies

Mount doors to perform as specified. Door framing must be continuously welded to the EM shield. The structural system supporting the door frame must provide proper support for doors and frames.

18.10 BOLTED CONSTRUCTION ENCLOSURE INSTALLATION

18.10.1 Enclosure Panel Installation

Install panels, without damage to the shielding steel, in accordance with the shielding manufacturer's recommendations. Exposed surfaces must be cleaned of dirt, finger marks, and foreign matter resulting from manufacturing processes, handling, and installation. Install electrical conduits as close to the EM shield as possible. Framing–joining system bolts must not be used to mount material and equipment. Material and equipment that penetrate the shielded enclosure must be seam welded or soldered to both shielding surfaces.

18.10.2 Surface Preparation

Clean and buff surfaces to ensure good electrical contact with shielding surface. Paint or other coverings on mating surfaces of special boxes (e.g., for fire alarm systems, buzzers, signal lights), including areas between the box and cover, box and wall, and box and conduit, must be removed. Remove insulating material to maintain a low-resistance ground system and to ensure firm mating of metal surfaces.

18.10.3 Floor Panel Setting

Place a polyethylene film vapor barrier that is 0.15 mm (6 mil) thick over the structural floor of the parent room before any other work is set thereon. Provide a 3-mm (1/8-in.) thick layer of hardboard over this film with joints loosely butted. Over this layer an additional layer of similar filler material of equal thickness as the projection of the framing–joining member from the bottom surface of the floor panel must be provided, leaving no more than 6 mm (1/4 in.) of space between the hardboard and the framing–joining member.

18.10.4 Framing–Joining System

Tighten screws with a calibrated adjustable torque wrench with equal torque set for each screw. Proper torque values must be in accordance with the manufacturer's requirements.

18.10.5   DOOR ASSEMBLIES

Mount the door to perform as specified. The door must be through-bolted to the EM shield.

18.10.6   FILTER INSTALLATION

Support filters independently of the wall shielding. Conduct inspections on filters provided under this specification to verify compliance with the specified requirements. Filters must be shipped after successful testing and must be examined prior to installation to determine if damage occurred during shipment. Damage, no matter how slight, must be reason for rejection of the filter.

18.11   WAVEGUIDE INSTALLATION

Penetrations of the EM shield must be treated with the appropriate waveguide method. Waveguides must be suitable for piping and for fluids or gases contained within, in accordance with specified requirements.

18.12   SHIELDING PENETRATION INSTALLATION

Penetrations must be installed in accordance with requirements of the penetration schedule and coordinated with system installation.

18.13   HEMP HARDNESS CRITICAL ITEM SCHEDULE

Hardness critical items must be identified during the detail drawing submittal period. These items are those components and/or construction features that singularly and collectively provide specified levels of HEMP protection (e.g., EM shield, surge arresters, EM shielded doors, shield welding, electrical filters, honeycomb waveguides, waveguides-below-cutoff).

18.13.1   PERFORMANCE TEST PLAN

A performance test plan details tests to be accomplished in three parts:

- In-progress
- Initial shielded enclosure effectiveness
- Final acceptance, shield enclosure effectiveness

The test plan includes the following:

- Equipment listings (including calibration dates and antenna factors)
- Specific test dates and durations during the overall construction period
- Separate testing schedule for the EM enclosure
- Proposed dates and duration of lowest and highest frequency tests

A test grid must be identified, and the plan for correlation of that grid to the structure must be provided.

18.13.2 Test Reports

Test reports include the method of testing, equipment used, personnel, location of tests, and test results. Daily reports of results of each test performed on each portion of the shielding system must be documented. Location of the area tested must be clearly identified. Leaks detected during testing must be identified with sufficient accuracy to permit relocation for testing in accordance with test procedures. Reports of testing will include required certification by the testing agency representative. The report should provide the following details:

- Shielded facility description
- Nomenclature of measurement equipment
- Serial numbers of measurement equipment
- Date of last calibration of measurement equipment
- Type of test performed
- Measured level of reference measurements and ambient level at each frequency and test point
- Measured level of attenuation in decibels at each frequency and test point
- Dynamic range at each test frequency and test point
- Test frequencies
- Location on the shielded enclosure of each test point
- Actual attenuation level at each test point

18.13.3 Field Testing

Reports must be provided of certified test results and results of all field and factory tests. Testing must be accomplished in three parts.

18.13.3.1 Part 1

Perform Part 1 as in-progress testing including inspection, visual seam inspection, and seam testing of all EM shielding materials and installation. In-progress testing consists of (1) welded shielding testing of the structural welds to be embedded prior to concrete placement by dye penetrant and magnetic particle testing and 100% testing of wall, ceiling, and floor shielding welds by the SELDS tests; and (2) bolted construction 100% testing of floor, wall, and ceiling shielding seams by SELDS testing. After successful completion of in-progress testing, including defect repairs and retest, placement of the embedment covering may be made to complete the structural systems. Submit an in-progress test report.

18.13.3.2 Part 2

Perform Part 2 initial testing, which consists of inspection, visual seam inspection, seam testing, and shielded enclosure effectiveness testing after shielding and shielding penetrations are completed but before the installation of finish materials over the shielding. Access to penetrations is required. All seam welds, including shielding and penetrations not tested in Part 1, must be SELDS tested. The initial shielded enclosure effectiveness acceptance test consists of a MIL-STD-188-125-1[*] test utilizing specified test frequencies for magnetic and plane wave. Testing must be conducted in accordance with the Section 18.15 (EM Shielding Effectiveness Testing). These tests must be performed with the number of shield penetrations limited to those required to support the test. After successful completion of Part 2 initial testing, including defect repairs and retest, placement of any covering may be made except in areas where penetrations are located. Submit an initial test report.

18.13.3.3 Part 3

Perform Part 3 final acceptance testing consisting of a visual inspection and a shielded enclosure effectiveness test of the EM shielding materials and installation. All seams welds, including shielding and penetrations not tested in Parts 1 and 2, must be SELDS tested. Part 3 testing must be performed upon completion of construction and when the building is ready for occupancy. Facilities requiring HEMP protection must be tested for shielding effectiveness in accordance with acceptance test procedures in MIL-STD-188-125-1. Corrective work must be accomplished immediately upon detection that any area has failed to meet the requirements specified. Retesting must be performed to verify that remedial work done to meet the required attenuation has been properly installed. A final acceptance test report must be provided.

18.13.4 WELD INSPECTION

The weld seams must be visually inspected by a qualified welder during the welding operation and after welding is completed. Completed welds must be inspected after the welds have been thoroughly cleaned by hand or power wire-brush. Welds must be inspected with magnifiers under bright light for surface cracking, porosity, slag inclusion, excessive roughness, unfilled craters, gas pockets, undercuts, overlaps, size, and insufficient throat and concavity. Defective welds must be ground out and replaced with sound welds.

18.14 SHIELDED ENCLOSURE LEAK DETECTION SYSTEM TESTING

The leak detection system must use a 95- to 105-kHz oscillator and a battery-operated, hand-held receiver. The receiver, or "sniffer," must have a ferrite loop probe capable of sensing leaks within 6 mm (1/4 in.) of the probe location with a dynamic range of 140 dB. Testing must be conducted in accordance with test equipment manufacturer's instructions. Test loops or leads must be placed under the shield floor or into inaccessible locations prior to installation to assist in the detection of seam leaks in the floor, ceiling, and walls. The loop or lead wires must be placed between the vapor barrier and the structural slab for the floor shield with the leads

---

[*] MIL-STD-188-125-1 (1998; Notice 1, 2005), Department of Defense Interface Standard: High-Altitude Electromagnetic Pulse (HEMP) Protection for Ground-Based C4I Facilities Performing Critical, Time-Urgent Missions (Part I—Fixed Facilities).

brought to an accessible location. The test leads must be insulated stranded copper conductors 2 to 2.5 mm (5/64 to 3/32 in.) in diameter and bonded to the shield only at the end. Test leads must be placed in plastic conduit for protection and must be ≤45 m (≤150 ft) in length.

The surface area of the shield will determine the number of test leads (drive points) that are required. Drive points must be installed on the shielding exterior and attached to two sets of diagonally opposing corners during construction. The distance between test lead connections on a shield surface must not be more than 20 m (66 ft). The maximum testing area must be 400 m² (4300 ft²). If the shield area exceeds this requirement, additional drive points must be provided. Bonding of the test leads to the shield is accomplished by brazing or high-temperature soldering. Test leads from the drive points must be run to a lockable test cabinet for connection to the SELDS oscillator.

If more than one test cabinet is required for a given area or building, test leads that would be common to different surface areas must be duplicated at each test cabinet to ensure that test point pairings are maintained. The location of the permanent test leads must be documented for permanent reference. Welds and seams must be 100% tested. Seams must be continuously probed with the test receiver set to detect abrupt changes of shielding level that are >10 dB on the shielding unit scale. Points having a change greater than 10 dB must be clearly marked and must have the weld repaired to meet the specified requirement. Each repaired point must be retested until no points on seams fail the test.

18.15   EM SHIELDING EFFECTIVENESS TESTING

The EM shielded enclosures must be tested. Test equipment used must have been calibrated within the last 12 months.

18.15.1   TEST PROCEDURE

The test procedures, equipment, and frequencies must be similar to those speci-fied in MIL-STD-188-125-1. Determine the test points (Table 18.1). Corner points of the grid should occur at the intersection of three planes (two wall surfaces and ceiling or two wall surfaces and floor). Measurement data at all test points must be recorded, and a grid map for each surface tested must be provided. For any test point where required attenuation is not provided, correct the discrepancy and retest. Both the results of the test failure and the successful results must be provided. The enclosure effectiveness test for magnetic attenuation must be performed with the antennas located directly opposite each other and separated by a distance of 600 mm (2 ft) plus the wall thickness. Plane wave attenuation tests must be performed with (1) the antennas located directly opposite each other; (2) the transmitting antenna placed 300 mm (1 ft) away from the enclosure wall; and (3) the receiving antenna set 300 mm (1 ft) from the wall for stationary measurements or 50 to 600 mm (2 in. to 2 ft) from the wall for swept measurements. The magnetic field test and the plane wave test must be performed as follows:

TABLE 18.1

Shielding Effectiveness Test Points

Testing Location	Test Points Spacing
Joints between steel panels for roof, walls, and floors	Test every 3 m (10 ft)[a] (minimum of one test point per side).
Corner seams for walls to floor, walls to roof, and wall to wall	Test every 3 m (10 ft)[a] (minimum of one test point per corner seam).
Corners (intersection of three surfaces)	Test at all corners in shield.
Single doors	Test at each corner; at midpoint of each side longer than 1.5 m (5 ft), and at center.
Double doors	Test each separately at same test point as single doors.
WBC vents and panels	Test in center (on-axis) for all sizes (including single); at all four corners if 300 by 300 mm (12 by 12 in.) or larger; and at the midpoint of each side longer than 1.5 m (5 ft).
At treated penetrations of shield (and entry panel and backshield)	Test as close to on-axis as possible, or orient for maximum signal.
All other shield joints, seams, or corners	Sweep all surfaces at one frequency in the range of 400 MHz to 1 GHz. Test every 3-m (10-ft) maximum plane wave.
Doors	Test door handles.
EM filter enclosures	Test at each seam corner and midpoint of each side longer than 1.5 m (5 ft) at center.
EM cabinets and enclosures	Test at each seam corner and each side 1.5 m (5 ft) on-center.

Source: NASA, *Electromagnetic (EM) Shielding*, Section 13 27 54.00 10, National Aeronautics and Space Administration, Washington, DC, 2007.

[a] Each point must be swept in space 600 mm (2 ft) around the point.

- Perform the calibrations at the beginning of each test day.
- Set up the test area for the 100- to 400-MHz stationary measurement on the two required polarizations.
- With the transmitter off, check the receiver sensitivity.
- Energize the transmitter and record the fixed measurement data.
- Remove the receiving antenna from the test stand and perform the swept measurement at the same frequency and transmitting antenna polarization.
- Rotate the transmitting antenna and perform the second 100- to 400-MHz stationary measurement.
- Perform the swept measurement for the second transmitting antenna polarization.
- Reconfigure the equipment for the 900- to 1000-MHz test frequency and repeat the series of four measurements.

To perform the swept measurement, remove the receiving antenna from the test stand and hold with a dielectric rod that is ≥300 mm (12 in.) in length. A dielectric spacer must be attached to the sweeping antenna to assist in maintaining the 50-mm

(2-in.) distance from the shield. A rapid sweep to locate hot spots must be made by rotating the polarization and waving the antenna through the specified volume. The final activity of each test day is to repeat the calibrations and verify the consistency with the previous calibration results. Test procedures include a definition of all test points, including (but not limited to) walls, door frames, accessible joints, and around filters and penetrations. Each EM door must be tested.

18.15.2 TEST POINTS

Additional test points must be measured in accordance with MIL-STD-188-125-1 for facilities requiring HEMP protection. Test points include the periphery of doors and covers, handles, latches, power filter penetrations, air vent filters, communications line filter penetrations, and points of penetration by pipes, tubes, and bolts.

18.15.3 TEST METHODOLOGY

Antennas must be oriented for maximum signal pickup. Each test point must be probed for area of maximum leakage (e.g., around door frames, accessible joints, filters, pipes, air ducts). The magnitude and location of maximum signal levels emanating from the enclosure must be determined for each accessible wall at a minimum of two locations on each wall, around doors, and at penetrations and seams of the enclosure. Measurement of attenuation must be accomplished.

18.15.4 TEST FREQUENCIES

The testing frequencies for shielded enclosures are shown in Table 18.2.

18.16 WELD TESTING

Structural welds to be embedded must be tested in accordance with AWS D1.1/D1.1M using magnetic particle inspection or dye penetrant inspection and 100% of the shielding seams by the SELDS testing prior to embedment.

TABLE 18.2
Fields, Waves, and Frequencies

Magnetic field	14 kHz, 400 kHz, and 10.1 MHz
Electric field	200 kHz and 16 MHz
Plane wave	100 MHz, 415 MHz, 1.29 to 18 GHz

MIL-STD-188-125-1 magnetic and plane wave frequencies are selected during the testing design phase.

Source: NASA, *Electromagnetic (EM) Shielding*, Section 13 27 54.00 10, National Aeronautics and Space Administration, Washington, DC, 2007.

18.17   GROUNDING

The contract drawings indicate the extent and general arrangement of the shielded enclosure grounding system. The grounding methods must be an equipotential grounding plane method in accordance with UL 1283,[*] NFPA 70,[†] NFPA 77,[‡] NFPA 780,[§] IEEE 142,[¶] MIL-STD-188-124,[**] and MIL-HDBK-419.[††]

REFERENCES

AWS. (2007). *Brazing Handbook*, 5th ed. Miami, FL: American Welding Society.

NFPA. (2005). *NFPA 70: National Electrical Code (NEC) Handbook*. Quincy, MA: National Fire Protection Association.

NFPA. (2012). *NFPA 70E: Standard for Electrical Safety in the Workplace®*. Quincy MA: National Fire Protection Association.

OSHA. (2007). Occupational Safety and Health Standards, 29 CFR, Part 1910, Subpart S, Electrical. Baltimore, MD: Occupational Safety and Health Administration.

---

[*] UL 1283 (2005), Electromagnetic Interference Filters.

[†] NFPA 70-2014 (AMD 1 2013; errata, 2013; AMD 2 2013), National Electrical Code.

[‡] NFPA 77-2014, Recommended Practice on Static Electricity.

[§] NFPA 780-2014, Standard for the Installation of Lightning Protection Systems.

[¶] IEEE 142-2007, Recommended Practice for Grounding of Industrial and Commercial Power Systems (IEEE Green Book).

[**] MIL-STD-188-124 (1998; Rev. B; Notice 2, 1998; Notice 3, 2000; Notice 4, 2013), Grounding, Bonding and Shielding for Common Long Haul/Tactical Communications Systems, Including Ground Based Communications—Electronics Facilities and Equipments.

[††] MIL-HDBK-419 (1987; Rev. A), Grounding, Bonding, and Shielding for Electronic Equipments and Facilities (Vol. I, Basic Theory; Vol. II, Applications).

19 Solid-State Technology

Martha J. Boss

CONTENTS

Use of solid-state technology will lessen the current and voltage differentials required to monitor and control circuits. Solid-state technology precludes open sparks, as all electrons are contained within the solid material. The equipment may be contained within potted units where circuitry is covered by a potting compound; however, this does not guarantee that no hot spots will exist.

19.1 SILICON-CONTROLLED RECTIFIER AND THYRISTOR FUNCTION

In a silicon-controlled rectifier (SCR) or thyristor, the outer two layers act as a *p-n* junction, and the inner two layers serve as an element to control that junction. The device has three external terminals: anode, cathode, and gate. Forward blocking occurs when an anode is positive with respect to the cathode (forward biased) and the gate is reverse biased in reference to the cathode. The balance of electrical charges exist in the four layers. Current flow is inhibited, and the SCR exhibits high resistance in both directions. If the gate is forward biased with respect to the cathode and the gate current upsets the electrical charge balance, anode-to-cathode current can flow. When this conduction starts, the gate loses all control. The gate can turn the thyristor on but not off. Gate control (or turning off the thyristor) is achieved by reverse-biasing the gate–cathode circuit, reducing the anode-to-cathode current essentially to zero. (*Note:* Voltages and currents are given with respect to cathode.) When the anode potential is positive, the gate is at cutoff (negative potential), with characteristics similar to those when the anode is negative (positive potential).

At breakover voltage (specific positive potential), the SCR will turn itself on and anode current increases considerably. The voltage drop is reduced substantially across the device. The breakover point can be altered by varying the gate current. A common technique for initiating conduction is to apply a current pulse at the gate. The pulse neutralizes the thyristor blocking action at a desired breakover voltage. The simplest thyristor application is half-wave rectifier. The circuit breakover is determined by gate current, and the average value of direct current (DC) through the load can be controlled by gate control pulse. In order to control alternating current (AC), the load must receive both sides of the AC waveform. Two single-phase circuits can be used as AC voltage controllers. A bidirectional arrangement uses two thyristors back to back. This arrangement gives a symmetrical output with appropriate firing-angle pulses for the two gates. A unidirectional circuit (with one thyristor and one diode) presents an asymmetrical waveform to load. The DC component created is a major disadvantage, because the source waveform must be ideal for the circuit to be of practical value. Voltage control for three-phase loads uses three single-phase controllers. The DC components are mostly canceled with three-phase waveform; however, unidirectional circuits introduce a higher harmonic content to line currents.

19.1.1   SIMPLE MOTOR CONTROL AND THYRISTORS

Commutation is the transfer of current from one device to another as voltage relations change. A commutation diode conducts armature current so the armature current will not be transferred to the thyristors; the diodes of the bridge and thyristors can then turn off at zero voltage. Without this diode, the thyristor current might not reduce to zero, and turning off would not be likely.

A chopper, or DC-to-DC converter, is a method of variable-speed DC motor control using thyristors that operate between the DC source and the load. The thyristors in chopper circuitry switch the DC source on and off, supplying unidirectional voltage pulses. This switching reduces the effective voltage to load.

A standard squirrel-cage induction motor can have variable-speed control by using a thyristor voltage control. The motor speed varies because the voltage is reduced and the motor allows greater percentage of slip. The controller then increases terminal voltage to the motor and allows lesser percentage of slip, so motor speed increases. Reduced-voltage operation causes more heat to be generated in the motor. Motor torque is reduced at slow speeds. The waveform produced by thyristors is rich in harmonics (frequencies other than line). These harmonics create even more motor heat. Squirrel-cage induction motors, however, do have effective starting if high-quality motors are used and reduced-voltage operation is restricted to a short time period.

19.1.2   CONTROL SYSTEMS

Control circuitry supplies the gate signal. The power circuits (thyristors) provide variable direct output or alternating voltage to the controlled system (any driven load). Two basic types of control systems are used:

- In open-loop systems, the controlling element is unaware of the effect on the controlled element.
- In closed-loop or feedback systems (regulators), the controlled element provides feedback. Appropriate feedback (current, voltage, temperature, speed) is supplied back to the controlling system. The controlling system responds to the feedback and input information. The controlling system supplies control signals to digital circuits. The digital circuits (acting as an additional transfer function) switch thyristors on and off.

19.1.3   PHYSICAL CHARACTERISTICS OF THYRISTORS

Thyristors that are stud-mount types are intended for smaller loads and have average forward-current ratings ≤150. Heat sinking is like that for diodes and, because thyristors have *p-n* junctions, the same heat dissipation theory applies to these devices. A larger thyristor (hockey puck, press pack, disk) can transfer junction heat on both sides and is clamped between two heat sinks. The heat-sink cooling is by either air or water. The single-disk thyristors have average forward-current ratings that are ≤7500 A.

19.2 THYRISTOR RATINGS AND PROTECTION

Thyristors are susceptible to damage from overvoltages, overcurrents, and rapid changes in voltage or current with respect to time. To protect thyristors against a high current change, proper design of the firing pulse is required. Thyristor ratings (oversizing based on ratings) and protective circuitry are used to protect thyristors. The thyristor is oversized to withstand most long-duration situations. These thyristors are used with thermal-overload sensors on thyristor heat sinks. When fuses are used, the fuse current limit is set near the failure point of the thyristor. A high rate of forward-voltage increase can turn a thyristor on even with a zero gate current. Inductive loads (induction motor belt drives) can cause voltage change with time problems. Either resistor–capacitor (RC) snubber networks across each pair of devices or metal oxide varistors (MOVs) across each device may be used to ensure that transient overvoltage does not destroy a thyristor.

19.3 ALTERNATING CURRENT APPLICATIONS

For an elementary inverter circuit, the incoming AC is rectified and then filtered by capacitance to provide high-quality DC. An inverter then converts the DC back to AC. Controlled switching of thyristors produces an alternating square-wave signal of the desired frequency. The signal can be used directly or filtered to provide a sinusoid. Three-phase AC output is obtained by using three inverters and firing inverters so that 120° timing is available.

19.3.1 STARTING CONSIDERATIONS

When starting an induction motor on a belt drive, consider the power system, motor, and mechanical equipment (speed reducers, belt). The motor should be started with as little current as practical to minimize overcurrent and undervoltage effects on the power system. Bring the motor up to full speed as quickly as possible to prevent insulation breakdown caused by rotor overheating. Smooth and easy acceleration of the belt reduces the wear and tear on the gears and the belt from excessive or uneven torque. Protection of the power system and mechanical components requires an extended low-level starting current, whereas protecting a motor from overheating requires a rapid-rise, high-level starting current. The area that is given priority determines which control system should be used.

A common design is the open-loop voltage ramp scheme without a current limit. The voltage starts at approximately 60% of line voltage, then gradually increases to 100% in a preset time limit. This scheme has no current-limiting capability and requires thermal overload sensing in the motor and simple circuitry for tripping the main circuit breaker when an overcurrent situation has exceeded a specified time limit. The open-loop voltage ramp scheme design allows for full locked-rotor torque and full locked-rotor current when needed by conveyor. Simplicity negates some of the advantages of soft starting (limiting starting currents and voltage drops). Conventional thermal overload relays do not provide adequate protection during long starting times caused by static control. Over-temperature detection devices installed in the motor provide thermal protection. Solid-state temperature sensors installed internally in motor windings, with a load-current detection backup, provide the best protection.

19.4 BELT STARTERS

Conveyor starters are an application of thyristor control of AC induction motors. Belt conveyor installations can call for high horsepower, resulting in a high starting current that can produce protective relaying problems and large voltage drops. Voltage drops can cause a motor torque decrease that can hamper belt conveyor acceleration. Controlled acceleration and limited starting current can be achieved using squirrel-cage induction motors with solid-state starters.

All three-phase solid-state belt starters use reduced-voltage motor starting. Remember that torque developed at the motor shaft is proportional to rotor current and square of terminal voltage. If terminal voltage is reduced to a given level to correspond with a desired torque value, then motor current is limited. Thyristors are used to control turning on power to the AC motor. Voltage is reduced by not firing the thyristors until some angle past the source voltage zero. This procedure effectively reduces the average voltage across the motor terminals to a value less than the source voltage. Firing control is obtained by either open-loop or closed-loop systems.

19.4.1 OPEN-LOOP SYSTEMS

Open-loop systems without feedback are simpler designs (voltage ramp or current ramp). The controller element brings the voltage or current up to a maximum in a predetermined time period that is independent of motor conditions. The maximum value of current in open-loop systems is limited by line voltage and motor characteristics, so control is limited to motors or power systems where starting current is not problematic.

19.4.2 CLOSED-LOOP SYSTEMS

In closed-loop systems, the reference signal or feedback from the motor is compared with a reference to adjust the controller output. The result is more control over motor starting characteristics and allowances for loading or operating conditions. For static belt-conveyor starters, the feedback signal corresponds to motor line current or motor speed. The line-current feedback control schemes are either current-limit or current-regulated types. Motor-speed feedback techniques are either linear acceleration or tachometer control.

In current-limiting schemes, the feedback control is used to compare the adjustable reference signal (current-limit setting) with a signal that represents motor current. Motor current is monitored by current transformers (CTs), preferably in all three lines. The torque being developed at the motor shaft is then computed by the starter circuitry from the current being used. The used current is limited to a preset value by thyristors. The resulting belt acceleration is a nonlinear (almost logarithmic) function of the belt load. When the belt is empty, a full-voltage start can occur.

In current-regulated schemes, the problem of a full-voltage start is removed by the addition of a second reference supplied to the control-system summing point. The second reference is an adjustable ramped reference signal, usually a linearly increasing voltage with time. The starter circuitry now restricts motor current to rise over a preselected time period (acceleration-time setting) to the current-limit setting. Motor acceleration is smoother than current-limit control but is still somewhat nonlinear.

19.4.3   Linear-Acceleration Starters

In basic linear-acceleration starters, the tachometer generator is placed on the motor shaft. The output of the tachometer generator provides a feedback signal proportional to either the motor speed (which is compared with a ramped reference) or acceleration (which is compared with constant preset reference). As a result, the motor voltage is adjusted to limit current, so the specified rate of acceleration is obtained. The acceleration is linear and constant regardless of load.

Some starters combine the linear-acceleration feature with an overriding current limit and current-regulation control. The starter tries to linearly accelerate the motor; however, current-regulation control keeps the rate of current rise within preset limits. The current-limit control keeps the maximum value of current below a certain level.

19.4.4   Start/Stop Circuitry

Start/stop circuitry is often relay controlled. When off, the relay contacts clamp the thyristor gates and the input of the control-system amplifier. When on (start command), the relay contacts sequentially unclamp the gates, allowing the control system to start the acceleration cycle. Thyristors do not physically disconnect the load. A circuit breaker is provided on the incoming line to physically disconnect the load and to provide short-circuit protection.

19.4.5   Thyristor Configuration

Thyristor configurations, unidirectional and bidirectional, are available for static belt conveyor starters. When in control mode, unidirectional configurations add higher harmonic content to line current. When applied to static induction motor starters, the harmonics tend to create excess heating during acceleration. The heating problem is reduced considerably if sufficient cooling time is allowed between starts and acceleration times are minimized. The main difficulty with unidirectional control is allowing DC to flow in the motor. Examples include

- During a motor ground fault, diodes can rectify the current and direct current can flow through the grounding resistor and grounding conductor without being detected by zero-sequence ground-fault relaying.

- Thyristor failures are rare and always occur in shorted mode. If thyristor fusing is not available, a low-impedance path for direct current exists through the motor windings and the other two diodes.

Bidirectional control is better, as two thyristors are always in series from line to line, with resultant less stress given per thyristor by transient and long-term overvoltages. Because of the symmetry, smoother control is provided under varying conditions. Motor windings and cable power conductors are near neutral potential when thyristors are off.

19.5   DIRECT CURRENT APPLICATIONS

Solid-state or static control can be applied to DC equipment, especially equipment with <100 connected hp. The solid-state or static control primarily uses chopper circuitry. Advantages include increased motor life by limiting armature current during acceleration, thus reducing brush and commutator problems, as well as significantly reduced drive-train maintenance. The general mechanical problems are reduced because motors have controlled acceleration even during plugging (worst shock-loading instance). Motors have controlled deceleration, reversal, and then acceleration. For battery-powered vehicles, battery life is increased because peak current demand is reduced and power is generally used more efficiently. High-amplitude voltage transients and significant electrical noise are one disadvantage.

Static control of the DC drive maximizes power-distribution efficiency, uses the traction advantages of DC series-wound motors, and has improved performance available from thyristor control. The transformer secondaries can be used to supply AC for DC systems that are using thyristor control of traction motors.

19.6   FIRING CIRCUITS

Sustained-pulse and multipulse are used. A pulse transformer is used in a multipulse system as isolation between the power circuit and control circuit. Numerous pulses each capable of turning the thyristor on are applied to a gate to keep the thyristor on during the thyristor's intended conduction period. The DC firing system is another technique that maintains a continuous gate current during the intended conduction period. The continuous gate current helps ensure (by continuous stimulation of the gate) that the thyristor will turn on and will turn on completely. Important parameters of the firing pulse are the level of current and the current duration. If the pulse is only of minimal current and duration, the pulse may cause conduction only in a limited area of the thyristor or no conduction. The small conduction area, coupled with a high rate of change of anode-to-cathode current, can result in concentrated heating and possible failure of the thyristor. To ensure conduction of the entire interface of the thyristor, hard firing is used in conjunction with DC or multipulse firing. This firing consists of a high-level initial pulse with a steep wavefront and enough duration to operate the thyristor at near the maximum input power level of the gate. The high initial gate current floods the conduction region of the thyristor and turns the thyristor on completely. After the thyristor is conducting, a sustained DC pulse or multipulse keeps the thyristor on even if the anode-to-cathode current approaches zero. This sustained pulse prevents thyristor turnoff during the critical end of the conduction cycle. Hard firing provides more consistent operation and reduced failures of thyristors.

19.7 SOLID-STATE/STATIC AND ELECTROMECHANICAL PROTECTIVE RELAYS

Solid-state/static relays replace the mechanical contact device with a solid-state device. The solid-state device is inserted in the trip-coil circuit and controlled by a sensor circuit. Solid-state devices include the transistor, thyristor, and triac. When unactuated, the device acts as very large resistance in the trip-coil circuit. This resistance limits the current through the trip coil to a very small value (leakage current). The leakage current is not capable of tripping the associated circuit breaker. When actuated, the device acts as a very small resistance in the trip-coil circuit. This resistance does not sufficiently limit the current through the trip coil. The result is ample current that is capable of tripping the associated circuit breaker.

19.7.1 SOLID-STATE RELAYS

Devices include continuous, digital-controlled continuous, and digital. In continuous relays, the power system is sensed continuously. The relaying becomes activated from sustained existence of a malfunction. As an example, a full-wave rectifier and CT continuously sense power-system operation. Particular relaying characteristics are provided by a function generator with a time delay provided by a linear-ramp generator. The pick-up level and ramp level detectors essentially function as digital control devices and operate in a go or no-go manner. The digital relays rely on discrete sampling of current and/or voltage waveforms, directing the data through an algorithm and decision process that ultimately decides whether the relay is to be actuated.

The ultimate in digital relays are microprocessors distributed throughout a power system to analyze circuit conditions on a real-time basis. Each microprocessor is responsible for its system portion, with control relinquished when demanded by a larger computer or when power-system conditions exist that are beyond the microprocessor's analytical capability. Static relays currently occupy only a small portion of protective relaying. The main benefits for the mining industry, for example, are decreased burden problems, increased sensitivity for zero-sequence relaying in resistance-grounded systems, and use as solid-state tripping elements in molded-case circuit breakers.

19.7.2 SOLID-STATE AND HYBRID RELAYS

In a solid-state relay, the load current flows through a common terminal of the semiconductor device that is common to trip (contact) and sensor (control) circuits. The transistor is a current-controlled device. If zero input (control) voltage is applied to the transistor base, no current will flow into the base and consequently no current will flow from the collector to the emitter. Thus, no current will pass through the load, and the contact-circuit voltage supply will not be dropped across the load. The voltage must, therefore, be across the collector–emitter terminals of the transistor base. With positive input, the voltage applied to a transistor base causes a positive current to flow into the base. This base current is the controlling factor that permits a large collector-to-emitter current to flow through the transistor base. With positive input, when sufficient base control current is supplied, the collector current will increase until all the contact-circuit voltage is across the trip coil (load) and switching characteristics are obtained. This circuit has a lack of electrical isolation between the control and the trip circuits.

The photon-coupler circuit transistor is an isolation method. This transistor is light controlled with a high-intensity light (i.e., LED) acting as a large base current. Light impinges on the phototransistor base region, allowing current to flow from the collector to emitter. Without an input to the LED, no light is produced and consequently no collector-to-emitter current flows.

The triac (triode AC semiconductor) is a bidirectional solid-state device that acts like two thyristors connected back to back. A triac provides full-wave voltage control in one solid-state structure with only one gate control. The load-voltage characteristics are like those exhibited by a thyristor. The single structure of the triac has heat dissipation limitations that restrict the triac to small current applications.

The output and input circuits of triac and thyristor relays have a common terminal between them. The hybrid relays are used when input/output isolation is required and a solid-state relay is needed. Hybrid relays use a low-power operating coil in either the input or output stage and solid-state device functioning in the other stage. An example hybrid relay has a reed relay in the input stage. Switching the reed relay activates a gate control input that fires the triac. An example hybrid circuit uses a solid-stage input stage that can react to power network conditions. When a predetermined threshold is reached, the solid-state stage sends a current through the relay coil that will pick up the relay contacts. *Note:* The best reed contacts are mercury wetted and provide bounce-free operation. The dry reed relays with heavy-duty silver contacts will carry 2 kW with a maximum of 30 A and can withstand 5 g (50 m/s^2) when mounted correctly.

Electromechanical and static relays are classified according to actuating quantity. Overcurrent, overvoltage, and undervoltage relays are used. The overcurrent or overvoltage relays trip when a predetermined current or voltage threshold value has been exceeded. The undervoltage relays trip whenever measured voltage falls short of a predetermined level. In an example static operation overcurrent relay, the alternating current from a CT is forced through a resistor. This resistor produces an alternating voltage across the input. The voltage is rectified by the bridge and applied to the series RC. When the capacitor voltage exceeds the battery voltage in base-emitter circuit by sufficient amount, the transistor begins to conduct. The current picks up a trip coil, tripping the circuit breaker. A time delay is achieved by the series RC time constant associated with the base-emitter circuit.

19.7.3 Arcing

Closing or opening of a contact occurs for only a small portion of the contact's service life; however, this movement is the essence of the difference between static and electromechanical relays. Potential arcing can occur due to the operation of electromechanical relays. Arcing does not occur in static relays (because of the nature of solid-state contact). High-inductive circuits will usually cause contact arcing during separation. Arcing can pit contact surfaces, giving a higher value of contact resistance. This arcing is particularly dangerous in explosive environments. Closing contacts are subject to contact bounce that can also cause arcing.

19.7.4 ELECTROMECHANICAL RELAYS

The physical separation of open contacts of an electromagnetic relay provides a very large resistance in the trip-coil circuit, which cannot be matched by a solid-state device. Electromechanical contacts when closed normally have only a few milliohms of resistance to control-power current. The input/output isolation resistance coupling of electromechanical relays provides a very large isolation resistance between input and output. Moving and stationary mechanical contacts are subject to nuisance tripping due to external vibrations (especially bothersome with portable power equipment used in mining). The mechanical nature and diverse configurations of electromechanical relays make their response difficult and costly to predict.

19.7.4.1 Power

Electromechanical relays are normally limited in voltage range and require different coils for AC and DC operation.

19.7.4.2 Temperature

Electromechanical relays require no heat sink, because in an underrated condition the contact resistance measures in the milliohm range. The maximum allowable continuous junction temperature for thyristors and triacs is 120°C. Electromechanical relays may be used in ambient temperatures that exceed 120°C. Temperature has little effect on electromechanical characteristics.

19.7.4.3 Transients

Trip circuit control-power transients may cause the false operation of an electromechanical relay that relies on control power to keep the relay closed under normal circuit conditions.

19.7.4.4 Ratcheting or Overtravel

Ratcheting is the accumulation of overtravel due to successive faults or sequential starting of motors. Electromechanical relays may trip erroneously due to ratcheting or overtravel. Mechanical inertia causes overtravel of induction-disk or induction-cup relays. Overtravel may occur even after the actuating quantity has been removed. If the relay is not set properly, overtravel may cause undesired tripping of a backup relay, thus isolating a sound part of power system being protected.

19.7.4.5 Breakdown Voltages

The dV/dt current is not a problem with electromechanical relays. Breakdown voltages of these relays are normally many times their operating voltage because of the air gap between mechanical contacts.

19.7.4.6 Exposure to Contaminants

Electromechanical units have contacts that close mechanically and electrically. Open electromechanical contacts are exposed to the surrounding atmospheric conditions. Settling dust can act as an abrasive agent and cause premature contact failure. Chemical contaminants promote oxide-layer production, which can also increase contact resistance.

19.7.5 COMPARISONS

Static relays are much faster than electromechanical relays because no mechanical motion is needed in solid-state devices. The solid-state device's high speed is advantageous. This speed gives static relays the capability to turn on at zero-voltage crossing. The input/output isolation resistance coupling of electromechanical relays provides a very large isolation resistance between input and output. Hybrid relays (static solid-state and electromechanical) can match the resistance of electromechanical arrays. Straight static devices can approach a value of between 10^{10} and 10^{11} ohms if a photon coupler isolator is used. Solid-state relaying is used for ground-check monitoring, especially when connected to portable equipment. Static protection relays have been applied successfully for common AC distribution areas:

- Line short circuit and overload
- Undervoltage
- Phase loss and phase sequence
- Zero-sequence ground fault
- Grounding resistor overvoltage

The replacement of electromechanical devices with static devices has led to

- Greater environmental resistance
- Easier adjustment and testing
- More sensitive ground-fault protection
- Lower maintenance

Solid-state relaying static overcurrent relays can provide safe, fast testing.

19.7.5.1 Radiofrequency Interference

The static relay eliminates much of the radiofrequency interference (RFI) caused by capacitance switching or prestrike (found with mechanical opening and closing of circuits with large capacitance values). Controlling RFI is important because digital circuits are easily affected by such interference.

19.7.5.2 Power

Static relays require far less input power than electromechanical relays. Input power delivered to any one static device may come from a wide range of voltages, including AC and DC. Static relays can use a range of AC and DC power supplies in their tripping circuit. Leakage current flowing through a solid-state device when in a high-impedance state can be 10 to 20 mA. Static relays should not be used where this current magnitude cannot be tolerated. When closed, a static relay's thyristors or triacs can provide a 10- to 15-V voltage drop.

19.7.5.3 Temperature

Heat production in solid-state relays results from a 1.0- to 1.5-V drop across the device. In the on position, power consumption is limited by the current required for the tripping operation. Static relays used for protective-relaying applications usually require only 5 A, which is insufficient to produce enough heat for concern. Power relays carrying between 5 and 25 A may require an auxiliary heat sink or special mounting arrangements. Above 25 A, the relay becomes a contactor even though its function does not change. A static relay above 25 A is considered a solid-state contactor.

Solid-state devices exhibit different characteristics at elevated temperatures. If subjected to these temperatures for long periods, the characteristics of the solid-state device may change permanently. Short-range thermal compensation is accomplished by use of thermistors in the design or using a differential-stage construction. The long-range effects of elevated temperature are guarded against by pre- and post-assembly heat soaking and testing.

If contacts become so degraded that contact resistance causes excessive heat generation, the contacts may weld closed. Moderate overloads can be tolerated for short periods. Solid-state relays underrated for a particular application will fail catastrophically.

19.7.5.4 Transients

Static relays are fast-responding devices that rarely trip on power-system transients, unless set to do so. Voltage spikes on the control-power circuit can cause unwanted tripping of static relays. If the transient is of sufficient magnitude, the solid-state device may break down, thus allowing conduction. This problem has been overcome by triacs with blocking voltages of ≥1000 V.

19.7.5.5 Gate Current

If the voltage spike has a high rate of rise, the inherent capacitance of the thyristor or triac will allow a momentary current to flow. With sufficient magnitude, this current will act as a gate current and turn on the device. This *dV/dt* problem is being overcome by including filters capable of absorbing energy contained in these transients.

19.7.5.6 Ratcheting or Overtravel

Static relays have negligible overtravel which precludes unneeded disconnects and makes possible closer time settings on backup relays to provide faster protection. Static relays do not exhibit ratcheting because of the negligible overtravel and short resetting time of the static relay.

19.7.5.7 Exposure to Contaminants and Seismic Issues

Static units have contacts that close electrically. Static relays are usually encapsulated in epoxy. Static relays are much less susceptible to seismic disturbances because of their epoxy resin covering, physical structure, and structural arrangement.

19.7.5.8 Cost

Additional contacts can only be added to static units by costly duplication of solid-state circuitry portions. Interlocks and remote indicators are more expensive in static relays (these tasks may be performed more economically by using digital logic on the control side).

19.7.5.9 Accuracy

Static relays are more precise than their counterparts. Common induction-disk relays are adjustable through 11 discrete tap settings in one range only. The typical static time–overcurrent relay can have an additive tap-block arrangement that permits numerous tap settings in more than one range. Static relay frequency response is uniform and should make the performance of static relays more predictable.

19.8 CURRENT TRANSFORMERS

Conventional thermal-magnetic trip elements in circuit breakers can be replaced by current transformers (CTs) and solid-state circuitry. The CTs proportionately reduce line currents to a level that can be used as input to solid-state circuitry. Breaker tripping is initiated when a low-power flux-transfer shunt trip or undervoltage release (UVR) is activated from the output of the solid-state circuitry. The overload rating can be altered by simply changing the small rating plug on the front of the breaker. The instantaneous-trip range is specified as a multiple of the overload-current setting.

19.8.1 BURDEN

Burden is an external load applied to the secondary of a current transformer. High CT burdens can cause core saturation due to the magnitude of voltage developed at the CT secondary terminals. Saturation causes a burden reduction because, under deep saturation, the burden approaches its DC resistive value. The burden change modifies relay characteristics and can cause incorrect operation. The possibility of CT saturation is of particular importance when low-ratio CTs cause large secondary currents (large secondary voltages) and when large burdens from additional secondary loads cause large secondary voltages from even moderate secondary currents. Relays presenting lower burdens are clearly advantageous and can be set on the minimum tap setting.

The static relay burden includes that of the relay current-measuring portion and the power-supply portion (because the relay receives its operating power from the CT burden). The burden of the current-measuring portion of the static relay is (1) much less than that for the electromechanical and (2) greater than the electromechanical burden at pick-up. The lower static burden offers the following advantages:

- More relays can be connected in series.
- The lowest tap of a multiratio CT can be used.
- Step-up auxiliary CTs can be used for greater sensitivity.
- More auxiliary burdens can be connected without CT saturation.

The low CT burden will permit use without saturation of a less expensive CT that has no relay classification.

19.9 SENSITIVE EARTH LEAKAGE SYSTEM

Sensitive earth leakage (SEL) is considered in order to reduce the dangers of incendive arcing from damaged cables and for ground-fault protection. SEL substantially reduces the chance of an electrocution by limiting ground-fault currents to extremely low values.

- Neutral-grounding impedance limits maximum ground-fault current to 750 mA.
- Fault detection is by a zero-sequence scheme. A solid-state amplifier increases sensitivity by ≤90 mA (compared to 4 to 6 A for typical electromechanical devices used on voltage on low-voltage systems in the United States).
- Currents less than 90 mA are allowed to flow continuously.
- Currents that are ≤750 mA are permitted for about 0.02 s. To provide selectivity, a 0.4-s time delay is introduced at the main circuit breaker.

The power factor of neutral grounding (earthing) impedance is normally specified from 0.65 to 0.75 (to avoid limiting current to a level that cannot be detected because of system capacitance). The circuit is a solid-state amplified, zero-sequence relay. Because of high sensitivity, grounded shields are placed over the CT to reduce electromagnetic interference that can cause nuisance tripping.

To prevent the circuit from being reset until a ground fault is cleared, the SEL system has an auxiliary circuit connected to a second winding on the CT. Upon a ground-fault pick-up, an auxiliary contact is closed by the relay, which in turn causes an auxiliary CT winding to be energized. This energized winding then induces a voltage on other CT windings that creates a lockout.

The multipoint SEL is another ground-fault method. A false-neutral transformer that is impedance grounded replaces the zero-sequence transformer. The source-transformer secondary is isolated from the ground across a spark gap. When a ground fault occurs, a potential is developed across the wye-connected impedances (false-neutral transformer). Current flows through the grounding impedance. Voltage developed across the impedance is amplified, causing the relay to pick up. An auxiliary changeover contact provides lockout until the fault is cleared.

The multipoint system has several disadvantages. The technique is indiscriminate and limits the number of units that can be utilized at a gate-end box (utilization center). The maximum number of units for a 550-V system is 37; for a 1100-V system, it is 18. The ground-fault current again is limited to ≤750 mA. However, all relays see fault current produced and will pick up (a definite drawback). The unfaulted units may be reset at once, and the faulted unit will be locked out. In the United Kingdom, automatic circuit-breaker resetting is allowed on this system to restore operation to the unaffected portion. The sensitivity of a multipoint system is excellent, with pick-up of ~3 mA on 550-V systems and ~6 mA on 1100-V systems. The high sensitivity is desirable in case there are two simultaneous ground faults on separate lines.

19.10   PHASE-SENSITIVE SHORT-CIRCUIT PROTECTION

Both motor starting and fault conditions result in large current flows of comparable magnitude. Standard relays must be adjusted to differentiate between starting and fault conditions so as to provide interruption only when a fault occurs. Phase-sensitive relaying can distinguish between starting and fault conditions by sensing the phase angle between current and voltage. The induction motor power factor at starting is ~0.5, whereas typical faults have power factors of about 0.9. Using phase-sensitive short-circuit protection, nuisance tripping on motor starting can be eliminated and short-circuit trip settings can be reduced. Two techniques can be used to provide protection: (1) diode bridge or (2) electronic comparator.

The circuit electronic phase-sensitive protection is tripped at much lower levels and in shorter times than standard short-circuit devices. The typical instantaneous relays may require a pick-up setting of 7 to 10 times normal full-load current to prevent nuisance tripping on motor start. The phase-sensitive relaying eliminates spurious tripping during motor start even when pick-up is as low as three times the full-load current (Morley, 1990).

REFERENCES

Morley, L.A. (1990). *Mine Power Systems*, Information Circular 9258, NTIS No. PB 91-241729. Philadelphia, PA: U.S. Bureau of Mines.

NFPA. (2005). *NFPA 70: National Electrical Code (NEC) Handbook*. Quincy, MA: National Fire Protection Association.

OSHA. (2007). Occupational Safety and Health Standards, 29 CFR, Part 1910, Subpart S, Electrical. Baltimore, MD: Occupational Safety and Health Administration.

20 Motors

Martha J. Boss

CONTENTS

Electric motors convert electrical power into mechanical force, which is used to produce rotation or torque. Motors include two concentric, cylindrical, laminated-iron cores separated by an air gap to carry magnetic flux, and two sets of windings wound or embedded in slots in iron cores with either or both excited by direct current (DC) or alternating current (AC). Inactive motor elements include the frame, end bells, and bearings. Motor speed is a function of applied power frequency. Speed control can be obtained by varying this power frequency. Load-speed and load-torque requirements are tied to the torque-speed characteristics. Torque developed at the motor shaft is proportional to rotor current and the square of the terminal voltage. Motor type classification is dependent upon how the stator or rotor windings are excited. The three general motor classes are induction alternating current, synchronous alternating current, and direct current. AC motors can be used effectively for the majority of motor applications, except when very high starting torques are required.

20.1 ALTERNATING CURRENT INDUCTION MOTOR

With induction motors, AC power is provided to the stator windings and induction occurs. This provides AC to the rotor winding. Full variable speed cannot be accomplished on AC induction motors with straight current or voltage control. Varying the power frequency, however, can create a variable-frequency AC drive.

20.1.1 STARTING

Most induction motors are started by directly connecting them to a power system. The system usually has enough impedance that protective devices can be set above inrush current to prevent nuisance tripping. For induction motors, full-voltage starting can usually be performed on 440- to 550-V motors up to 1600 hp with National Electrical Manufacturers Association (NEMA) standard magnetic starters. However, many systems (e.g., conveyor belt drives) cannot take the shock of full-voltage starting.

20.1.2 REDUCED-VOLTAGE STARTING

Above 1600 hp, full-voltage starting becomes impractical even when the load is connected to a motor that can withstand stress. A common method for starting these large induction motors is reduced-voltage starting. Autotransformers are used to start the motor at reduced voltage (50 to 80% of rated), thus limiting starting current and torque. When almost at full speed, contactors quickly change the motor from autotransformer to full-voltage supply.

20.1.3 WOUND-ROTOR MOTORS AND VARIABLE SPEED

A wound-rotor motor has insulated windings like a stator with the same number of poles and windings placed in the rotor slots. The starting and running characteristics of the induction motor may be adjusted by varying the resistance-to-reactance ratio of the rotor conductors (instead of rotor bars and end rings). External resistance can be used to vary speed-torque characteristics by changing the rotor resistance-to-reactance ratio. Wound-rotor induction motors can be considered variable-speed machines. Applications for wound-rotor motors include the following:

- Loads that require constant-torque variable-speed drives
- Where a sequence of slow-speed steps is required to limit motor current during acceleration for high-inertia or high-torque loads

These motors are suited to high-torque loads and are used in crushers, grinders, ball and roller mills, conveyor belt drives, and hoists.

20.1.4 SQUIRREL-CAGE MOTORS WITH SOLID-STATE STARTERS

Squirrel-cage induction motors have no slip rings, commutators, or brushes to wear out and use the simplest kind of starting equipment. *Note:* Wound-rotor induction motors are being replaced by squirrel-cage induction motors equipped with solid-state starters. This change is to eliminate the failures inherent with brushes, slip rings, and relay contacts.

20.2 THREE-PHASE ALTERNATING CURRENT SYNCHRONOUS MOTORS

Synchronous motors correspond to three-phase generators. AC power is provided to the stator armature windings. An external DC source (exciter) excites the field windings in the rotor. The rotor consists of field poles connected in series, parallel, or series–parallel combinations, which are terminated at slip rings. The number of field winding poles equals the number of magnetic poles present in the stator. Rotor field excitation is often supplied from a small DC generator mounted on the same rotor shaft. Alternatively, DC supply is obtained from a three-phase, full-wave bridge rectifier or by a separate motor–generator set.

20.2.1 General Method Used for Starting

The general starting sequence is as follows (of course, many variations can occur):

1. The start button is pressed to energize the control relay, which
   - Closes one set of contacts.
   - Electrically locks in the start sequence (terminated by pressing the stop button).
   - Energizes the motor relay so that the motor relay contacts close.
2. Three-phase power is applied to the stator winding. The machine accelerates as an induction motor. The motor is allowed to accelerate to maximum induction speed, where slip between the stator rotating field and rotor is very small.
   - The resistor and inductor are placed across the field winding with an FR relay across the inductor. Inductance of the relay coil is selected to be much lower than that of the inductor.
   - Immediately after starting commences, a high-frequency potential is induced in the field winding. The majority of the current flows through the resistor relay coil because the inductance exhibits a high reactance.
   - The motor further accelerates and the frequency of induced current decreases. At close to synchronous speed, the frequency has decreased to the point where most current flows through the inductor.
6. A switch is closed to apply direct current to the rotor field winding. The steady rotor electromagnetic field established can lock in step with the rotating field of the stator motor, which will turn at synchronous speed.
   - Voltage is reduced to the point where the FR relay cannot hold its contacts open.
   - The FR contacts close, the FS relay is energized, DC excitation is applied to the field winding, and the resistor is removed from the circuit.

20.2.2 Automatic Excitation of DC Field

If direct current is applied before the maximum induction speed is achieved, the rotor may not pull into synchronization. Severe vibration can occur, caused by repulsion every time the rotating pole passes the stator pole. Consequently, most synchronous-motor starters do not rely on manual control and instead automatically excite the rotor field at the appropriate time. Thus, synchronization is based on the frequency technique, which uses voltage induced in the field winding during acceleration and before direct current is applied.

20.2.3 Load-Delivered Torque

If a load delivers torque to the motor shaft, the rotor produces generated voltage, which may exceed supply power delivered back or may be regenerated into the line. If the supply voltage is then removed, the load acts as a prime mover. The voltage must be generated as long as field excitation exists and until the load dissipates its energy. This consideration is important when supply voltage is lost (short circuit), as the synchronous motor can deliver significant current to malfunction. Due to the possibility of load-delivered torque, machines with high rotational speeds should always be analyzed to determine if power generation is likely. If so, during lockout/tagout and hazardous energy control procedures thereof, the time required for the dissipation of this energy must be allowed prior to any maintenance activities that require a zero-energy state.

20.2.4 Pure Synchronous

Pure synchronous motors are not self-starting and are generally accelerated in the same manner as inductor motors. Salient-pole rotors commonly have squirrel-cage windings to produce the necessary induction motor action.

20.2.5 Cylindrical Rotors

Low-speed cylindrical rotors closely resemble the wound-rotor induction motor but with five slip rings (three rings are used for the wound-rotor circuit and the other two for the DC field). Cylindrical-rotor motors can provide high starting torque to accelerate high-inertia loads. Some large synchronous motors are accelerated by a small induction motor mounted on the synchronous-motor shaft.

20.2.6 Motor Static Capacitors and Silicon Rectifiers

Elementary motor static capacitors have replaced synchronous motors for power-factor improvement, flexibility, and ease of installation. Silicon rectifiers have replaced motor–generator sets for power conversion. Due to the possibility of capacitance essentially storing energy, machines should always be analyzed to determine if capacitance feedback Is probable. If so, determine how to discharge this capacitance. During lockout/tagout, the methods used to discharge this capacitance must be included as required steps in the machine-specific hazardous energy control procedures.

20.3 DIRECT CURRENT MOTORS

Essential motor parts include armature (rotor), commutator, main field frame, and field windings (stator). Actual DC industrial motors have many commutator segments, armature conductors, and main field poles. Elementary two-pole DC motors have a one-coil armature.

1. The main field is supplied by direct current through field windings (stators), with energy expended in these windings forming a power (I^2R) heat loss.
2. With armature (rotor) current flow, the armature reaction (distortion of the main electromagnetic field by the rotating armature field) produces forces on the conductors. The main field provides the necessary medium for the armature windings to push against when developing rotary motion.
3. Torque results in clockwise rotation.
4. The commutator acts as a switch to reverse the armature current each time the conductors pass the neutral plane. To reverse the armature rotation, armature current flow is reversed.

20.3.1 Motor Control

One method of speed control in DC motors is by adjusting the DC voltage to the machine. This method applies to loaded series-wound motors. Shunt and compound DC motors can be controlled by two direct means: field voltage or armature voltage. For field voltage/control, field excitation is usually considered constant; however, motor speed can be adjusted inversely with the field current.

20.3.2 Interpoles or Commutating Poles

Interpoles (or commutating poles) are mounted between field pole windings and are connected in series with the armature. These interpole windings produce a small electromagnetic field. The electromagnetic field produced opposes the main field (armature reaction) in the same plane as the brushes and thus reduces the electromagnetic field cut by armature conductors undergoing commutation (current reversal). The electromagnetic field reduction leads to reduced brush sparking.

20.3.3 Compensating and Stabilizing Windings

The armature reaction can be further neutralized through use of compensating or stabilizing windings. These windings are placed in slots on the ends of the main field poles next to the armature and are connected in series with armature windings. This arrangement is especially useful in motors intended for variable-speed or reversing operation.

20.3.4 Shunt Motors

Exactly like DC generators, DC motors can be connected as separately excited shunt, series, and compound. The performance of a separately excited motor is similar to that of the shunt. In shunt motors, the main field winding is connected in parallel with the armature field winding. Because the field winding is connected across the supply, resistance must be rather high; however, because of space constraints, armature windings have much smaller resistance.

1. When the motor is energized, the armature current is limited only by the armature's winding resistance; consequently, the armature current is much higher than the field winding current.
2. As soon as the armature begins to rotate, the armature's conductors cut the main field.
3. The magnetic flux is counter-electromotive force (cemf) generated in the windings. This cemf opposes the applied armature voltage and begins to limit armature current.
4. As the armature accelerates, the cemf rises and the armature current drops. The motor torque then decreases.
5. Final speed is reached when the cemf is almost equal in magnitude to the supply voltage.

Under motor loading, the armature slows down, cemf decreases, and current enters. The armature speed of the shunt motor remains relatively constant from no-load conditions up to 100% rated. Speed can be adjusted by changing the resistance in series with the field winding. Weakening field flux by decreasing the field current increases the motor speed. For a constant field flux, torque varies linearly with the armature current.

20.3.4.1   Starting and Controlling

If across-the-line starting was attempted with a shunt motor, the cemf would not build up fast enough to limit armature current and safeguard the commutator brushes and armature winding. For DC motors (except fractional horsepowers), starting resistance used in series with the armature is selected to limit armature current from 150 to 250% of the rated current. Shunt winding is always connected across full line voltage when starting. Use of fixed-resistance starting has widespread application as the starting resistance remains in series with the armature for running. Although this resistance gives poor speed regulation, the motor can be started unattended.

20.3.4.2   Braking

If a shunt motor is unloaded, the difference between the terminal voltage and the cemf will allow only enough armature current to overcome the friction, winding, and core losses. If a shunt motor is running under load and the armature circuit is opened, inertia of that load will drive the machine as a DC generator. Dynamic braking connects resistance across the armature to dissipate available energy and decelerate the load. Braking action is most effective at high armature speeds, becoming negligible at low speeds. Due to the possibility of inertia driving the machine as a DC generator, machines should always be analyzed to determine if power generation is probable. If so, the time required for the dissipation of this energy via braking must be considered prior to any maintenance activities that require a zero energy state.

20.3.5   Series Motor

The armature and main field winding are connected in series and both carry load current. Magnetic flux is produced in the main field winding proportional to the armature current. The torque varies as the square of the armature current. The main field strength will change with the load, causing a speed decrease with increased load.

- When a series motor is started, the cemf builds up as armature speed increases. During initial acceleration, the cemf is small and the armature and field current are high.
- After the power-source contacts close, the motor is accelerated with both resistances in series with the armature. The same control that activated the power source contacts simultaneously energizes a definite-time relay after the preset time (about 1.0 s). This relay closes its contacts, which creates a path that shunts the starting resistors.

At light loads, the motor speeds may become excessively high. Series motors must be directly connected to loads that cannot be removed freely or the motor may race to destruction. DC series-wound motors are used due to the speed–torque characteristics of these motors. A series machine has the best traction or starting characteristics and is the most-used motor for traction purposes in vehicles (diesel-electric trucks).

20.3.6 COMPOUND MOTORS

Cumulative compounding combines the characteristics of both series and shunt motors. Compound motors have both shunt and series field windings installed on the same poles. Series windings may be differentially or cumulatively compounded. Subtracting from or adding to magnetizing force of the shunt field causes either reduced or increased armature speed with load. Cumulative compounding gives greater torque than possible with a simple shunt motor because of the greater amount of main field flux available. Increased flux causes the speed to drop off more rapidly than for the shunt motor but not as much as for the series motor. The cumulative compound motor will develop high torque with any sudden increase in load. At light loads, the motor will not run away because the shunt field provides constant field flux. Cumulative compound motors are often applied to loads requiring high starting torque and fairly constant operating speed under normal conditions.

A differential compound motor produces torque that is always lower than that of the shunt motor. The amount of series winding can be adjusted to offset any drop in speed as loading increases or to give a slightly higher speed than normal at full load. A flat compounded motor has a constant speed from full load to no load. An overcompounded motor has a slightly higher speed than normal.

20.4 DIRECT CURRENT SINGLE-PHASE MOTORS

Direct current single-phase motors find widespread use for auxiliary functions. Single-phase induction motors have one running speed and require a separate means for starting rotation. Usually a separate stator or starting winding motor is classified by its starting method: split phase or capacitor start.

20.4.1 ROTATING STATOR FIELD

When single-phase AC voltage is applied to one stator winding, current flow produces an electromagnetic field with a resultant direction that alternates online. If a squirrel-cage rotor winding is in the stator field, voltage must be induced in the rotor conductors. Current produced will create an electromagnetic field that coincides with the stator field. As no magnetic interaction occurs, no torque is developed and the rotor remains stationary.

If the rotor is moved by some means, the rotor conductors cut the stator electromagnetic field. Induced voltages are in phase with the current through the stator winding. Rotor winding impedance appears as almost pure inductance. Rotor current will lag induced voltage by almost 90°. The rotor electromagnetic field is now

90° from the stator field and is termed a *cross-magnetizing field*. The rotor and stator fields combine to produce a resultant field that rotates at synchronous speed. Cross-field strength is proportional to rotor speed and about equal to the stator field strength at synchronous speed. The same operational principles for three-phase induction motors hold for single-phase motors. Slip must always exist between the rotating field and rotor because of the cross field. Slower rotor speed causes the rotating field to pulsate. With single-phase induction motors, vibration and noise are inherent.

20.5 SPLIT-PHASE STARTING

Split-phase motors have two stator windings connected in parallel. The impedance of each winding is such that currents through the windings are out of phase. One winding (auxiliary or starting winding) is usually constructed of small-gauge wire and has high resistance and low reactance. The other winding (running or main winding) has a heavier gauge conductor, low resistance, and high inductance. When energized, the phase angle between the currents through the two windings is only about 30° (enough to produce a rotating electromagnetic field). The rotating field pulsates and the starting torque is small. Once the motor is started, a rotor cross field is produced. The starting winding is no longer needed and is usually disconnected when the rotor speed reaches 70 to 80% of synchronous speed.

20.6 CAPACITOR-START MOTORS

The capacitor-start motor also has two stator windings. The main winding is arranged for direct connection to the power source. The auxiliary winding is connected in series with the capacitor. Currents through the two windings can be as high as 90° out of phase. The starting torque can approach 100% of rated. The starting winding is disconnected at 70 to 80% of synchronous speed using a centrifugal switch or relay sensing current through the main stator winding. Apart from high starting torque, the operation of the capacitor-start motor is basically the same as that of the split-phase motor. Popular split-phase motors have an upper power limit of 1/3 hp, while capacitor-start machines are available at up to 10 hp (Morley, 1990).

20.7 PREDICTIVE TESTING AND INSPECTION SUSTAINABILITY AND STEWARDSHIP—MOTORS

Determining the proper functioning of electrically energized motors is an element of machine sustainability assessment. Motor analysis consists of various technologies to determine the condition of both the motor and motor circuitry. The total impedance of a motor is the sum of its resistance, capacitance, and inductance. Any impedance imbalances in a motor will result in a voltage imbalance. Voltage imbalances result in higher operating current and temperatures, which will weaken the insulation and shorten the motor's life. For resistance, the maximum allowable percentage imbalance is 2%. For inductance, the maximum allowable percentage is 10%. The percent imbalance is calculated as follows:

$$R_{avg} = (R_1 + R_2 + R_3)/3$$

$$\text{Percent voltage imbalance} = [(\text{High} - R_{avg})/R_{avg}] \times 100\%$$

The formula is the same for impedance imbalance. The capacitance value is used for trending purposes and indicates wet or dirty windings.

20.7.1   REQUIRED EQUIPMENT INFORMATION

- Equipment foundation data (if applicable)
- Bearing layout
- Design analysis and calculations
- Bearing data

20.7.2   REQUIRED TESTING RESULTS

- Insulation resistance test results
- Airborne ultrasonic test results
- Power factor test results
- High-voltage test results
- Infrared thermography (IRT) test results
- Test point locations
- Vibration analysis test results
- Motor circuit evaluation test results
- Start-up test results

20.7.3   MOTORS (TIP UP)

A motor should be tested using a grounded specimen test (GST). Ideally, each phase should be tested individually, so this test must be accomplished prior to terminating the motor at the terminal box.

20.7.4   MOTOR CIRCUIT ANALYSIS

Motor circuit analysis is a technology designed to monitor the condition of the complete motor circuit. The test device measures the basic electrical characteristics, conductor phase resistance, conductor phase inductance, resistance to ground, and capacitance to ground. The root causes of motor failure in AC three-phase motor stators are frequently found outside the motor. The most common cause is found after the power supply enters the facility, including (in order of decreasing frequency):

1. Distribution transformer defects
2. Uneven loads on individual phases
3. Deterioration of individual conductor paths

Checking three-phase motor circuits for resistive balance can be important to ensuring long motor operating life. In each three-phase motor circuit the resistance of each conductor path should be as close to equal as possible. Small values of resistive imbalance (up to 2 to 3%) can be tolerated for short periods without much loss. However, resistive imbalance beyond 5% will begin to reduce life expectancy radically. Similarly, inductive balance should be checked. At some point above 15% inductive imbalance, electrically induced mechanical vibrations will occur. When a rotor has a defect, (e.g., broken bars, end rings), highly variable imbalances will occur as the rotor position changes relative to the stator windings. During equipment acceptance, the lower the initial imbalance the longer will be the expected motor life from both electrical and mechanical standpoints.

Motor circuit analysis also evaluates motor circuit capacitance and resistance to ground. The capacitance indicates the amount of dirt and moisture present on the outside of the motor winding insulation, which (1) has the effect of helping current leak through defective insulation to ground through the motor frame, and (2) interferes with the motor's designed capability to dissipate heat. From a baseline condition of low capacitance (to ground) this parameter can be trended. Test equipment is portable and computer based, which provides for automated test performance and data collection.

20.7.5 Motor Current Signature Analysis

Motor current signature analysis (MCSA) detects the following:

- Broken rotor bars
- Defective shorting rings
- Rotor porosity
- Air gap eccentricity

The main indicator of rotor problems is the difference in height, measured in decibels, between the spike at the line frequency and the spike at the pole pass frequency of the motor current signature. As the difference decreases, increased severity of damage to the rotor bars and end rings that make up the rotor cage is indicated. MCSA takes advantage of the fact that an electrical motor is a reliable transducer of mechanically induced loads. Variations in motor load modulate the current flowing through the motor stator windings. These variations in the motor load are due to the non-symmetrical magnetic field in the motor and mechanical feedback due to variations in system response. Spikes are present at or near the ones at the pole pass frequency that are induced by other mechanical faults, mainly those in the machinery that the motor is driving. The motor current is filtered to remove line frequency harmonics and transformed to the frequency domain by performing a fast Fourier transform.

20.7.6 Electrical Signal Analysis

Electrical signature analysis is an online technology where all currents and voltages of a motor circuit are measured, conditioned, and displayed as time domain and frequency spectral data. The technology allows for the diagnosis of conditions of the power system, motor, and driven component. The data collected include voltage and current balance, power quality, impedance balance, and current sequence data.

20.7.7 ELECTRICAL ACCEPTANCE TESTS

Perform the high-voltage test to verify the insulation in a motor and to ensure that no excessive leakage current is present. This is a potentially destructive test. The minimum acceptance criterion is performance better than or equal to the manufacturer's specifications.

Perform the motor circuit evaluation test to determine the condition of the complete motor circuit. The motor should be loaded at 75% or greater. Minimum acceptance results are

- Resistive imbalance < 2%
- Inductive imbalance < 10%
- Resistance, inductance, and capacitance within ±5% of the manufacturer's specifications

Use the start-up test to determine the coast-down time and the peak starting current.

20.7.8 FLUX ANALYSIS

Flux analysis is a diagnostic technique involving measurement and analysis of the magnetic leakage flux field around a motor. The technology is designed to detect faults in rotor bars, stator turn-to-turn shorts, and phase-to-phase faults. The analysis is similar to that of motor current signature analysis (NASA, 2004).

20.7.9 MOTORS 200 HP AND ABOVE

On motors 200 hp and above, the insulation resistance test and polarization index test should be required on each phase of the motor prior to terminating the motor to the feeder cable. On motors where the phases cannot be separated, all three phases are tested together.

20.7.9.1 Insulation Resistance

Insulation tests on motors up to 2400 volts are conducted using 2500 volts DC. Insulation tests on motors rated above 2400 volts are conducted using 5000 volts DC. Data recorded will be megohm readings relative to time. Readings are taken at 15, 30, or 45 seconds and in 1-minute increments thereafter for up to 10 minutes. Megohm readings should not be less than 25 megohms for each phase, and each phase reading should be within 10% of the other two.

20.7.9.2 Polarization Index

The polarization index of each phase is calculated by dividing the 10-minute reading by the 1-minute reading. The polarization index should be greater than 1.25. Any values lower than 1.25 should be rejected and the motor not accepted. The dielectric absorption ratio (DAR) of each phase is calculated by dividing the 30-second reading by the 1-minute reading. The DAR should be greater than 1.25. Any values lower than 1.25 should be rejected and the motor not accepted (NASA, 2004).

20.7.10 MONITORING MOTOR VIBRATION

Use a vibration data collector that has

- A minimum of 400 lines of resolution
- A dynamic range greater than 70 dB
- A frequency response of 5 Hz to 10 kHz (300 to 600,000 cpm)
- The capability to perform ensemble averaging
- The use of a Hanning window
- Autoranging frequency
- A minimum amplitude accuracy over the selected frequency range of ±20% or ±1.5 dB
- Sensor frequency response

20.7.10.1 Vibration Criteria for Electric Motors

All motor vibration spectra should be analyzed at the following forcing frequencies:

- One times running speed (1×) for imbalance
- Two times running speed (2×) for misalignment
- Multiples of running speed (N×) for looseness, resonance, and plain bearing defects
- Electric line frequency and harmonics (60 or 120 Hz for AC motors) for stator and rotor problems

Rolling element bearing frequencies are

- Outer race defect frequency
- Inner race defect frequency
- Ball defect (ball spin frequency)
- Fundamental train frequency

Plain or journal bearings indicate faults at harmonics of running speed and at the frequency corresponding to 0.4 to 0.5× running speed. Other sources of vibration in motors are dependent on the number of motor rotor bars and stator slots, the number of cooling fan blades, the number of commutator bars and brushes, and the silicon-controlled rectifier (SCR) firing frequencies for variable speed motors. Broken rotor bars will often produce sidebands spaced at two times the slip frequency. The presence of broken rotor bars can be confirmed through the use of electrical testing.

20.7.10.2 Vibration Standards for Electric Motors

Alternating current motors must be tested at the rated voltage and frequency and no load. Single-speed alternating current motors must be tested at synchronous speed. A multi-speed AC motor must be tested at all of its rated synchronous speeds, and DC motors must be tested at their highest rated speed. Series and universal motors must be tested at operating speed. All electrical motors, according to size, should meet the following requirements:

- The velocity amplitude (in./s peak) of any line of resolution, measured at bearing locations in any direction, should not exceed the line-amplitude band limit values specified in Figures 2-9 and 2-10 of *Reliability Centered Building and Equipment Acceptance Guide* (NASA, 2004) for small and large motors, respectively.
- The acceleration overall amplitude (*g* peak) at bearing locations in any direction should not exceed the band-limited overall amplitude acceptance limit appropriate for the motor being tested.

Data for electrical motor certification include the following:

- Identifiers (motor serial number, frame number, model number, horse-power, and synchronous speed)
- Amplitude of vibration at bearing locations in any direction
- Vibration signatures of velocity and acceleration
- Listing of the maximum peak velocity in each band for vibration measurements taken at position 1 horizontal, at position 2 vertical, and at position 3 axial
- Vibration amplitude at the fundamental rotational frequency or one times (1×) running speed (this is a narrow-band limit; an overall reading is not acceptable)

All testing should be conducted at normal operating speed under full load conditions.

20.7.11 REWOUND ELECTRIC MOTORS

Due to the potential of both rotor and/or stator damage being incurred during the motor rewinding process (usually resulting from bake-out of the old insulation and subsequent distortion of the pole pieces) a rewound electrical motor should be checked both electrically and mechanically. The mechanical check consists of post-overhaul vibration measurements at the same location as for new motors. The vibration level at each measurement point should not exceed the reference spectrum for that motor by more than 10%. In addition, vibration amplitudes associated with electrical faults (e.g., slip, rotor bar, stator slot) should be noted for any deviation from the reference spectrum.

20.7.11.1 Required Equipment Information

- Motor type
- Motor specifications
  - Bearing information
  - Frame size
  - Motor class
  - Full load and locked rotor current
  - Winding resistance
  - Winding inductance
  - Cooling fan blades

- Number of rotor bars
- Number of stator slots
- SCR firing sequence

20.7.11.2 Balance Measurement

Use balance measurement to verify that rotating shafts are balanced and mass and rotational centerlines are coincident. Minimum acceptance criterion is performance better than or equal to the manufacturer's specifications.

20.7.11.3 Using Laser and Infrared Thermography

Alignments should be measured using laser alignment or other equivalent alignment methods. Infrared thermography (IRT) should be used to detect abnormal hot spots that may indicate flaws in the stator winding.

REFERENCES

Morley, L.A. (1990). *Mine Power Systems*, Information Circular 9258, NTIS No. PB 91-241729. Philadelphia, PA: U.S. Bureau of Mines.

NASA. (2004). *Reliability Centered Building and Equipment Acceptance Guide*, National Aeronautics and Space Administration, Washington, DC, 2004.

NFPA. (2005). *NFPA 70: National Electrical Code (NEC) Handbook*. Quincy, MA: National Fire Protection Association.

OSHA. (2007). Occupational Safety and Health Standards, 29 CFR, Part 1910, Subpart S, Electrical. Baltimore, MD: Occupational Safety and Health Administration.

21 Hazardous Energy Control and OSHA

Martha J. Boss and Dennis Day

CONTENTS

The terms *hazardous energy* and *lockout/tagout* (LO/TO) are used interchangeably in this chapter. LO/TO may be employed for work on machines or work on the electrical equipment proper. Electrical work is defined herein as work performed near or on electrical power systems (bus, branch, distribution network) employed to prevent electric shock or other injuries resulting from either direct or indirect electrical contacts, when work is performed near or on equipment or circuits that are or may be energized. When equipment is not locked out, performing work on that machine or piece of equipment would be considered live work; however, when LO/TO has isolated the energy source, the work can be done in a non-live-work situation. The lockout or tagout devices either isolate energy sources or otherwise disable machines and equipment to prevent unexpected energization, start up, or release of stored energy.

21.1   DEFINITIONS

Affected employee: An employee whose job requires him/her (1) to operate or use a machine or piece of equipment on which servicing or maintenance is being performed under LO/TO, or (2) to work in an area in which such servicing or maintenance is being performed.

Authorized employee: A person who locks out or tags out machines or equipment in order to perform servicing or maintenance on those machines or equipment.

Capable of being locked out: An energy-isolating device is capable of being locked out (1) if it has a hasp or other means of attachment to which, or through which, a lock can be affixed, or it has a built-in locking mechanism; or (2) if lockout can be achieved without the need to dismantle, rebuild, or replace the energy-isolating device or permanently alter its energy control capability.

De-energized parts: Live parts to which an employee may be exposed. Per 29 CFR 1910.333, equipment is to be de-energized before the employee works on or near it, unless the employer can demonstrate that de-energizing (1) introduces additional or increased hazards, or (2) is infeasible due to equipment design or operational limitations. Live parts that operate at less than 50 volts to ground need not be de-energized if no increased exposure to electrical burns or to explosion due to electric arcs will occur during the work.

Energized: Refers to being connected to an energy source or containing residual or stored energy.

Energized parts: If the exposed live parts are not de-energized (i.e., for reasons of increased or additional hazards or infeasibility), then (1) other safety-related work practices must be used to protect employees who may be exposed to the electrical hazards involved, and (2) work practices must protect employees against contact with energized circuit parts directly with any part of their body or indirectly through some other conductive object.

Energy-isolating device: A mechanical device that physically prevents the transmission or release of energy, including but not limited to the following: (1) manually operated electrical circuit breaker; (2) motor disconnect switch; (3) manually operated switch by which the conductors of a circuit can be disconnected from all ungrounded supply conductors and, in addition, no pole can be operated independently; (4) line valve; and (5) block or any similar device used to block or isolate energy. Push buttons, selector switches, and other control circuit types of devices are not energy-isolating devices.

Energy sources: Any source of electrical, mechanical, hydraulic, pneumatic, chemical, thermal, or other energy.

Fixed equipment: Equipment fastened in place or connected by permanent wiring methods.

Hot tap: A procedure used in repair, maintenance, and service activities that involves welding on a piece of equipment (pipelines, vessels, or tanks) under pressure to install connections or appurtenances; used to replace or add sections of pipeline without the interruption of service for air, gas, water, steam, and petrochemical distribution systems.

Isolation device: Any device that presents a physical barrier that interrupts and stops the exchange of energy in a machine.

Lockout: The placement of a lockout device on an energy-isolating device, in accordance with the site-specific LO/TO procedures, ensures that the energy-isolating device and the equipment being controlled cannot be operated until the lockout device is removed by the authorized employee.

Lockout device: A device that utilizes a positive means (e.g., lock, key, or combination type; blank flanges; bolted slip blinds) to hold an energy-isolating device in the safe position and prevent the energizing of a machine or piece of equipment. These devices must have the following attributes:

Durable—Capable of withstanding the environment for the maximum period of time that exposure is expected. Tagout devices must be constructed and printed so that exposure to weather conditions or wet, corrosive, and/or damp locations will not cause the tag to deteriorate or the message on the tag to become illegible.

Identifiable—Each lockout device or tagout device must indicate the identity of the employee applying the device.

Standardized—Lockout and tagout devices must be standardized within the facility in at least one of the following criteria: color, shape, or size. In the case of tagout devices, print and format must also be standardized.

Substantial—Lockout devices must be substantial enough to prevent removal without the use of excessive force or unusual techniques (e.g., use of bolt cutters or other metal-cutting tools). Tagout devices, including their means of attachment, must prevent inadvertent or accidental removal. Tagout device attachment means must be of a nonreusable type, attachable by hand, self-locking, and non-releasable with a minimum unlocking strength of no less than 50 pounds and having the general design and basic characteristics of being at least equivalent to a one-piece, all-environment-tolerant nylon cable tie.

Normal production operations: The utilization of a machine or equipment to perform its intended production function.

Other employee: Refers to an employee whose current work activities are not in an area where energy control procedures are required; thus, this employee is not an affected employee either due to temporary removal from a work location during hazardous energy control procedure implementation or ongoing work assignments in areas where hazardous energy control is never required.

Protective materials and hardware: Protective materials and hardware provided by the employer for isolating, securing, or blocking machines and equipment from energy sources; may include locks, tags, chains, wedges, key blocks, adapter pins, self-locking fasteners, or other hardware specified in the site-specific LO/TO procedures.

Service and/or maintenance: Any workplace activity, including constructing, installing, setting up, adjusting, inspecting, modifying, maintaining, and servicing machines and equipment. These activities include lubrication, cleaning, unjamming of machines or equipment, and making adjustments or tool changes where the employee may be exposed to the unexpected energization or startup of the equipment or release of hazardous energy.

Setting up: Any work performed only to prepare a machine or piece of equipment for normal production operations.

Tagout: The placement of a tagout device on an energy-isolating device, in accordance with an established procedure, to indicate that the energy-isolating device and the equipment being controlled may not be operated until the tagout device is removed.

Tag/tagout device: A prominent warning device (i.e., a tag and a means of attachment) that can be securely fastened to an energy-isolating device, to indicate that the energy-isolating device and the equipment being controlled may not be operated until the tagout device is removed. Tagout devices will warn against hazardous conditions if the machine or equipment is energized and will include a legend such as *Do Not Start. Do Not Open. Do Not Close. Do Not Energize. Do Not Operate.*

21.2 LOCKOUT/TAGOUT DEVICES AND CONDITIONS

A uniquely keyed lock, singularly identifiable as a lock used for the control of energy sources by an individual, must be issued to each authorized employee. Each authorized employee should be issued personalized tags to be attached to the lock to identify the authorized employee applying the devices. The lockout device is under the exclusive control of the authorized employee performing the servicing or maintenance. Each lockout device should only be removed by the employee who applied the device (an exception being group LO/TO). Each authorized employee issued a tag, lock, and lockout is required to return them upon retirement, resignation, termination, posting to a different job, or on request of management.

21.2.1 ILLUMINATION

Employees may not enter spaces containing exposed energized parts, unless illumination is provided that enables the employees to perform the work safely. Where lack of illumination or an obstruction precludes observation of the work to be performed, employees may not perform tasks near exposed energized parts. Employees may not reach blindly into areas that may contain energized parts.

21.2.2 CONFINED OR ENCLOSED WORK SPACES

When an employee works in a confined or enclosed space (i.e., manhole or vault) that contains exposed energized parts, protective shields, protective barriers, or insulating materials must be used as necessary to avoid inadvertent contact with these parts. Doors and hinged panels must be secured to prevent them from swinging into an employee and causing the employee to contact exposed energized parts.

21.2.3 CONDUCTIVE MATERIALS AND EQUIPMENT

Conductive materials and equipment that are in contact with any part of an employee's body must be handled in a manner that will prevent the employee from contacting exposed energized conductors or circuit parts. If an employee must handle long-dimensional conductive objects (e.g., ducts, pipes) in areas with exposed live parts, work practices such as the use of insulation, guarding, and material handling techniques must be used to minimize the hazard.

21.2.4 PORTABLE LADDERS

Portable ladders must have non-conductive side rails if used where the employee or the ladder could contact exposed energized parts.

21.2.5 CONDUCTIVE APPAREL

Conductive articles of jewelry and clothing (e.g., watch bands, bracelets, rings, key chains, necklaces, metalized aprons, cloth with conductive thread, metal headgear) may not be worn.

21.2.6 HOUSEKEEPING

Where live parts present an electrical contact hazard, employees may not perform housekeeping duties at such close distances to the parts that a possibility of contact exists. Adequate safeguards (e.g., insulating equipment, barriers) may prevent contact and where necessary may be required. Electrically conductive cleaning materials (e.g., steel wool, metalized cloth, silicon carbide, conductive liquid solutions) may not be used in proximity to energized parts unless procedures are followed to prevent electrical contact.

21.2.7 INTERLOCKS

Only a qualified person may defeat an electrical safety interlock and then only temporarily while he or she is working on the equipment. The interlock system should be returned to its operable condition when work is completed.

21.3 CORD- AND PLUG-CONNECTED EQUIPMENT

Cord- and plug-connected equipment operating at 120 volts or less that can be completely de-energized by unplugging the device must be unplugged and are not subject to LO/TO procedures; however, when unplugging the device could cause an additional hazard, site-specific procedures must be used to unplug the machine or piece of equipment. To guarantee the plug will not be reinstated either a lockable cover must be placed over the plug or control of the plug must be maintained at all times.

21.4 LOCKOUT/TAGOUT PROCEDURES

21.4.1 NOTIFICATION

Affected employees should be notified by the authorized employee of the application and removal of lockout devices or tagout devices. Notification is to be given before the controls are applied and after these controls are removed from the machine or piece of equipment.

21.4.2 PREPARATION FOR SHUTDOWN

An initial survey must be made to locate and identify all energy-isolating devices (e.g., switch, valve, other energy-isolating devices) to be applied to the machine or piece of equipment to be locked out.

21.4.3 SHUTDOWN

Prior to the application of LO/TO devices and controls, all affected employees should be notified that LO/TO procedures are going to be implemented. Before an authorized or affected employee turns off a machine or piece of equipment, the authorized employee must have knowledge of the type and magnitude of the energy, the hazards of the energy to be controlled, and the method or means to control the energy as detailed in the site's machine and equipment LO/TO procedures. The operator of the machine or piece of equipment should use the normal stopping procedure to shut it off. An orderly shutdown must be utilized to avoid any additional or increased hazards to employees as a result of the equipment stoppage. If normal on/off procedures are not sufficient (due to an increased hazard present), an authorized employee should consult the site's machine and equipment LO/TO procedures. The authorized employee can then turn off the machine or equipment.

21.4.4 ISOLATION AND LOCKOUT APPLICATIONS

Isolation and lockout applications should occur as specified in the site's machine and equipment LO/TO procedures. In general, the following procedure should be conducted by the authorized employee:

- Physically locate and operate the electrical disconnect, valve, or other energy-isolating devices so the equipment is isolated from its energy sources.
- Disconnect the circuits and equipment to be worked on from all electric energy sources.
  - Do not use control circuit devices (e.g., push buttons, selector switches, interlocks) as the sole means for de-energizing circuits or equipment.
- Do not use interlocks for electric equipment as a substitute for lockout and tagging procedures.
- Apply the LO/TO device and lock using the assigned individual locks. Attach the lock so as to prevent persons from operating the disconnecting means, and verify that these devices prevent the reinitiation of power sources.
- Dissipate all potentially hazardous stored or residual energy in rams, rotating machine parts, flywheels, springs, pneumatic, thermal systems, hydraulic systems, or other components.
  - Disconnect, restrain, and otherwise render safe all energy stores.
  - Release stored electric energy.
  - If the stored electric energy might endanger personnel, be sure that capacitors are discharged and high-capacitance elements are short-circuited and grounded.
  - Treat capacitors or associated equipment handled in meeting this requirement as energized.
  - Prevent or control energy reaccumulation/reinitiation.
- Prevent reaccumulation of stored energy by disconnecting all power sources and/or by applying mechanical stops (e.g., ram stops/blocks). If energy reinitiation is possible, stored non-electrical energy in devices that could re-energize electric circuit parts must be blocked or relieved to the extent that the circuit parts could not be accidentally energized by the device. Verification of isolation must be continued until the servicing or maintenance is completed or until the possibility of energy reaccumulation no longer exists.
- Use the normal operating (on) controls to verify that the machine or equipment has been de-energized and will not operate.

- Return the operating controls to the neutral or off position after the verification test.
- Apply LO/TO devices. All energy-isolating devices that are required to control the energy to the machine or equipment must be physically located and operated in such a manner as to isolate the machine or equipment from the energy sources.

Servicing or maintenance can then be conducted.

21.4.5 SPECIAL PROVISIONS FOR ELECTRICAL EQUIPMENT

Before any circuits or equipment can be considered and worked as de-energized,

- A qualified person must operate the equipment operating controls or otherwise verify that the equipment cannot be restarted.
- A qualified person must use test equipment to test the circuit elements and electrical parts of equipment to which employees will be exposed and verify that the circuit elements and equipment parts are de-energized.
  - The test must also determine if any energized condition exists as a result of inadvertently induced voltage or unrelated voltage backfeed even though specific parts of the circuit have been de-energized and presumed to be safe.
  - If the circuit to be tested is over 600 volts, nominal, the test equipment must be checked for proper operation immediately after this test.

21.4.6 LOCKOUT DEVICE REMOVAL

Lockout device removal should occur as specified in the machine or equipment LO/TO procedures. In general, before circuits or equipment are re-energized, even temporarily, the authorized employee performing the service/maintenance must

- Check that the machine/equipment/component is operationally intact, tools and other nonessential items have been removed, and guards have been replaced.
- Check to be sure that all affected employees are safely positioned.

A qualified person must conduct tests and visual inspections to verify that all tools, electrical jumpers, shorts, grounds, and other such devices have been removed, so that the circuits and equipment can be safely energized. Employees exposed to the hazards associated with re-energizing the circuit or equipment must be warned to stay clear of circuits and equipment. All affected employees must be notified that locks or tags are going to be removed and the machine is ready for operation. After all lockout devices have been removed, all energy to the machine or equipment can be sequentially restored.

21.4.7 TESTING AND POSITIONING PROCEDURE FOR TEMPORARY LO/TO DEVICE REMOVAL

When lockout devices must be temporarily removed so the machine or equipment can be energized to test or position the machine, equipment, or component, the authorized employee should apply the following procedure:

- Clear the machine or piece of equipment of all tools and materials.
- Remove affected employees from the vicinity of the machine or piece of equipment.
- Follow the site's machine and equipment LO/TO procedure for re-energizing the machine or piece of equipment.
  - Remove the lockout device.
  - Energize and proceed with testing or positioning.
  - De-energize all machine, equipment, or component systems.
- Follow the machine and equipment LO/TO procedure to reinstitute LO/TO.
  - De-energize all systems.
  - Reapply energy control measures.

21.4.8 REMOVAL OF ABSENT AUTHORIZED EMPLOYEE'S LOCK

If the authorized employee who applied the lock is not available, the plant manager should contact the maintenance department supervisor, who will

- Verify (by a visual determination) that the authorized employee who locked out the equipment is not on the premises.
- Attempt to contact the authorized employee to inform that person that his or her lock will be removed from the machine or piece of equipment.
- Secure the master keys from the plant manager and remove the lock.
- Return the master key to the plant manager.
- Return the lock and tag to the appropriate employee immediately upon his or her return to work.
- Document that the authorized employee has been informed of the above prior to resumption of his or her work at that facility.

21.5 TAGOUT FOR TAGS WITHOUT LOCKS

The preferred LO/TO method is to lock out the energy source; however, when complete lockout is not feasible, tagout can be used upon approval by the plant manager. A tag used without a lock should be supplemented by at least one additional safety measure that provides a level of safety equivalent to that obtained by the use of a lock. Examples of additional safety measures include removing an isolating circuit element, blocking a controlling switch, or opening an extra disconnecting device. Tagout devices must clearly indicate that the operation or movement of energy-isolating devices from the safe or off position is prohibited. Where a tag cannot be affixed directly to the energy-isolating device, the tag must be located as close as safely possible to the device, in a position that will be immediately obvious to anyone attempting to operate the device.

When a tagout device is used on an energy-isolating device, the tagout device must be attached at the same location where the lockout device would have been attached, and the employer should demonstrate that the tagout program will provide a level of safety equivalent to that obtained by using a lockout program. Tag removal requires authorization by the authorized person who applied the tag. Tags are never to be bypassed, ignored, or otherwise defeated. Tags must be legible and understandable by all authorized employees, affected employees, and all other employees whose work operations are or may be in the area. Each tag must contain a statement prohibiting unauthorized operation of the disconnecting means and removal of the tag.

21.6 LOCKS WITHOUT TAGS

A lock may be placed without a tag only under the following conditions:

- Only one circuit or piece of equipment is de-energized.
- The lockout period does not extend beyond the work shift.
- Employees exposed to the hazards associated with re-energizing the circuit or equipment must be familiar with the procedures to be used.

21.7 GROUP LO/TO

When servicing and/or maintenance is performed by a group, the group must utilize a procedure that affords employees a level of protection equivalent to that provided by the implementation of a personal LO/TO device. Group LO/TO devices must be used in accordance the site's machine and equipment requirements with the following understanding:

- Primary responsibility is vested in an authorized employee for a set number of employees working under the protection of a group LO/TO device (e.g., an operations lock).
- Provisions for the authorized employee to ascertain the exposure status of individual group members with regard to the LO/TO of the machine, equipment, or component must be documented in an attachment to the site's machine and equipment LO/TO procedures.
- When more than one group is involved, overall job-associated LO/TO control responsibility is assigned to an authorized employee designated to coordinate affected work forces and ensure continuity of protection.
- Each authorized employee should affix a personal LO/TO device to the group lockout device, group lockbox, or comparable mechanism when he or she begins work and must remove those devices when he or she stops working on the machine or piece of equipment being serviced or maintained.

21.8   SHIFT OR PERSONNEL CHANGES

Specific procedures should be utilized during shift or personnel changes to ensure the continuity of LO/TO protection, including providing for the orderly transfer of LO/TO device protection between off-going and oncoming employees.

21.9   OUTSIDE PERSONNEL (CONTRACTORS)

Whenever outside servicing personnel are to be engaged in LO/TO, plant management and the outside employer should inform each other of their respective LO/TO procedures. Outside contractors must be informed and supplied a copy of the company's lockout procedures by the employee awarding the contract. The contractor's procedures must be reviewed and approved by the manufacturing engineer prior to beginning work. Notification of affected and other employees of the contractor's lockout procedures must be completed prior to beginning work. The plant or site manager should ensure that his or her employees understand and comply with the restrictions and prohibitions of the outside employer's energy control program.

21.10   SPECIAL PROVISIONS FOR LOCKOUT/TAGOUT

21.10.1   WORKING ON OR NEAR EXPOSED ENERGIZED PARTS

Work performed on or near exposed live parts (involving either direct contact or by means of tools or materials) may expose workers to electrical hazards; consequently, LO/TO is required. Note that, until the power is disconnected, the parts are considered live and that a normally operated on/off switch is not a power supply disconnect.

21.10.2   WORK ON ENERGIZED EQUIPMENT

Only qualified persons may work on electric circuit parts or equipment that has not been de-energized. Such persons must be capable of working safely on energized circuits and be familiar with the proper use of special precautionary techniques, personal protective equipment, insulating and shielding materials, and insulated tools.

21.10.3   OVERHEAD LINES

If work is to be performed near overhead lines, the lines must be de-energized and grounded or other protective measures must be provided before work is started. If the lines are to be de-energized, arrangements must be made with the person or organization that operates or controls the electric circuits involved to de-energize and ground them. If protective measures (e.g., guarding, isolating, insulating) are provided, these precautions must prevent employees from contacting such lines directly with any part of their body or indirectly through conductive materials, tools, or equipment.

21.10.4 UNQUALIFIED PERSONS

When an unqualified person is working in an elevated position near overhead lines, the location must be such that the person and the longest conductive object that he or she may contact cannot come closer to any unguarded, energized overhead line than the following distances:

For voltages to ground 50 kV or below	305 cm (10 ft)
For voltages to ground over 50 kV	305 cm (10 ft) plus 10 cm (4 in.) for every 10 kV over 50 kV

When an unqualified person is working on the ground in the vicinity of overhead lines, the person may not bring any conductive object closer to unguarded, energized overhead lines than the distances shown above. *Note:* For voltages normally encountered with overhead power lines, objects that do not have an insulating rating for the voltage involved are considered to be conductive.

21.10.5 QUALIFIED PERSONS

When a qualified person is working in the vicinity of overhead lines, whether in an elevated position or on the ground, the person may not approach or take any conductive object without an approved insulating handle closer to exposed energized parts than shown in Table 21.1, unless

- The person is insulated from the energized part. Gloves, with sleeves if necessary, rated for the voltage involved are considered to be insulation of the person from the energized part on which work is performed.
- The energized part is insulated from all other conductive objects at a different potential and from the person.
- The person is insulated from all conductive objects at a potential different from that of the energized part.

TABLE 21.1
Approach Distances for Qualified Employees—Alternating Current

Voltage Range (Phase to Phase)	Minimum Approach Distance
300 V and less	Avoid contact
Over 300 V but not over 750 V	30.5 cm (1 ft, 0 in.)
Over 750 V but not over 2 kV	46 cm (1 ft, 6 in.)
Over 2 kV but not over 15 kV	61 cm (2 ft, 0 in.)
Over 15 kV but not over 37 kV	91 cm (3 ft, 0 in.)
Over 37 kV but not over 87.5 kV	107 cm (3 ft, 6 in.)
Over 87.5 kV but not over 121 kV	122 cm (4 ft, 0 in.)
Over 121 kV but not over 140 kV	137 cm (4 ft, 6 in.)

21.10.6 VEHICULAR AND MECHANICAL EQUIPMENT

Any vehicle or mechanical equipment capable of having parts of its structure elevated near energized overhead lines should be operated so that a clearance of 305 cm (10 ft) is maintained. If the voltage is higher than 50 kV, the clearance must be increased 10 cm (4 in.) for every 10 kV over that voltage. However, under any of the following conditions, the clearance may be reduced:

- If the vehicle is in transit with its structure lowered, the clearance may be reduced to 122 cm (4 ft). If the voltage is higher than 50 kV, the clearance may be increased 10 cm (4 in.) for every 10 kV over that voltage.
- If insulating barriers are installed to prevent contact with the lines, and if the barriers are rated for the voltage of the line being guarded and are not a part of or an attachment to the vehicle or its raised structure, the clearance may be reduced to a distance within the designed working dimensions of the insulating barrier.
- If the equipment is an aerial lift insulated for the voltage involved, and if the work is performed by a qualified person, the clearance (between the uninsulated portion of the aerial lift and the power line) may be reduced to the distance given in Table 21.1.

Employees standing on the ground may not contact the vehicle or mechanical equipment or any of its attachments, unless

- The employee is using protective equipment rated for the voltage.
- The equipment is located so that no uninsulated part of its structure (that portion of the structure that provides a conductive path to employees on the ground) can come closer to the line than permitted and shown below:

For voltages to ground 50 kV or below   305 cm (10 ft)
For voltages to ground over 50 kV   305 cm (10 ft) plus 10 cm (4 in.) for every 10 kV over 50 kV

If any vehicle or mechanical equipment capable of having parts of its structure elevated near energized overhead lines is intentionally grounded, then employees working on the ground near the point of grounding may not stand at the grounding location whenever the potential for overhead line contact is present. Additional precautions (e.g., use of barricades or insulation) must be taken to protect employees from hazardous ground potentials, depending on earth resistivity and fault currents, which can develop within the first few feet or more outward from the grounding point.

21.11 CERTIFICATION

The employer must certify that employee training has been accomplished and is being kept up to date. The certification should contain each employee's name and dates of training.

21.12   PERIODIC INSPECTION

A periodic inspection of the LO/TO procedure is to be conducted annually. The periodic inspection should be performed by an authorized employee other than the ones utilizing the energy control procedure being inspected. The inspector must be an authorized person and may not review their own use of lockout procedures. The periodic inspection should be conducted to correct any deviations or inadequacies identified. It should include a review, between the inspector and each authorized employee, of that employee's responsibilities under the LO/TO procedure being inspected. Energy control procedures for each machine or type of machine must be inspected, and the employer must certify that the periodic inspections have been performed. The certification should identify the following:

- Machine or piece of equipment on which the energy control procedure was being utilized
- Inspection date
- Employees included in the inspection
- Person performing the inspection

REFERENCES

OSHA. (2007). Occupational Safety and Health Standards, 29 CFR, Part 1910, Subpart S, Electrical. Baltimore, MD: Occupational Safety and Health Administration.

Glossary

Acceleration: The time rate of change of velocity. Typical units are ft/s^2 and g (1 g = 32.17 ft/s^2 = 386 in./s^2 = 9.81 m/s^2). Acceleration measurements are made with accelerometers.

Accelerometer: Transducer where the output is directly proportional to acceleration. Most commonly used are mass loaded piezoelectric crystals to produce an output proportional to acceleration.

Accepted: An installation is accepted if it has been inspected and found by a nationally recognized testing laboratory to conform to specified plans or to procedures of applicable codes.

Accessible: The ability to fully reach, adjust, and maintain the equipment. Consideration should be given to confined space restrictions and removing guards, bushing plates, hydraulic lines, lubrication lines, and electric lines. Also, on a broader scale, refers to the ability to gain access to the equipment due to security, safety, and other restrictions.

Accessible (as applied to equipment): Allowing close approach; not guarded by locked doors, elevation, or other effective means.

Accessible (as applied to wiring methods): Capable of being removed or exposed without damaging the building structure or finish, or not permanently closed in by the structure or finish of the building.

Age exploration: The process of determining the most effective intervals for maintenance tasks. Often associated with identifying age-related maintenance actions (e.g., overhaul and discard tasks) and then extending the interval between tasks.

Alignment target specifications: Desired intentional offset and angularity at coupling center to compensate for thermal growth and/or dynamic loads. Most properly specified as an *offset* and an angle in two perpendicular planes, horizontal and vertical.

Ampacity: The current, in amperes, that a conductor can carry continuously under the conditions of use without exceeding the conductor's temperature rating.

Amplitude: A measure of the severity of vibration. Amplitude is expressed in terms of peak-to-peak, zero-to-peak (peak), or root-mean-square (rms). For pure sine waves only:

$$\text{Peak} = 1.414 \times \text{rms}$$

$$\text{Peak-to-peak} = 2 \times \text{Zero-to-peak (peak)}$$

Amplitude limits: The total vibration level in a band should not exceed the overall amplitude acceptance limit specified for the band.

Angular error: A misalignment condition characterized by the angular error between the desired centerline and the actual centerline. This misalignment condition may exist in planes both horizontal and vertical to the axis of rotation.

Angularity: The angle between the rotational centerlines of two shafts. Angularity is a slope expressed in terms of a rise (millimeters or thousandths of an inch) over a run (meter or inches).

Anti-aliasing filter: A low-pass filter designed to filter out frequencies higher than one-half the sample rate in order to prevent aliasing.

Appliances: Utilization equipment, generally other than industrial, normally built in standardized sizes or types that is installed or connected as a unit to perform one or more functions.

Approved: Acceptable to the authority enforcing this subpart. The authority enforcing this subpart is the Assistant Secretary of Labor for Occupational Safety and Health. The definition of acceptable indicates what is acceptable to the Assistant Secretary of Labor and therefore approved within the meaning of 29 CFR 1926.

Armored cable (type AC): A fabricated assembly of insulated conductors in a flexible metallic enclosure.

Askarel: A generic term for a group of nonflammable synthetic chlorinated hydrocarbons used as electrical insulating media. Askarels of various compositional types are used. Under arcing conditions, the gases produced, while consisting predominantly of non-combustible hydrogen chloride, can include varying amounts of combustible gases depending upon the askarel type.

Attachment plug (plug cap; cap): A device that, by insertion in a receptacle, establishes a connection between the conductors of the attached flexible cord and the conductors connected permanently to the receptacle.

Automatic: Self-acting, operating by its own mechanism when actuated by some impersonal influence (e.g., change in current strength, pressure, temperature, mechanical configuration).

Availability: (1) Informally, the time a machine or system is available for use. (2) From the overall equipment effectiveness calculation, the actual run time of a machine or system divided by the scheduled run time. Note that *availability* differs slightly from *asset utilization (uptime)* in that scheduled runtime varies between facilities and is changed by scheduled maintenance actions, logistics, or administrative delays.

Axial play: Shaft axial movement along the shaft's centerline caused by axial forces, thermal expansion, or contraction and permitted by journal bearings, sleeve bearings, and/or looseness. Also known as *axial float* or *end float*.

Balance: When the mass center line and rotational center line of a rotor are coincident.

Balancing: A procedure for adjusting the radial mass distribution of a rotor by adding or removing weight so the mass centerline approaches the rotor geometric centerline, achieving less vibration amplitude at rotational speed.

Band-limited overall amplitude: For vibration level limits specified in terms of *band-limited overall reading*.

Band-limited overall reading: The vibration severity amplitude measured over a frequency range defined by a F_{min} and a F_{max}.

Barrier: A physical obstruction that is intended to prevent contact with equipment or live parts or to prevent unauthorized access to a work area.

Beat frequency: The absolute value of the difference in frequency of two oscillations of slightly different frequencies.

Beats: Periodic variations in the amplitude of an oscillation resulting from the combination of two oscillations of slightly different frequencies. The beats occur at the difference frequency.

Blade-pass frequency: A potential vibration frequency on any bladed machine (turbine, axial compressor, fan, pump); represented by the number of fan blades or pump vanes times shaft rotating frequency. Also known as *pumping frequency*.

Bonding (bonded): The permanent joining of metallic parts to form an electrically conductive path that ensures electrical continuity and the capacity to conduct safely any current likely to be imposed.

Bonding jumper: A conductor that provides the necessary electrical conductivity between metal parts required to be electrically connected.

Borescoping: A condition-monitoring technique to verify that the internal parts are in satisfactory shape and, if potential failures exist, to what extent. A borescope, or fiberscope, is an optical instrument that allows internal inspection of machinery through port holes without equipment disassembly.

Branch circuit: The circuit conductors between the final overcurrent device protecting the circuit and the outlets.

Building: A structure that stands alone or is cut off from adjoining structures by fire walls with all openings therein protected by approved fire doors.

Building commissioning: The systematic process for achieving, verifying, and documenting that the performance of facilities and collateral equipment meets the design intent. The process extends through all phases of a project and culminates with occupancy and operation. The process includes the testing and accepting of new or repaired building, system, or component parts to verify proper installation.

Cabinet: An enclosure designed for either surface or flush mounting and provided with a frame, mat, or trim in which a swinging door or doors are or can be hung.

Cable tray system: A unit or assembly of units or sections and associated fittings forming a rigid structural system used to securely fasten or support cables and raceways. Cable tray systems include ladders, troughs, channels, solid bottom trays, and other similar structures.

Cablebus: An assembly of insulated conductors with fittings and conductor terminations in a completely enclosed, ventilated, protective metal housing.

Capacitor: An electrical component used to store energy in an electric field.

Calibration: A test to verify the accuracy of measurement instruments. For vibration, a transducer is subjected to a known motion, usually on a shaker table, and the output readings are verified or adjusted.

Cell line: An assembly of electrically interconnected electrolytic cells supplied by a source of direct current power.

Cell line attachments and auxiliary equipment: Cell line attachments and auxiliary equipment include, but are not limited to, auxiliary tanks, process piping, ductwork, structural supports, exposed cell line conductors, conduits and other raceways, pumps, positioning equipment, and cell cutout or bypass electrical devices. Auxiliary equipment also includes tools, welding machines, crucibles, and other portable equipment used for operation and maintenance within the electrolytic cell line working zone. In the cell line working zone, auxiliary equipment includes the exposed conductive surfaces of ungrounded cranes and crane-mounted cell-servicing equipment.

Center pivot irrigation machine: A multi-motored irrigation machine that revolves around a central pivot and employs alignment switches or similar devices to control individual motors.

Certified: Equipment is considered to be certified if it bears a label, tag, or other record of certification stating that the equipment (1) has been tested and found by a nationally recognized testing laboratory to meet nationally recognized standards or to be safe for use in a specified manner; or (2) is of a kind whose production is periodically inspected by a nationally recognized testing laboratory and is accepted by the laboratory as safe for its intended use.

Circuit breaker: A device designed to open and close a circuit by nonautomatic means and to open the circuit automatically on a predetermined overcurrent without damage to itself when properly applied within the circuit breaker's rating.

Class I locations: Class I locations are those in which flammable gases or vapors are or may be present in the air in quantities sufficient to produce explosive or ignitable mixtures. Class I locations include the following:

1.   Class I, Division 1—A location (a) in which ignitable concentrations of flammable gases or vapors may exist under normal operating conditions; (b) in which ignitable concentrations of such gases or vapors may exist frequently because of repair or maintenance operations or because of leakage; or (c) in which breakdown or faulty operation of equipment or processes might release ignitable concentrations of flammable gases or vapors and might also cause simultaneous failure of electric equipment. *Note:* This classification usually includes locations where volatile flammable liquids or liquefied flammable gases are transferred from one container to another; interiors of spray booths and areas in the vicinity of spraying and painting operations where volatile flammable solvents are used; locations containing open tanks or vats of volatile flammable liquids; drying rooms or compartments for the evaporation of flammable solvents; locations containing fat and oil extraction equipment using volatile flammable solvents; portions of cleaning and dyeing plants where flammable liquids are used; gas generator rooms and other portions of gas manufacturing plants where flammable gas may escape; inadequately ventilated pump rooms for flammable gas or for volatile flammable liquids; the interiors of refrigerators and freezers in which volatile flammable materials are stored in open, lightly stoppered, or easily ruptured containers; and all other locations where ignitable concentrations of flammable vapors or gases are likely to occur in the course of normal operations.

2. Class I, Division 2—A location (a) in which volatile flammable liquids or flammable gases are handled, processed, or used but in which the hazardous liquids, vapors, or gases will normally be confined within closed containers or closed systems from which escape occurs only in the event of accidental rupture or breakdown of such containers or systems, or as a result of abnormal operation of equipment; (b) in which ignitable concentrations of gases or vapors are normally prevented by positive mechanical ventilation and which might become hazardous through failure or abnormal operations of the ventilating equipment; or (c) that is adjacent to a Class I, Division 1, location and to which ignitable concentrations of gases or vapors might occasionally be communicated unless such communication is prevented by adequate positive-pressure ventilation from a source of clean air and effective safeguards against ventilation failure are provided. *Note:* This classification usually includes locations where volatile flammable liquids or flammable gases or vapors are used but which would become hazardous only in case of an accident or of some unusual operating condition. The quantity of flammable accident, the adequacy of ventilating equipment, the total area involved, and the record of the industry or business with respect to explosions or fires are all factors that merit consideration in determining the classification and extent of each location. Piping without valves, checks, meters, and similar devices would not ordinarily introduce a hazardous condition even though used for flammable liquids or gases. Locations used for the storage of flammable liquids or liquefied or compressed gases in sealed containers would not normally be considered hazardous unless also subject to other hazardous conditions. Electrical conduits and their associated enclosures separated from process fluids by a single seal or barrier are classed as a Division 2 location if the outside of the conduit and enclosures is a non-hazardous location.

3. Class I, Zone 0—A location in which one of the following conditions exists: (a) ignitable concentrations of flammable gases or vapors are present continuously; or (b) ignitable concentrations of flammable gases or vapors are present for long periods of time. *Note:* As a guide in determining when flammable gases or vapors are present continuously or for long periods of time, refer to API RP 505 (1997; R2013)—Recommended Practice for Classification of Locations for Electrical Installations of Petroleum Facilities Classified as Class I, Zone 0, Zone 1, or Zone 2; IEC 79 10 (1995)—Electrical Apparatus for Explosive Gas Atmospheres; Institute for Petroleum Model Code, Part 15—Area Classification Code for Petroleum Installations; and ISA S12.24.01-1997—Electrical Apparatus for Explosive Gas Atmospheres, Classifications of Hazardous (Classified) Locations.

4. Class I, Zone 1—A location in which one of the following conditions exists: (a) ignitable concentrations of flammable gases or vapors are likely to exist under normal operating conditions; (b) ignitable

concentrations of flammable gases or vapors may exist frequently because of repair or maintenance operations or because of leakage; (c) equipment is operated or processes are carried on of such a nature that equipment breakdown or faulty operations could result in the release of ignitable concentrations of flammable gases or vapors and also cause simultaneous failure of electric equipment in a manner that would cause the electric equipment to become a source of ignition; or (d) location is adjacent to a Class I, Zone 0, location from which ignitable concentrations of vapors could be communicated, unless communication is prevented by adequate positive-pressure ventilation from a source of clean air and effective safeguards against ventilation failure are provided.

5. Class I, Zone 2—A location in which one of the following conditions exists: (a) ignitable concentrations of flammable gases or vapors are not likely to occur in normal operation and if occurring will exist only for a short period; (b) volatile flammable liquids, flammable gases, or flammable vapors are handled, processed, or used but the liquids, gases, or vapors are normally confined within closed containers or closed systems from which escape only occurs as a result of accidental rupture or breakdown of the containers or system or as the result of the abnormal operation of the equipment with which the liquids or gases are handled, processed, or used; (c) ignitable concentrations of flammable gases or vapors normally are prevented by positive mechanical ventilation but may become hazardous as the result of failure or abnormal operation of the ventilation equipment; or (d) location is adjacent to a Class I, Zone 1, location, from which ignitable concentrations of flammable gases or vapors could be communicated, unless such communication is prevented by adequate positive-pressure ventilation from a source of clean air and effective safeguards against ventilation failure are provided.

Class II locations: Class II locations are those that are hazardous because of the presence of combustible dust. Class II locations include the following:

1. Class II, Division 1—A location: (a) in which combustible dust is or may be in suspension in the air under normal operating conditions, in quantities sufficient to produce explosive or ignitable mixtures; (b) where mechanical failure or abnormal operation of machinery or equipment might cause such explosive or ignitable mixtures to be produced and might also provide a source of ignition through simultaneous failure of electric equipment, through operation of protection devices, or from other causes; or (c) in which combustible dusts of an electrically conductive nature may be present. *Note:* This classification may include areas of grain handling and processing plants, starch plants, sugar-pulverizing plants, malting plants, hay-grinding plants, coal pulverizing plants, areas where metal dusts and powders are produced or processed, and other similar locations that contain dust-producing machinery and equipment (except where the equipment is

dust-tight or vented to the outside). These areas would have combustible dust in the air, under normal operating conditions, in quantities sufficient to produce explosive or ignitable mixtures. Combustible dusts that are electrically non-conductive include dusts produced in the handling and processing of grain and grain products, pulverized sugar and cocoa, dried egg and milk powders, pulverized spices, starch and pastes, potato and wood flour, oil meal from beans and seed, dried hay, and other organic materials that may produce combustible dusts when processed or handled. Dusts containing magnesium or aluminum are particularly hazardous, and the use of extreme caution is necessary to avoid ignition and explosion.

2. Class II, Division 2—A location where (a) combustible dust will not normally be in suspension in the air in quantities sufficient to produce explosive or ignitable mixtures, and dust accumulations will normally be insufficient to interfere with the normal operation of electric equipment or other apparatus, but combustible dust may be in suspension in the air as a result of infrequent malfunctioning of handling or processing equipment; and (b) resulting combustible dust accumulations on, in, or in the vicinity of the electric equipment may be sufficient to interfere with the safe dissipation of heat from electric equipment or may be ignitable by abnormal operation or failure of electric equipment. *Note:* This classification includes locations where dangerous concentrations of suspended dust would not be likely but where dust accumulations might form on or in the vicinity of electric equipment. These areas may contain equipment from which appreciable quantities of dust would escape under abnormal operating conditions or be adjacent to a Class II, Division 1, location, as described above, into which an explosive or ignitable concentration of dust may be put into suspension under abnormal operating conditions.

Class III locations: Class III locations are those that are hazardous because of the presence of easily ignitable fibers or flyings, but in which such fibers or flyings are not likely to be in suspension in the air in quantities sufficient to produce ignitable mixtures. Class III locations include the following:

1. Class III, Division 1—A location in which easily ignitable fibers or materials producing combustible flyings are handled, manufactured, or used. *Note:* Such locations usually include some parts of rayon, cotton, and other textile mills; combustible fiber manufacturing and processing plants; cotton gins and cotton-seed mills; flax-processing plants; clothing manufacturing plants; woodworking plants; and establishments and industries involving similar hazardous processes or conditions. Easily ignitable fibers and flyings include rayon, cotton (including cotton linters and cotton waste), sisal or henequen, istle, jute, hemp, tow, cocoa fiber, oakum, baled waste kapok, Spanish moss, excelsior, and other materials of similar nature.

2. Class III, Division 2—A location in which easily ignitable fibers are stored or handled, other than in the process of manufacture.

Colinear: Two lines that are positioned as if they were one line. Colinear as used in alignment means two or more centerlines of rotation with no offset or angularity between them. Two or more lines are colinear when no offset or angularity exists between them (i.e., the lines follow the same path).

Collector ring: An assembly of slip rings for transferring electric energy from a stationary to a rotating member.

Competent person: One who is capable of identifying existing and predictable hazards in the surroundings or working conditions that are unsanitary, hazardous, or dangerous to employees and who has authorization to take prompt corrective measures to eliminate them.

Complete machine: A complete machine is defined as the entire assembly of components, subcomponents, and structure that is purchased to perform a specific task. On a complete machine assembly with all individual components operating in their normal operating condition, mode, and sequence, the component vibration level limits for the complete machine acceptance are the same as when the component is tested individually.

Concealed: Rendered inaccessible by the structure or finish of the building. Wires in concealed raceways are considered concealed even though accessible upon withdrawal from the raceways.

Conductor, bare: A conductor having no covering or electrical insulation whatsoever.

Conductor, covered: A conductor encased within material of composition or thickness that is not considered to be electrical insulation.

Conductor, insulated: A conductor encased within material of such composition and thickness that it is considered to be electrical insulation.

Conduit body: A separate portion of a conduit or tubing system that provides access through one or more removable covers to the interior of the system at a junction of two or more sections of the system or at a terminal point of the system. Boxes such as FS and FD or larger cast or sheet metal boxes are not classified as conduit bodies.

Controller: A device or group of devices that serves to govern, in some predetermined manner, the electric power delivered to the apparatus to which the device is connected.

Coplanar: The condition of two or more surfaces having all elements in one plane (per ANSI Y14.5).

Cost effective: Economic determination of the maintenance approach that entails the evaluation of maintenance costs, support costs, and consequences of failure.

Coupling point: The phrase "coupling point" in the definition of shaft alignment is an acknowledgment that vibration due to misalignment originates at the point of power transmission, the coupling. The shafts are being aligned and the coupling center is just the measuring point.

Critical failure: A failure involving a loss of function or secondary damage that could have a direct adverse effect on operating safety or mission or could have significant economic impact.

Critical failure mode: A failure mode that has significant mission, safety, or maintenance effects that warrant the selection of maintenance tasks to prevent the critical failure mode from occurring.

Critical speed: The speed of a rotating system corresponding to a system resonance frequency.

Current: The number of electrons passing a given location in a specified unit of time, usually Coulombs per second.

Cutout (over 600 volts, nominal): An assembly of a fuse support with a fuseholder, fuse carrier, or disconnecting blade. The fuseholder or fuse carrier may include a conducting element (fuse link) or may act as the disconnecting blade by the inclusion of a nonfusible member.

Cutout box: An enclosure designed for surface mounting and having swinging doors or covers secured directly to and telescoping with the walls of the box proper.

Dead front: Without live parts exposed to a person on the operating side of the equipment.

Decibel (Db): A logarithmic representation of amplitude ratio, defined as 20 times the base-10 logarithm of the ratio of the measured amplitude to a reference; for example, decibel volt (dBV) readings are referenced to 1 volt rms. The dB amplitude scales are required to display the full dynamic range of an F analyzer.

De-energized: Free from any electrical connection to a source of potential difference and from electrical charge; not having a potential different from that of the earth.

Device: A unit of an electrical system that is intended to carry but not utilize electric energy.

Dielectric heating: The heating of a nominally insulating material due to its own dielectric losses when the material is placed in a varying electric field.

Disconnecting means: A device, or group of devices, or other means by which the conductors of a circuit can be disconnected from their source of supply.

Disconnecting (or isolating) switch (over 600 volts, nominal): A mechanical switching device used for isolating a circuit or equipment from a source of power.

Displacement: The distance traveled by a vibrating object. For purposes of this document, displacement represents the total distance traveled by a vibrating part or surface from the maximum position of travel in one direction to the maximum position of travel in the opposite direction (peak-to-peak) and is measured in the unit mil (1 mil = 0.001 inch).

Dominant failure mode: A single failure mode that accounts for a significant portion of the failures of a complex item.

Dynamic mass: To determine if the mass of the transducer is affecting the measurement, perform the following steps: (1) Make the desired measurement with an accelerometer. (2) Place a mass equivalent to the mass of the accelerometer adjacent to the measuring accelerometer. (3) Repeat the measurement. (4) Compare data from steps (1) and (3). If any differences (i.e., shift in frequencies) between steps (1) and (3) exist, then a less massive transducer should be used.

Dynamic range: The difference between the highest measurable signal level and the lowest measurable signal level that is detectable for a given amplitude range setting. Dynamic range is usually expressed in decibels, typically 60 to 90 dB for modern instruments.

Electrolytic cell line working zone: The space envelope wherein operation or maintenance is normally performed on or in the vicinity of exposed energized surfaces of electrolytic cell lines or their attachments.

Electrolytic cells: A tank or vat in which electrochemical reactions are caused by applying energy for the purpose of refining or producing usable materials.

Electromagnetic field: A field composed of both an electric wave and a magnetic wave, propagating at 90 degrees to each other. Electromagnetic fields are produced when electricity is produced.

Electromagnetic interference: A disturbance caused by electromagnetic radiation from an external source.

Enclosed: Surrounded by a case, housing, fence, or walls that will prevent persons from accidentally contacting energized parts.

Enclosure: The case or housing of apparatus or the fence or walls surrounding an installation to prevent personnel from accidentally contacting energized parts or to protect the equipment from physical damage.

Energized: Electrically connected to a source of potential difference.

Equipment: A general term including material, fittings, devices, appliances, fixtures, apparatus, and the like used as a part of, or in connection with, an electrical installation.

Equipment grounding conductor: See *Grounding conductor, equipment*.

Explosion-proof apparatus: Apparatus enclosed in a case that is capable of withstanding an explosion of a specified gas or vapor that may occur within the case and of preventing the ignition of a specified gas or vapor surrounding the enclosure by sparks, flashes, or explosion of the gas or vapor within, and that operates at such an external temperature that the apparatus (internal circuitry, etc.) will not ignite a surrounding flammable atmosphere.

Exposed: Where the circuit is in such a position that, in case of failure of supports or insulation, contact with another circuit may result (29 CFR 1910.308(e)).

Exposed (as applied to live parts): Capable of being inadvertently touched or approached nearer than a safe distance by a person. The term is applied to parts not suitably guarded, isolated, or insulated.

Exposed (as applied to wiring methods): On or attached to the surface, or behind panels designed to allow access.

Externally operable: Capable of being operated without exposing the operator to contact with live parts.

Failure: A cessation of proper function or performance; the inability to meet a standard; nonperformance of what is requested or expected.

Failure effect: The consequences of failure.

Failure mode, motor: The manner of failure (e.g., motor stops) and the reason why the motor failed (e.g., motor bearing seized).

Failure modes and effects analysis (FMEA): Analysis used to determine what parts fail, why they usually fail, and what effect their failure has on the systems in total.

Feeder: All circuit conductors between the service equipment, the source of a separate derived system, or other power supply source and the final branch-circuit overcurrent device.

FFT (fast Fourier transform): A calculation procedure that converts a time-domain signal into a frequency-domain display.

FFT analyzer: Vibration analyzer that uses the fast Fourier transform to display vibration frequency components.

Fitting: An accessory (e.g., locknut, bushing, other part of a wiring system) that is intended primarily to perform a mechanical rather than an electrical function.

F_{max}: Maximum frequency limit of the spectrum being evaluated.

F_{min}: Minimum frequency limit of the spectrum being evaluated.

Fountain: Fountains, ornamental pools, display pools, and reflection pools. *Note:* This definition does not include drinking fountains.

Frequency: The repetition rate of a periodic event, usually expressed in cycles per second (hertz, HZ), cycles per minute (CPM), or multiples of rotational speed (orders). Orders are commonly referred to as 1× for rotational speed and 2× for twice rotational speed. Frequency is the reciprocal of the period. *Note:* Vibration frequencies are expressed in Hz (cycles per second) or CPM (cycles per minute). Rotational speed (running speed) is expressed in revolutions per minute (RPM).

Frequency domain: Presentation of a signal whose amplitude is measured on the y-axis and the frequency is measured on the x-axis.

Frequency resolution (ΔF): $\Delta F = (F_{max} - F_{min})$/number of lines of resolution. ΔF represents the minimum spacing between data points in the spectrum.

Frequency response: Portion of the frequency spectrum that can be covered within specified frequency limits.

Function: A defined performance standard. Usually quantitative in nature (flow rate, cooling capacity).

Fuse (over 600 volts, nominal): An overcurrent protective device with a circuit-opening fusible part that is heated and severed by the passage of overcurrent through the fuse. A fuse is comprised of all of the parts that form a unit capable of performing the prescribed functions. The fuse may or may not be the complete device necessary to connect the fuse into an electrical circuit.

Gear mesh frequency: A potential vibration frequency on any machine that contains gears; equal to the number of teeth multiplied by the rotational frequency of the gear.

Ground: A conducting connection, whether intentional or accidental, between an electric circuit or equipment and the earth or some conducting body that serves in place of the earth.

Ground-fault circuit interrupter: A device intended for the protection of personnel that functions to de-energize a circuit or a portion of a circuit within an established period of time when a current to ground exceeds some predetermined value that is less than that required to operate the overcurrent protective device of the supply circuit.

Grounded: Connected to the earth or to some conducting body that serves in place of the earth.

Grounded conductor: A system or circuit conductor that is intentionally grounded.

Grounded, effectively: Intentionally connected to earth through a ground connection or connections of sufficiently low impedance and having sufficient current-carrying capacity to prevent the buildup of voltages that may result in undue hazards to connected equipment or to persons.

Grounding conductor: A conductor used to connect equipment or the grounded circuit of a wiring system to a grounding electrode or electrodes.

Grounding conductor, equipment: The conductor used to connect the non-current-carrying metal parts of equipment, raceways, and other enclosures to the system grounded conductor, the grounding electrode conductor, or to both, at the service equipment or at the source of a separately derived system.

Grounding electrode conductor: The conductor used to connect the grounding electrode to the equipment grounding conductor, to the grounded conductor, or to both, of the circuits at the service equipment or at the source of a separately derived system.

Guarded: Covered, shielded, fenced, enclosed, or otherwise protected by means of suitable covers, casings, barriers, rails, screens, mats, or platforms to remove the likelihood of approach to a point of danger or contact by persons or objects.

Hanning window: A digital signal analysis (DSA) window function that provides better frequency resolution than the flat-top window but with reduced amplitude.

Harmonic: Frequency component at a frequency that is an integer (whole number, such as 2x, 3x, or 4x) multiple of the fundamental (reference) frequency.

Heating equipment: Includes any equipment used for heating purposes if heat is generated by induction or dielectric methods (29 CFR 1910.306(g)).

Hertz (Hz): The unit of frequency represented by cycles per second.

High band-pass filter: A device that separates the components of a signal and allows only those components above a selected frequency to be amplified.

Hoistway: Any shaftway, hatchway, well hole, or other vertical opening or space that is designed for the operation of an elevator or dumbwaiter.

Horizontal: Parallel to the mounting surface.

Identified (as applied to equipment): Approved as suitable for the specific purpose, function, use, environment, or application described in a particular requirement. *Note:* Some examples of ways to determine suitability of equipment for a specific purpose, environment, or application include investigations by a nationally recognized testing laboratory (through listing and labeling), inspection agency, or other organization recognized under the definition of *acceptable.*

Imbalance: Unequal radial weight distribution of a rotor system; a shaft condition such that the mass and shaft geometric centerlines do not coincide.

Induction heating: The heating of a nominally conductive material due to its own I^2R losses when the material is placed in a varying electromagnetic field.

Infrared thermography: A predictive technique that uses infrared imaging to identify defects in electrical and electromechanical devices (e.g., fuse boxes, circuit breakers, switchgear). Thermography also can be used effectively in a non-predictive manner to detect thermal cavities and leaks in walls, ceilings, and rooftops, the correction of which can result in sizeable reductions in heating and air conditioning expenses. Thermal imaging is extremely sensitive and evaluates the heat an object emits. Emittance and reflective factors of the object and environment must be considered.

Inspection: A time- or cycle-based action performed to identify hidden failure or potential failure.

Insulated: Separated from other conducting surfaces by a dielectric (including air space) offering a high resistance to the passage of current.

Interrupter switch (over 600 volts, nominal): A switch capable of making, carrying, and interrupting specified currents.

Irrigation machine: An electrically driven or controlled machine, with one or more motors, that is not hand portable and is used primarily to transport and distribute water for agricultural purposes.

Isolated (as applied to location): Not readily accessible to persons unless special means for access are used.

Isolated power system: A system comprised of an isolating transformer or its equivalent, a line isolation monitor, and its ungrounded circuit conductors.

Jackbolts, jackscrews: Positioning bolts on the machine base that are located at each foot of the machine and are used to adjust the position of the machines.

Labeled: Equipment is considered to be labeled if it has attached to it a label, symbol, or other identifying mark of a nationally recognized testing laboratory that makes periodic inspections of the production of such equipment and indicates compliance with nationally recognized standards or tests to determine safe use in a specified manner.

Level: Parallel to a reference plane or a reference line established by a laser.

Lighting outlet: An outlet intended for the direct connection of a lampholder, a lighting fixture, or a pendant cord terminating in a lampholder.

Line amplitude limit: The maximum amplitude of any line of resolution contained within a band should not exceed the line amplitude acceptance limit for the band.

Line-clearance tree trimming: The pruning, trimming, repairing, maintaining, removing, or clearing of trees or cutting of brush that is within 305 cm (10 ft) of electric supply lines and equipment.

Line of resolution: A single data point from a spectrum that contains vibration amplitude information. The line of resolution amplitude is the band overall amplitude of the frequencies contained in the ΔF frequency resolution.

Linear non-overlapping average: An averaging process where each time block sample used in the averaging process contains data not contained in other time blocks (i.e., non-overlapping) used in the averaging. Linear averaging is performed in the frequency domain, and each of the samples is weighted equally.

Listed: Equipment is considered to be listed if it is of a kind mentioned in a list that is published by a nationally recognized laboratory that makes periodic inspection of the production of such equipment and states that such equipment meets nationally recognized standards or has been tested and found safe for use in a specified manner.

Live parts: Energized conductive components.

Location, damp: Partially protected locations under canopies, marquees, roofed open porches, and like locations, and interior locations subject to moderate degrees of moisture.

Location, dry: A location not normally subject to dampness or wetness. A location classified as dry may be temporarily subject to dampness or wetness, as in the case of a building under construction.

Location, wet: Locations that are underground or in concrete slabs or masonry in direct contact with the earth, subject to saturation with water or other liquids (e.g., vehicle-washing areas), or unprotected and exposed to weather.

Machine: The total entity made up of individual machine components (e.g., motors, pumps, spindles, fixtures).

Machine base: The structure that supports the machine or machine components under consideration.

Machine component: An individual unit (e.g., motor, pump, spindle, fixture) often referred to as a machine in its own context.

Maintainability: The ability to retain or restore function within a specified period of time, when provided with an identified level of tools, training, and procedures. Maintainability factors include machine and systems access, visibility, simplicity, ease of monitoring or testing, special training requirements, special tools, and capability of local work force.

Maintenance: Action taken to retain function (i.e., prevent failure). Actions include preventive maintenance, predictive testing and inspection, lubrication and minor repair (e.g., replacing belts and filters), and inspection for failure.

Measurement point: A location on a machine or component at which vibration measurements are made.

Medium-voltage cable (type MV): A single or multiconductor solid dielectric insulated cable rated 2001 volts or higher.

Metal-clad cable (type MC): A factory assembly of one or more insulated circuit conductors with or without optical fiber members enclosed in an armor of interlocking metal tape or a smooth or corrugated metallic sheath.

Micrometer (micron, μm): One millionth (0.000001) of a meter (1 micron = 1×10^{-6} meters = 0.04 mils).

mil: One thousandth (0.001) of an inch (1 mil = 25.4 microns).

Mineral-insulated, metal-sheathed cable (type MI): Type MI cable is a factory assembly of one or more conductors insulated with a highly compressed refractory mineral insulation and enclosed in a liquid-tight and gas-tight continuous copper or alloy steel sheath.

Mobile X-ray: X-ray equipment mounted on a permanent base with wheels or casters or both for moving while completely assembled.

Motor circuit analysis (MCA): A predictive technique whereby the static characteristics (i.e., impedance, capacitance to ground, and inductance) of a motor or generator are measured as indicators of equipment condition.

Motor control center: An assembly of one or more enclosed sections having a common power bus and principally containing motor control units.

Motor current spectrum analysis (MCSA): A predictive technique whereby motor current signatures provide information on the electromechanical condition of AC induction motors. MCSA detects faults (e.g., broken rotor bars, high resistance joints, cracked rotor end rings) by collecting motor current spectrums with clamp-on sensors and analyzing the data.

Natural frequency: The frequency of free vibration of a system when excited with an impact force (bump test).

Non-metallic-sheathed cable (types NM, NMC, and NMS): A factory assembly of two or more insulated conductors having an outer sheath of moisture-resistant, flame-retardant, non-metallic material.

Offset: The distance (in 1/1000 of an inch or in millimeters) between the rotational centerlines of two parallel shafts.

Oil (filled) cutout (over 600 volts, nominal): A cutout in which all or part of the fuse support and its fuse link or disconnecting blade are mounted in oil with complete immersion of the contacts and the fusible portion of the conducting element (fuse link), so that arc interruption by severing of the fuse link or by opening of the contacts will occur under oil.

Open wiring on insulators: An exposed wiring method that utilizes cleats, knobs, tubes, and flexible tubing for the protection and support of single insulated conductors that are run in or on buildings and not concealed by the building structure.

Order: A unit of frequency unique to rotating machinery where the first order is equal to rotational speed.

Outlet: A point on the wiring system at which current is taken to supply utilization equipment.

Outline lighting: An arrangement of incandescent lamps or electric discharge lighting to outline or call attention to certain features (e.g., shape of a building, decoration of a window).

Overcurrent: Any current in excess of the rated current of equipment or the ampacity of a conductor; may result from overload, short circuit, or ground fault.

Overhaul: Perform a major replacement, modification, repair, or rehabilitation similar to that involved when a new building or facility is built, a new wing is added, or an entire floor is renovated.

Overload: Operation of equipment in excess of normal, full-load rating or of a conductor in excess of rated ampacity that, when such operation persists for a sufficient length of time, would cause damage or dangerous overheating. A fault, such as a short circuit or ground fault, is not an overload.

Panelboard: A single panel or group of panel units designed for assembly in the form of a single panel, including buses and automatic overcurrent devices, with or without switches for the control of light, heat, or power circuits; designed to be placed in a cabinet or cutout box placed in or against a wall or partition and accessible only from the front.

Pass frequency: A potential vibration frequency on any bladed machine (e.g., turbine, axial compressor, fan, pump). Represented by the number of fan blades or pump vanes times shaft rotating frequency. Also known as *pumping frequency*.

Peak: Refers to the maximum of the units being measured (e.g., peak velocity, peak acceleration, peak displacement).

Peak-to-peak: Refers to the displacement from one travel extreme to the other travel extreme. In English units, this is measured in mils (0.001 inch) and in metric units is expressed in micrometers (μm; 0.000001 meters).

Period: The amount of time, usually expressed in seconds or minutes, required to complete one cycle of motion of a vibrating machine or machine part. The reciprocal of the period is the frequency of vibration.

Permanently installed decorative fountains and reflection pools: Pools constructed in the ground, on the ground, or in a building in such a manner that the fountain or pool cannot be readily disassembled for storage, whether or not served by electrical circuits of any nature. These units are primarily constructed for their aesthetic value and are not intended for swimming or wading.

Permanently installed swimming, wading, and therapeutic pools: Pools that are constructed in the ground or partially in the ground and all others capable of holding water at a depth greater than 1.07 m (42 in.). The definition also applies to all pools installed inside of a building, regardless of water depth, whether or not served by electric circuits of any nature.

Phase (phase angle): The relative position, measured in degrees, of a vibrating part at any instant in time to a fixed point or another vibrating part. The phase angle (usually expressed in degrees) is the angle between the instantaneous position of a vibrating part and the reference position. This angle represents the portion of the vibration cycle through which the part has moved relative to the reference position.

Pitch: An angular misalignment in the vertical plane.

Portable x-ray: X-ray equipment designed to be hand carried.

Position error (centerline/offset misalignment): A misalignment condition that exists when the shaft centerline is parallel but not in line with (not coincidental with) the desired alignment centerline.

Potential failure: An identifiable condition indicating that a failure is imminent.

Power and control tray cable (type TC): A factory assembly of two or more insulated conductors with or without associated bare or covered grounding conductors under a non-metallic sheath approved for installation in cable trays, in raceways, or where supported by a messenger wire.

Power fuse (over 600 volts, nominal): See *Fuse.*

Power-limited tray cable (type PLTC): A factory assembly of two or more insulated conductors under a non-metallic jacket.

Power outlet: An enclosed assembly (which may include receptacles, circuit breakers, fuseholders, fused switches, buses, and watt-hour meter mounting means) that is intended to supply and control power to mobile homes, recreational vehicles, or boats or to serve as a means for distributing power needed to operate mobile or temporarily installed equipment.

Predictive maintenance: See *Predictive testing and inspection (PT&I).*

Predictive testing and inspection (PT&I): The use of advanced technology to assess machinery condition. The PT&I data obtained allow for planning and scheduling preventive maintenance or repairs in advance of failure.

Also known as *condition monitoring, predictive maintenance,* and *condition-based maintenance.*

Premises wiring (system): The interior and exterior wiring, including power, lighting, control, and signal circuit wiring, together with all of their associated hardware, fittings, and wiring devices, both permanently and temporarily installed, that extend from the service point of utility conductors or source of power (e.g., battery; solar photovoltaic system; generator, transformer, or converter) to the outlets. Such wiring does not include wiring internal to appliances, fixtures, motors, controllers, motor control centers, and similar equipment.

Preventive maintenance: Time- or cycle-based actions performed to prevent failure, monitor condition, or inspect for failure.

Proactive maintenance: The collection of efforts to identify, monitor, and control future failure with an emphasis on the understanding and elimination of the cause of failure. Proactive maintenance activities include the development of design specifications to incorporate maintenance lessons learned and to ensure future maintainability and supportability, the development of repair specifications to eliminate underlining causes of failure, and performing root-cause failure analysis to understand why in-service systems failed.

Qualified person: One who has received training in and has demonstrated skills and knowledge in the construction and operation of electric equipment and installations and the hazards involved. Whether an employee is considered to be a qualified person depends on various circumstances in the workplace; for example, an individual may be considered qualified with regard to certain equipment in the workplace but unqualified with regard to other equipment (see 29 CFR 1910.332(b)(3) for training requirements that specifically apply to qualified persons). An employee who is undergoing on-the-job training and who, in the course of such training, has demonstrated an ability to perform duties safely at his or her level of training and who is under the direct supervision of a qualified person is considered to be a qualified person for the performance of those duties.

Raceway: An enclosed channel of metal or non-metallic materials designed expressly for holding wires, cables, or busbars, with additional functions as permitted in this standard. Raceways include, but are not limited to, rigid metal conduit, rigid non-metallic conduit, intermediate metal conduit, liquid-tight flexible conduit, flexible metallic tubing, flexible metal conduit, electrical metallic tubing, electrical non-metallic tubing, underfloor raceways, cellular concrete floor raceways, cellular metal floor raceways, surface raceways, wireways, and busways.

Radial measurement: Measurements taken perpendicular to the axis of rotation.

Radial vibration: Shaft dynamic motion or casing vibration, which is in a direction perpendicular to the shaft centerline.

Readily accessible: Capable of being reached quickly for operation, renewal, or inspections, so that those needing ready access do not have to climb over or remove obstacles or to resort to portable ladders or chairs.

Receptacle: A contact device installed at the outlet for the connection of an attachment plug. A single receptacle is a single contact device with no other contact device on the same yoke. A multiple receptacle is two or more contact devices on the same yoke.

Receptacle outlet: An outlet where one or more receptacles are installed.

Reliability: (1) The dependability constituent or dependability characteristic of design. (2) The duration or probability of failure-free performance under stated conditions; the probability that an item can perform its intended function for a specified interval under stated conditions (MIL-STD-721C).

Reliability-centered maintenance (RCM): The process used to determine the most effective approach to maintenance; it involves identifying actions that, when taken, will reduce the probability of failure and are the most cost effective, and it seeks the optimal mix of condition-based actions, other time- or cycle-based actions, or run-to-failure approach.

Remote-control circuit: Any electric circuit that controls any other circuit through a relay or an equivalent device.

Repair: Facility work required to restore a facility or component thereof, including collateral equipment, to a condition substantially equivalent to its originally intended and designed capacity, efficiency, or capability. Repair includes the substantially equivalent replacement of utility systems and collateral equipment necessitated by incipient or actual breakdown. Also, the restoration of function, usually after failure.

Repeatability: The consistency of readings and results between consecutive sets of measurements.

Resistor: An electrical component that resists the flow of electrons in a circuit.

Resonance: The condition of vibration amplitude and phase change response caused by a corresponding system sensitivity to a particular forcing frequency. A resonance is typically identified by a substantial amplitude increase and related phase shift.

Rolling element bearing: Bearing whose low friction qualities derive from rolling elements (balls or rollers), with little lubrication.

Root-cause failure analysis (RCFA): A process of exploring, in increasing detail, all possible causes related to a machine failure. Failure causes are grouped into general categories for further analysis. Causes can be related to machinery, people, methods, materials, policies, environment, or measurement error.

Rotational speed: The number of times an object completes one complete revolution per unit of time (e.g., 1800 RPM).

Sealable equipment: Equipment enclosed in a case or cabinet that is provided with a means of sealing or locking so that live parts cannot be made accessible without opening the enclosure. The equipment may or may not be operable without opening the enclosure.

Separately derived system: A premises wiring system whose power is derived from a battery, a solar photovoltaic system, or from a generator, a transformer, or converter windings and that has no direct electrical connection, including a solidly connected grounded circuit conductor, to supply conductors originating in another system.

Service: The conductors and equipment for delivering electric energy from the serving utility to the wiring system of the premises served.

Service cable: Service conductors made up in the form of a cable.

Service conductors: The conductors from the service point to the service disconnecting means.

Service drop: The overhead service conductors from the last pole or other aerial support to and including the splices, if any, connecting to the service-entrance conductors at the building or other structure.

Service-entrance cable: A single conductor or multiconductor assembly provided with or without an overall covering, primarily used for services, and is of the following types: (1) type SE, which has a flame-retardant, moisture-resistant covering; and (2) type USE, which is identified for underground use and has a moisture-resistant covering but is not required to have a flame-retardant covering. Cabled, single-conductor, type USE constructions recognized for underground use may have a bare copper conductor cabled with the assembly. Type USE single, parallel, or cable conductor assemblies recognized for underground use may have a bare copper concentric conductor applied. These constructions do not require an outer overall covering.

Service-entrance conductors, overhead system: The service conductors between the terminals of the service equipment and a point usually outside the building, clear of building walls, where joined by tap or splice to the service drop.

Service-entrance conductors, underground system: The service conductors between the terminals of the service equipment and the point of connection to the service lateral.

Service equipment: The necessary equipment, usually consisting of one or more circuit breakers or switches and fuses, and their accessories, connected to the load end of service conductors to a building or other structure, or an otherwise designated area, and intended to constitute the main control and cutoff of the supply.

Service point: The point of connection between the facilities of the serving utility and the premises wiring.

Shaft alignment: Positioning two or more machines (e.g., a motor driving a hydraulic pump) so the rotational centerlines of their shafts are collinear at the coupling center under operating conditions.

Shielded non-metallic-sheathed cable (type SNM): A factory assembly of two or more insulated conductors in an extruded core of moisture-resistant, flame-resistant, non-metallic material, covered with an overlapping spiral metal tape and wire shield and jacketed with an extruded moisture-, flame-, oil-, corrosion-, fungus-, and sunlight-resistant non-metallic material.

Show window: Any window used or designed to be used for the display of goods or advertising material, whether fully or partly enclosed or entirely open at the rear and whether or not having a platform raised higher than the street floor level.

Side band: Equals the frequency of interest plus or minus one times the frequency of the exciting force.

Signaling circuit: Any electric circuit that energizes signaling equipment.

Signature (spectrum): Term usually applied to the vibration frequency spectrum that is distinctive and special to a machine or component, system, or subsystem at a specific point in time and under specific machine operating conditions. Usually presented as a plot of vibration amplitude (displacement, velocity, or acceleration) vs. time or frequency. When the amplitude is plotted against time, the amplitude is usually referred to as the *time wave form.*

Soft foot: A condition that exists when the bottom of all of the feet of the machinery components are not on the same plane (can be compared to a chair with one short leg). Soft foot is present if the machine frame distorts when a foot bolt is loosened or tightened. A soft foot condition must be corrected before the machine is actually aligned.

SpecsIntact: Short for "Specifications Kept Intact," a specifications processing system that uses standard master specifications (Master Text) issued by three government agencies (NASA, Army, and Navy) for the preparation of facility projects.

Storable swimming or wading pool: A pool that is constructed on or above the ground and is capable of holding water to a maximum depth of 1.07 m (42 in.), or a pool with non-metallic, molded polymeric walls or inflatable fabric walls regardless of dimension.

Stress-free condition: The condition that exists when no forces are acting on the structure of a machine, machine component, or machine base that would cause distortion in the structure (e.g., bending, twisting).

Switch, general-use: A switch intended for use in general distribution and branch circuits. Rated in amperes and capable of interrupting its rated current at its rated voltage.

Switch, general-use snap: General-use switch constructed to be installed in device boxes or on box covers or otherwise used in conjunction with wiring systems.

Switch, isolating: A switch intended for isolating an electric circuit from the source of power. It has no interrupting rating and is intended to be operated only after the circuit has been opened by some other means.

Switch, motor-circuit: A switch, rated in horsepower, capable of interrupting the maximum operating overload current of a motor of the same horsepower rating as the switch at the rated voltage.

Switchboard: A large, single panel, frame, or assembly of panels on which are mounted (on the face or back, or both) switches, overcurrent and other protective devices, buses, and (usually) instruments. Switchboards are generally accessible from the rear as well as from the front and are not intended to be installed in cabinets.

Switching devices (over 600 volts, nominal): Devices designed to close and open one or more electric circuits. Included in this category are circuit breakers, cutouts, disconnecting (or isolating) switches, disconnecting means, interrupter switches, and oil (filled) cutouts.

Thermal effects (growth or shrinkage): This term is used to describe displacement of shaft axes due to machinery temperature changes (or dynamic loading effects) during start-up.

Time-/cycle-based actions: Maintenance activities performed from time-to-time that have proven to be effective in preventing failure (e.g., lubrication, restoration of wear, time- or cycle-based area inspection, condition monitoring, predictive testing, inspection).

Time domain: Presentation of a signal whose amplitude is measured on the *y*-axis and the time period is measured on the *x*-axis.

Tolerance: An area where all misalignment forces sum to a negligible amount and no further improvement in alignment will reduce significantly the vibration of the machine or improve efficiency. Also known as *deadband, envelope,* or *window.*

Tolerance values: Maximum allowable deviation from the desired values, whether such values are zero or non-zero.

Transducer (pickup): A device that converts shock or vibratory motion into an electrical signal that is proportional to a parameter of the vibration measured. Transducer selection is related to the frequencies of vibration, which are important to the analysis of the specific machine being evaluated or analyzed.

Transportable x-ray: X-ray equipment installed in a vehicle or that may readily be disassembled for transport in a vehicle.

Utilization equipment: Equipment that utilizes electric energy for electronic, electromechanical, chemical, heating, lighting, or similar purposes.

Velocity: The time rate of change of displacement with respect to some reference position. For purposes of this definition, velocity is measured in the units of inches per second (peak).

Ventilated: Provided with a means to permit circulation of air sufficient to remove an excess of heat, fumes, or vapors.

Vertical: Perpendicular to the horizontal plane.

Vibration analysis: Type of analysis that uses noise or vibration created by mechanical equipment to determine the equipment's actual condition by using transducers to translate a vibration amplitude and frequency into electronic signals. When measurements of both amplitude and frequency are available, diagnostic methods can be used to determine both the problem's magnitude and probable cause. Vibration techniques most often used include broadband trending (looks at the overall machine condition), narrowband trending (looks at the condition of a specific component), and signature analysis (visual comparison of current vs. normal condition). Vibration analysis most often reveals problems in machines involving mechanical imbalance, electrical imbalance, misalignment, looseness, and degenerative problems.

Volatile flammable liquid: A flammable liquid having a flash point below 38°C (100°F), a flammable liquid whose temperature is above its flash point, or a Class II combustible liquid having a vapor pressure not exceeding 276 kPa (40 psia) at 38°C (100°F) and whose temperature is above its flash point.

Voltage, circuit: The greatest root-mean-square (rms) (effective) difference of potential between any two conductors of the circuit concerned.

Voltage, nominal: A nominal value assigned to a circuit or system for the purpose of conveniently designating the circuit's or system's voltage class (as 120/240 volts, 480Y/277 volts, 600 volts). The actual voltage at which a circuit operates can vary from the nominal within a range that permits satisfactory operation of equipment.

Voltage to ground: For grounded circuits, the voltage between the given conductor and that point or conductor of the circuit that is grounded; for ungrounded circuits, the greatest voltage between the given conductor and any other conductor of the circuit.

Watertight: So constructed that moisture will not enter the enclosure.

Weatherproof: So constructed or protected that exposure to the weather will not interfere with successful operation. Rainproof, raintight, or watertight equipment can fulfill the requirements for weatherproof where varying weather conditions other than wetness (e.g., snow, ice, dust, temperature extremes) are not a factor.

Wireways: Sheet-metal troughs with hinged or removable covers for housing and protecting electric wires and cable and in which conductors are laid in place after the wireway has been installed as a complete system.

Yaw misalignment: An angular misalignment in the horizontal plane.

Appendix. RCM Evaluation Tables

The tables on the following pages are provided as an example of reliability centered maintenance (RCM) evaluation tables that can be made site specific. The tables were originally presented in *Reliability Centered Building and Equipment Acceptance Guide* (National Aeronautics and Space Administration, 2004).

TABLE A.1
Airborne Ultrasonic Criteria

Item	Date of Inspection	Acceptable Limit	Actual Value	Inspector Initials
Warble tone generator				
Differential pressure method				
Contact probe method				
Frequency range set				
Sensitivity level set				
Scale set				
System loading (%)				

Results:

TABLE A.2
Balance—Keyed Shaft Criteria

Item	Date of Inspection	Acceptable Limit	Actual Value	Inspector Initials
Key length A				
Key length B				
Final key length				

Results:

TABLE A.3
Alignment Criteria

Item	Date of Inspection	Acceptable Limit	Actual Value	Inspector Initials
Alignment				
RPM				
Soft foot actual (in.)				
Soft foot tolerance				
Vertical angularity at coupling (actual)				
Vertical offset at coupling (actual)				
Vertical angularity at coupling (actual)				
Vertical offset at coupling (actual)				
Axial shaft play				
Shims				
Shim type				
Shim condition				
Number of shims in pack				
Thickness				
Sheaves				
True to shaft				
Runout (in.)				

Results:

TABLE A.4
Battery Impedance Test Criteria

Item	Date of Inspection	Acceptable Limit	Actual Value	Inspector Initials
Battery age				
Applied voltage				
Read voltage				
Delta				
Impedance				
Compare vs. previous reading (<5% difference)				
Compare vs. similar batteries (<10% difference)				

Results:

TABLE A.5
Breaker Timing Test Criteria

Item	Date of Inspection	Acceptable Limit	Actual Value	Inspector Initials
Voltage applied				
C1, Phase A				
C2, Phase B				
C3, Phase C				

Results:

TABLE A.6
Contact Resistance Test Criteria

Item	Date of Inspection	Acceptable Limit	Actual Value	Inspector Initials
DC current applied				
Measured voltage				
Calculated resistance				
Manufacturer resistance				

Results:

TABLE A.7
High-Voltage Test Criteria

Item	Date of Inspection	Acceptable Limit	Actual Value	Inspector Initials
DC high-voltage applied				
Leakage current				

Results:

TABLE A.8
Hydraulic Oil Test Criteria

Item	Date of Inspection	Acceptable Limit	Actual Value	Inspector Initials
System sensitivity				
Actual particle level (part/100 mL)				

Results:

TABLE A.9
Infrared Thermography (IRT) Test Criteria

Item	Date of Inspection	Acceptable Limit	Actual Value	Inspector Initials
Sensitivity				
Accuracy				

Results:

TABLE A.10
Insulation Oil Test Criteria

Item	Date of Inspection	Acceptable Limit	Actual Value	Inspector Initials
Dissolved gas analysis				
Nitrogen (N_2)		<100 ppm		
Oxygen (O_2)		<10 ppm		
Carbon dioxide (CO_2)		<10 ppm		
Carbon monoxide (CO)		<100 ppm		
Methane (CH_4)		None		
Ethane (C_2H_6)		None		
Ethylene (C_2H_4)		None		
Hydrogen (H_2)		None		
Acetylene (C_2H_2)		None		
Karl Fisher (@ 20°C)		<25 ppm		
Dielectric breakdown strength		>30 kV		
Neutralization number		<0.05 mg/g		
Interfacial tension		>40 dynes/cm		
Color (ASTM D-1524)		<3.0		
Sediment		Clear		
Power factor		<0.05%		
Sediment/visual examination		Clear		

Results:

TABLE A.11
Insulation-Resistance Test Criteria

Item	Date of Inspection	Acceptable Limit	Actual Value	Inspector Initials
Capacitance charging current				
Dielectric absorption current				
Leakage current				
Temperature correction factor				
Polarization index				
Dielectric absorption ratio				
Minimum resistance value				

Results:

TABLE A.12
Turns Ratio Test Criteria

Item	Date of Inspection	Acceptable Limit	Actual Value	Inspector Initials
Voltage applied				
Induced voltage				
Calculated ratio				
Nameplate data ratio				
Certification				
Other				
Submitted				

Results:

TABLE A.13
Lubrication Oil Test Criteria

Item	Date of Inspection	Acceptable Limit	Actual Value	Inspector Initials
Liquids				
Viscosity grade (ISO units)				
AGMA/SAE classification				
Additives				
Grease				
Type of base stock				
NLGI number				
Type/% of thickener				
Dropping point				
Base oil viscosity (SUS)				
Total acid number				
Visual observation (cloudiness)				
IR spectral analysis, metal count				
Particle count				
Water content				
Viscosity				

Results:

TABLE A.14
Motor Circuit Evaluation Test Criteria

Item	Date of Inspection	Acceptable Limit	Actual Value	Inspector Initials
Resistance				
Resistance imbalance		<2%		
Inductance				
Inductance imbalance		<10%		
Capacitance				
Total impedance				
Impedance imbalance				
Motor current analysis				

Results:

TABLE A.15
Power Factor Test Criteria

Item	Date of Inspection	Acceptable Limit	Actual Value	Inspector Initials
Grounded specimen test (GST)				
Ungrounded specimen test (UST)				
GST with guard				
Environment humidity				
Environment temperature				
Surface cleanliness				
Phase I				
Applied voltage				
Total current				
Capacitive current				
Dissipation factor				
Power factor				
Normal power factor				
Phase II				
Applied voltage				
Total current				
Capacitive current				
Dissipation factor				
Power factor				
Normal power factor				
Phase III				
Applied voltage				
Total current				
Capacitive current				
Dissipation factor				
Power factor				
Normal power factor				

Results:

TABLE A.16
Vibration Analysis Test Criteria

Item	Date of Inspection	Acceptable Limit	Actual Value	Inspector Initials
Test instrumentation				
FFT analyzer				
Type				
Model				
Serial number				
Last calibration date				
Line resolution bandwidth				
Dynamic range				
Hanning window				
Linear non-overlap averaging				
Anti-aliasing filters				
Amplitude accuracy				
Sound disk thickness				
Adhesive (hard/soft)				
Vibration readings				
1H				
1V				
1A				
2H				
2V				
2A				
Velocity amplitude (in./s-peak)				
Running speed order				
Acceleration overall amp (g peak)				
Vibration signatures (H, V, A)				
Frequency (CPM)				
Balanced condition				
Balance weight type				

Results:

TABLE A.17
Borescope/Fiberscope Test Criteria

Item	Date of Inspection	Acceptable Limit	Actual Value	Inspector Initials
Equipment Information				
Model number				
Scope serial number				
Manufacturer				
Date purchased				
Diameter				
Working length				
Test information/results				
Direction of view				
Degrees direct				
Degrees foreoblique				
Degrees side				
Degrees retro				
Field of view				
Degrees focusing				
Degrees fixed focus				
Depth of field (distance)				
Minimum (cm)				
Maximum (cm)				

Special Instructions/Notes:

Results:

Bibliography

GENERAL BIBLIOGRAPHY

ACGIH. (1995). *Industrial Ventilation: A Manual of Recommended Practice*, 22nd ed. Cincinnati, OH: American Conference of Governmental Industrial Hygienists.

Aldridge, M.D. (1973). *Analysis of Communication Systems in Coal Mines*, U.S. Bureau of Mines Contract No. J0101702, NTIS No. PB/225-862. Morgantown: West Virginia University.

Allen, P. (2009). Risk control hierarchy clarifies electrical safety. *EHS Today*, 2(10): 69–72.

Cawley, J.C. and Homce, G.T. (2003). Occupational electrical injuries in the United States, 1992–1998, and recommendations for safety research. *Journal of Safety Research*, 34(3): 241–248.

Cawley, J.C. and Homce, G.T. (2008). Trends in electrical injury in the U.S., 1992–2002. *EE Transactions on Industrial Applications*, 44(4): 962–972.

CFR. (2006). Title 30, Mineral Resources, Parts 1 to 199. Washington, DC: U.S. Government Printing Office (http://www.msha.gov/30CFR/CFRINTRO.HTM).

Clarke, S.L. (2009). *Explosion Risk Assessment for "EPL" Selection*. Chichester, UK: Intertek (http://aiche.confex.com/aiche/CCPS09/webprogrampreliminary/Paper171698.html).

CMAA. (1994). *Specifications for Electric Overhead Traveling Cranes*, CMAA Specification 1B61. Charlotte, NC: Crane Manufacturer's Association of America.

Doan, D.R., Hoagland, E., and Neal, T. (2010). Update of field analysis of arc flash incidents, PPE protective performance and related work injuries. In: *Proceedings of Electrical Safety Workshop (ESW)*, 2010 IEEE IAS, Memphis, TN, February 1–5.

DOD. (2000). *Standard Practice for System Safety*, MIL-STD-882D. Washington, DC: Department of Defense.

Fesak, G., Helfrich, W., Vilcheck, W., and Deutsch, D. (1981). Instantaneous circuit breaker settings for the short circuit protection of three phase 480, 600, and 1040 V trailing cables. *IEEE Transaction of Industrial Applications*, 17(4): 369–376.

Gardner, J. and Alanis, J. (2008). A Comparison of International Wiring and Grounding Systems with North American Methods, paper presented at IEEE Petroleum and Chemical Industry Technical Conference, Cincinnati, OH, September 22–24.

Gordon, L.B. and Cartelli, L. (2009). *A Complete Electrical Hazard Classification System and Its Application*, paper presented at IEEE Electrical Safety Workshop, St. Louis, MO, February 2–6.

Gregory, G. Preventing arc flash incidents in the workplace. *EC&M*, June 1.

Homce, G.T., Cawley, J.C., Sacks, H.K., and Yenchek, M.R. (2002). Heavy equipment near overhead power lines? *Engineering and Mine Journal*, 203(4): 36–39.

Homce, G.T., Cawley, J.C., and Yenchek, M.R. (2005). Avoid the shock. *Water Well Journal*, 59(8): 12–14.

HSA. (2002). *Guidance for Carrying Out Risk Assessment at Surface Mining Operations*, Document No. 5995/2/98-EN. Dublin, Ireland: Health and Safety Authority, Committee on Surface Workings (http://www.hsa.ie/eng/Publications_and_Forms/Publications/Mines_and_Quarries/Guidance_For_Carrying_Out_Risk_Assessment_at_Surface_Mining_Operations_.html).

Iannacchione, A., Varley, F., and Brady, T. (2008). *The Application of Major Hazard Risk Assessment (MHRA) to Eliminate Multiple Fatality Occurrences in the U.S. Minerals Industry*, Information Circular 9508. Spokane, WA: Department of Health and Human Services.

IEC. (2002). *Effects of Current on Human Beings and Livestock. Part 1. General Aspects*, IEC/TS 60479-1. Geneva, Switzerland: International Electrotechnical Commission.

ITT. (2008). Meeting Summary Focus Group Meeting, Defense Information Systems Agency Joint Spectrum Center, Annapolis, MD, August 27.

Jensen, P. (1987). Electric injury causing ventricular arrhythmias. *British Heart Journal*, 57(3): 279–283.

Kohler, J.L., Sottile, J., and Trutt, F.C. (1999). Condition-based maintenance of electrical machines. In: *Proceedings of the IEEE Industry Applications Conference*, 34th IAS Annual Meeting, Phoenix, AZ, October 3–7.

Kowalski-Trakofler, K. and Barrett, E. (2007). Reducing non-contact electric arc injuries: an investigation of behavioral and organizational issues. *Journal of Safety Resources*, 38(5): 597–608.

L-3 Global Security & Engineering Solutions and NIOSH. (2009). *Wireless Mesh Communications System: Phase 3 Final Report*, Contract Number 200-2007-20388. Chantilly, VA: L-3 Global Security & Engineering Solutions.

Littelfuse. (2007). *Misconceptions About Arc-Flash Hazard Assessments*. Chicago, IL: Littelfuse.

Mastrullo, K.G., Jones, R.A., and Jones, J.G. (2003). *The Electrical Safety Program Book*. Quincy, MA: National Fire Protection Association.

Morley, L.A. (1990). *Mine Power Systems*, Information Circular 9258, NTIS No. PB 91-241729. Pittsburgh, PA: U.S. Department of the Interior, Bureau of Mines.

Morse, M.S. and Weiss, D. (1993). An Evaluation Protocol for Electric Shock Injury Supported by Minimal Diagnostic Evidence, paper presented at 15th Annual International Conference of the IEEE Engineering in Medicine and Biology Society (EMBS), San Diego, CA, October 28–31.

NEMA. (1962). *Requirements for Electric Arc Welding Apparatus*, NEMA EW-1. Arlington, VA: National Electrical Manufacturer's Association.

NETA ATS. (2003). *Acceptance Testing Specifications for Electrical Power Distribution Equipment and Systems*. Morrison, CO: International Electrical Testing Association.

NFPA. (2005). *NFPA 70: National Electrical Code (NEC) Handbook*. Quincy, MA: National Fire Protection Association.

NSW DoPI. (2005). *DC Systems Used Underground*, Decision Sheet 6.1. Albury, NSW: NSW Department of Primary Industries, Electrical Engineering Safety.

OMSHR. (2012). *Mining Topic: Electrical Safety*, Atlanta, GA: Office of Mine Safety and Health Research, Centers for Disease Control and Prevention (http://www.cdc.gov/niosh/mining/topics/ElectricalAccidents.html).

OSHA. (2007). Occupational Safety and Health Standards, 29 CFR, Part 1910, Subpart S, Electrical. Baltimore, MD: Occupational Safety and Health Administration.

Regan, R. (2008). Department of Primary Industries Official Notices: Coal Mine Health and Safety Act 2002: types of electrical plant used in hazardous zone. *New South Wales Government Gazette*, January 25, pp. 181–182.

Sacks, H.K., Cawley, J.C., Homce, G.T., and Yenchek, M.R. (2001). Feasibility study to reduce injuries and fatalities caused by contact of cranes, drill rigs, and haul trucks with high-tension lines. *IEEE Transactions on Industry Applications*, 37(3): 914–919.

State of Queensland. (1999). *Recognised Standard 01: Underground Electrical Equipment and Electrical Installations*, Coal Mining Safety and Health Act 1999 (http://mines.industry.qld.gov.au/assets/mines-safety-health/recognised_standard01.pdf).

State of Queensland. (2013a). *Coal Mining Safety and Health Regulation 2001*, current as at 20 December 2013 (https://www.legislation.qld.gov.au/legisltn/current/c/coalminshr01.pdf).

State of Queensland. (2013b) *Coal Mining Safety and Health Act 1999*, current as at 31 March 2013 (https://www.legislation.qld.gov.au/LEGISLTN/CURRENT/C/CoalMinSHA99. pdf).

Tajali, R. (2012). *Arc Flash Analysis: IEEE Method versus the NFPA 70E Tables*, January 2012/1910DB1201. Palatine, IL: Schneider Electric Engineering Services.

Thurnherr, P. and Schwarz, G. (2009). Selection of electrical equipment for hazardous areas. *IEEE Industry Applications*, January/February, 50–55.

Trutt, F.C., Sottile, J., and Kohler, J.L. (2001). Detection of AC machine winding deterioration using electrically excited vibrations. *IEEE Transactions on Industry Applications*, 37(1): 10–14.

U.S. Army Corps of Engineers. (2008). *Safety and Health Requirements Manual*, EM 385-1-1. Washington, DC: U.S. Army Corps of Engineers.

Walls, G. (2005). *Understanding Arc Flash Requirements*, Revision 3. Virginia Beach, VA: Professional Power Systems.

Wang, M.Y. and Chang, C.H. (2002). Applications of a multi-generation diffusion model to pagers and mobile phones. In: *Proceedings of IEEE International Engineering Management Conference (IEMC '02)*, St. John's College, Cambridge, UK, August 18–20.

WPSAC. (2014). *Train-the-Trainers Guide to Electrical Safety for General Industry*. Washington, DC: Workplace Safety Awareness Council.

Yenchek, M.R., Cawley, J.C., Brautigam, A.L., and Peterson, J.S. (2002). Distinguishing motor starts from short circuits through phase-angle measurements. *IEEE Transactions on Industry Applications*, 38(1): 195–202.

IEEE STANDARDS

IEEE SA—141-1993. IEEE Recommended Practice for Electric Power Distribution for Industrial Plants (http://standards.ieee.org).

IEEE SA—1584-2002. IEEE Guide for Performing Arc Flash Hazard Calculations (http://standards.ieee.org).

IEEE SA—1584a-2004. IEEE Guide for Performing Arc-Flash Hazard Calculations—Amendment 1 (http://standards.ieee.org).

IEEE SA—1584b-2011. IEEE Guide for Performing Arc-Flash Hazard Calculations—Amendment 2: Changes to Clause 4 (http://standards.ieee.org).

IEEE SA—3007.3-2012. IEEE Recommended Practice for Electrical Safety in Industrial and Commercial Power Systems (http://standards.ieee.org).

ANSI STANDARDS INCORPORATED BY OSHA IN 29 CFR 1910.6

ANSI A11.1-65 (R 70). Practice for Industrial Lighting.

ANSI A14.2-56. Safety Code for Portable Metal Ladders, Supplemented by ANSI A14.2a-77.

ANSI A17.1-65. Safety Code for Elevators, Dumbwaiters and Moving Walks, Including Supplements A17.1a (1967); A17.1b (1968); A17.1c (1969); A17.1d (1970).

ANSI A17.2-60. Practice for the Inspection of Elevators, Including Supplements A17.2a (1965), A17.2b (1967).

ANSI A90.1-69. Safety Standard for Manlifts.

ANSI A92.2-69. Standard for Vehicle Mounted Elevating and Rotating Work Platforms.

ANSI A120.1-70. Safety Code for Powered Platforms for Exterior Building Maintenance.

ANSI B15.1-53 (R 58). Safety Code for Mechanical Power Transmission Apparatus.

ANSI B20.1-57. Safety Code for Conveyors, Cableways, and Related Equipment.

ANSI B30.2.0-67. Safety Code for Overhead and Gantry Cranes.

ANSI B30.2-43 (R 52). Safety Code for Cranes, Derricks, and Hoists.

ANSI B56.1-69. Safety Standard for Powered Industrial Trucks.
ANSI C33.2-56. Safety Standard for Transformer-Type Arc Welding Machines.
ANSI H38.7-69. Specification for Aluminum Alloy Seamless Pipe and Seamless Extruded Tube.
ANSI J6.4-71. Standard Specification for Rubber Insulating Blankets.
ANSI J6.6-71. Standard Specification for Rubber Insulating Gloves.
ANSI Z9.2-60. Fundamentals Governing the Design and Operation of Local Exhaust Systems.
ANSI Z9.2-79. Fundamentals Governing the Design and Operation of Local Exhaust Systems.
ANSI Z35.1-68. Specifications for Accident Prevention Signs.
ANSI Z41-99. American National Standard for Personal Protection—Protective Footwear.
ANSI Z41.1-67. Men's Safety Toe Footwear.
ANSI Z54.1-63. Safety Standard for Non-Medical X-Ray and Sealed Gamma Ray Sources.
ANSI Z87.1-68. Practice of Occupational and Educational Eye and Face Protection.
ANSI Z87.1-03. American National Standard Practice for Occupational and Educational Eye
    and Face Protection.
ANSI Z89.1-69. Safety Requirements for Industrial Head Protection.
ANSI Z89.1-09. American National Standard for Industrial Head Protection.
ANSI Z89.2-71. Safety Requirements for Industrial Protective Helmets for Electrical Workers,
    Class B.
ANSI Z535.1-06 (R2011). Safety Colors, reaffirmed July 19, 2011.
ANSI Z535.2-11. Environmental and Facility Safety Signs.

ASTM STANDARDS INCORPORATED BY OSHA IN 29 CFR 1910.6

ASTM D93-71. Test for Flash Point by Pensky Martens.
ASTM D93-08. Standard Test Methods for Flash Point by Pensky-Martens Closed Cup Tester.
ASTM D120-87. Specification for Rubber Insulating Gloves.
ASTM D178-93 (D 178-88). Specification for Rubber Insulating Matting.
ASTM D240-02 (Reapproved 2007). Standard Test Method for Heat of Combustion of Liquid
    Hydrocarbon Fuels by Bomb Calorimeter.
ASTM D323-68. Standard Test Method of Test for Vapor Pressure of Petroleum Products
    (Reid Method).
ASTM D1048-93 (D 1048-88a). Specification for Rubber Insulating Blankets.
ASTM D1049-93 (D 1049-88). Specification for Rubber Insulating Covers.
ASTM D1050-90. Specification for Rubber Insulating Line Hose.
ASTM D1051-87. Specification for Rubber Insulating Sleeves.
ASTM D3278-96 (Reapproved 2004). E1, Standard Test Methods for Flash Point of Liquids
    by Small Scale Closed-Cup Apparatus.
ASTM D3828-07a. Standard Test Methods for Flash Point by Small Scale Closed Cup Tester.
ASTM F478-92. Specification for In-Service Care of Insulating Line Hose and Covers.
ASTM F479-93. Specification for In-Service Care of Insulating Blankets.
ASTM F496-93b. Specification for In-Service Care of Insulating Gloves and Sleeves.
ASTM F2412-05. Standard Test Methods for Foot Protection.
ASTM F2413-05. Standard Specification for Performance Requirements for Protective Footwear.

NFPA STANDARDS INCORPORATED BY OSHA IN 29 CFR 1910.6

NFPA 30-1969. Flammable and Combustible Liquids Code.
NFPA 33-1969. Standard for Spray Finishing Using Flammable and Combustible Material.
NFPA 51B-1962. Standard for Fire Protection in Use of Cutting and Welding Processes.
NFPA 68-1954. Guide for Explosion Venting.
NFPA 86A-1969. Standard for Oven and Furnaces Design, Location and Equipment.

NFPA 91-1961. Standard for the Installation of Blower and Exhaust Systems for Dust, Stock, and Vapor Removal or Conveying (ANSI Z33.1-61).

NFPA 101-1970. Code for Life Safety From Fire in Buildings and Structures.

NFPA 101-2009. Life Safety Code, IBR Approved for §§1910.34, 1910.35, 1910.36, and 1910.37.

NFPA 385-1966. Recommended Regulatory Standard for Tank Vehicles for Flammable and Combustible Liquids.

NFPA 496-1967. Standard for Purged Enclosures for Electrical Equipment in Hazardous Locations.

NFPA 505-1969. Standard for Type Designations, Areas of Use, Maintenance, and Operation of Powered Industrial Trucks.

NFPA 566-1965. Standard for the Installation of Bulk Oxygen Systems at Consumer Sites, IBR Approved for §§1910.253(b)(4)(iv) and (c)(2)(v).

Index

385